# POLYCYCLIC AROMATIC HYDROCARBONS IN THE AQUATIC ENVIRONMENT

## Sources, Fates and Biological Effects

# POLYCYCLIC AROMATIC HYDROCARBONS IN THE AQUATIC ENVIRONMENT

## Sources, Fates and Biological Effects

JERRY M. NEFF, Ph.D.

*Department of Biology, Texas A & M University,*
*College Station, Texas* 77843, *USA*

APPLIED SCIENCE PUBLISHERS LTD
LONDON

APPLIED SCIENCE PUBLISHERS LTD
RIPPLE ROAD, BARKING, ESSEX, ENGLAND

**British Library Cataloguing in Publication Data**

Neff, Jerry M
Polycyclic aromatic hydrocarbons in the aquatic environment.
1. Aquatic ecology 2. Water—Pollution
3. Hydrocarbons—Environmental aspects
4. Aromatic compounds—Environmental aspects
5. Polycyclic compounds—Environmental aspects
I. Title
574.5′263 QH541.5.W3

ISBN 0-85334-832-4

WITH 89 TABLES AND 30 ILLUSTRATIONS



Printed in Great Britain by Galliard (Printers) Ltd, Great Yarmouth

# *Acknowledgements*

The preparation of this review was supported in part by a contract from the American Petroleum Institute, Washington, D.C. I should like to express my appreciation to Margaret Weirich, Lawrence A. Reitsema, and William J. Rue, Jr., for their assistance in the literature search and to Becky Schult, Kay Barker, and Gail Snook for their assistance in typing the manuscript. Michelle M. Michot provided invaluable assistance in preparing the figures and in checking the manuscript. I should also like to express my appreciation to Dr Geraldine V. Cox, Dr L. W. Pollock, Dr Winston K. Robbins, and Dr Fred T. Weiss for reviewing a preliminary draft of the review and for offering many valuable suggestions for its improvement.

# Preface

It has been known for many years that some polycyclic aromatic hydrocarbons (PAH) can cause cancer in laboratory mammals. In addition, the incidence of certain types of human cancer has been correlated with occupational or other exposure to these compounds. Concern over the hazards to human and animal health from exposure to carcinogenic PAH has lead to an accumulation over the last twenty years of a very large body of scientific literature dealing with the sources, distribution in the environment, and biological effects of PAH. With improvements in chemical methods for analysis of complex PAH assemblages in environmental samples has come the realization that PAH are nearly ubiquitous trace contaminants of marine and freshwater ecosystems. Considerable discussion has arisen in recent years about the possible sources of these aquatic PAH, their effects on aquatic organisms, and the possible health hazards they pose to man. In this volume, I have attempted to critically review and bring together the available scientific literature on the subject with the objective of drawing some general conclusions about the relative rates of entry of PAH from different sources into the aquatic environment, the rates and pathways of degradation of aquatic PAH by abiotic and biotic processes, the biological effects, including cancer, of PAH on aquatic organisms, and the potential health hazards to man of aquatic PAH. The major emphasis is on the biological aspects of the aquatic PAH problem.

It is my sincere hope that this review will be of use to scientists, technologists, and students in several environmental disciplines because it not only provides a broad summary and synthesis of our current state of knowledge of the aquatic PAH problem but also reveals important areas where information is scanty or completely lacking. It is hoped that this review will encourage environmental scientists to conduct the research necessary to fill in these gaps in our knowledge.

Information on the sources and distribution of PAH in the aquatic environment has come from many countries. Russian scientists have provided much of the available data on PAH concentrations in industrial wastewaters, while information on emissions of airborne PAH from industrial, domestic, and internal combustion engine sources has come mainly from the United States and Canada. British and German scientists have published much of the data on PAH in fresh waters, while much of the research on PAH in marine sediments and organisms has been conducted by French, American, and Canadian scientists. To gain a better understanding of the cycle of PAH in the environment, a single study should be conducted using the most modern analytical techniques in which selected PAH from a single point source are monitored simultaneously in the wastewater, gaseous, and solid effluents from the source, and in the air, water, and aquatic biota at different distances downstream from the source.

The question of the possible biosynthesis of PAH by bacteria, fungi and higher plants has not been adequately resolved. Much of the research in this area was performed before the advent of modern high-resolution techniques for PAH analysis and without full realization of the magnitude of the problem of sample contamination from exogenous PAH sources. Carefully designed experiments utilizing modern analytical methodologies could perhaps resolve this important question.

Much of our concern about environmental PAH stems from the knowledge that a few of them are carcinogenic. However, the vast majority of PAH detected in the environment have not been demonstrated to be carcinogenic or mutagenic to laboratory mammals or aquatic organisms. Relatively little is known about the toxicity to and sublethal biological effects on aquatic organisms of those PAH which are not known carcinogens. More research is needed on the carcinogenicity of known or suspect PAH carcinogens and on the chronic toxicity of non-carcinogenic PAH to aquatic organisms.

Although the scientific literature on the metabolism of PAH by aquatic microorganisms and animals is growing rapidly, more information is needed on the chemical identity of the PAH metabolites (are they the same as or different from those produced by well-characterized mammalian systems?) and on the influence of physical environmental variables such as temperature, salinity, and dissolved oxygen on the rate of uptake, metabolism, and release of PAH by aquatic organisms.

Finally, there are several other important topics which warrant further careful investigation. These include: rates of photochemical oxidation of PAH in the water column and surface sediments under different conditions of salinity, temperature, and oxygen concentration; bioavailability of sediment-adsorbed and food-adsorbed PAH to aquatic organisms; biomagnification of PAH in aquatic food chains; induction of cancer in aquatic organisms by exposure to realistic levels of PAH in the water, sediment and food; and induction of cancer in laboratory mammals by consumption of PAH-contaminated water and aquatic foods.

JERRY M. NEFF

*January*, 1979 *College Station, Texas*

# *Contents*

*CHAPTER* 1

# *Introduction*

There has been a growing concern in recent years regarding possible harmful effects to man and other living organisms of pollutant organic chemicals released into the environment as a consequence of man's activities. Polycyclic aromatic hydrocarbons (PAH) are one of the more significant classes of pollutant chemicals which give rise to this concern. This concern arises primarily from the fact that a small fraction of the PAH generated and released to the environment by human activities has been shown to be carcinogenic to mammals. The environmental impact of the majority of PAH which are not carcinogenic is largely unknown. The purpose of this review is to summarize and synthesize recent data on the sources, fates, and effects of PAH in the aquatic environment with special emphasis on the biological aspects of this problem.

PAH are composed of two or more fused aromatic (benzene) rings. Two aromatic rings are said to be fused when a pair of carbon atoms is shared. The resulting structure is a molecule with all carbon and hydrogen atoms lying in a single plane. Naphthalene ($C_{10}H_8$), which consists of two fused aromatic rings, is the lowest molecular weight member. The ultimate fused-ring aromatic system is graphite, an allotropic form of elemental carbon. Of primary environmental concern are mobile compounds ranging in molecular weight from naphthalene ($C_{10}H_8$, MW 128·16) to coronene ($C_{24}H_{12}$, MW 300·36). Within this range is an extremely large number of PAH differing in the number and positions of aromatic rings and in the number, chemistry, and position of substituents on the basic ring system.

Physical and chemical characteristics of PAH vary in a more or less regular fashion with molecular weight. Resistance to oxidation and reduction tends to decrease with increasing molecular weight. Vapor pressure and aqueous solubility decrease almost logarithmically with increasing molecular weight. As a consequence of these differences, PAH of different molecular weight vary substantially in their behavior and distribution in the environment and their effects on biological systems. Two molecular weight classes of PAH can be distinguished on the basis of physical, chemical, and biological properties. These are the lower molecular weight 2–3 ring aromatics (naphthalenes, fluorenes, phenanthrenes, and anthracenes) and the higher molecular weight 4–7 ring aromatics (chrysene–coronene). The low molecular weight PAH have significant acute toxicity to aquatic organisms, whereas the high molecular weight PAH do not. On the other hand, all of the 20–30 proven PAH carcinogens are in the high molecular weight PAH group.

Most methods designed for the collection, extraction, and analysis of PAH in environmental samples do not efficiently handle the low molecular weight PAH. Thus, the majority of published values for 'total PAH' in environmental samples do not include the low molecular weight PAH.

Several systems of nomenclature have been used to describe PAH ring structures. Nomenclature used in this review is that adopted by the International Union of Pure and Applied Chemistry (IUPAC) and described in detail in *The Ring Index* (Patterson *et al.*, 1960). The most important rules used in this system are the following:

1. The structural diagram is written to present the greatest possible number of rings in a horizontal row.
2. Horizontal and vertical axes are drawn through the center of the horizontal row and the molecule is oriented in such a way as to place the maximal number of rings in the upper right quadrant and the minimal number of rings in the lower left quadrant.
3. Carbon atoms are numbered in a clockwise direction starting with the carbon atom that is not part of another ring and is in the most counterclockwise position of the uppermost ring farthest to the right; carbon atoms common to two or more rings are not numbered.
4. Ring faces (except those common to two rings) are lettered in alphabetical order with the side between carbon atoms 1 and 2 designated '*a*'. Alphabetical order is continued clockwise around the molecule.
5. In naming a compound formed by the addition of a component, numbers and letters are enclosed in square brackets and placed immediately after the name of the added component to describe where a substituent group is attached or where a ring is fused to the face of the molecule. If a ring is fused to more than one face of the

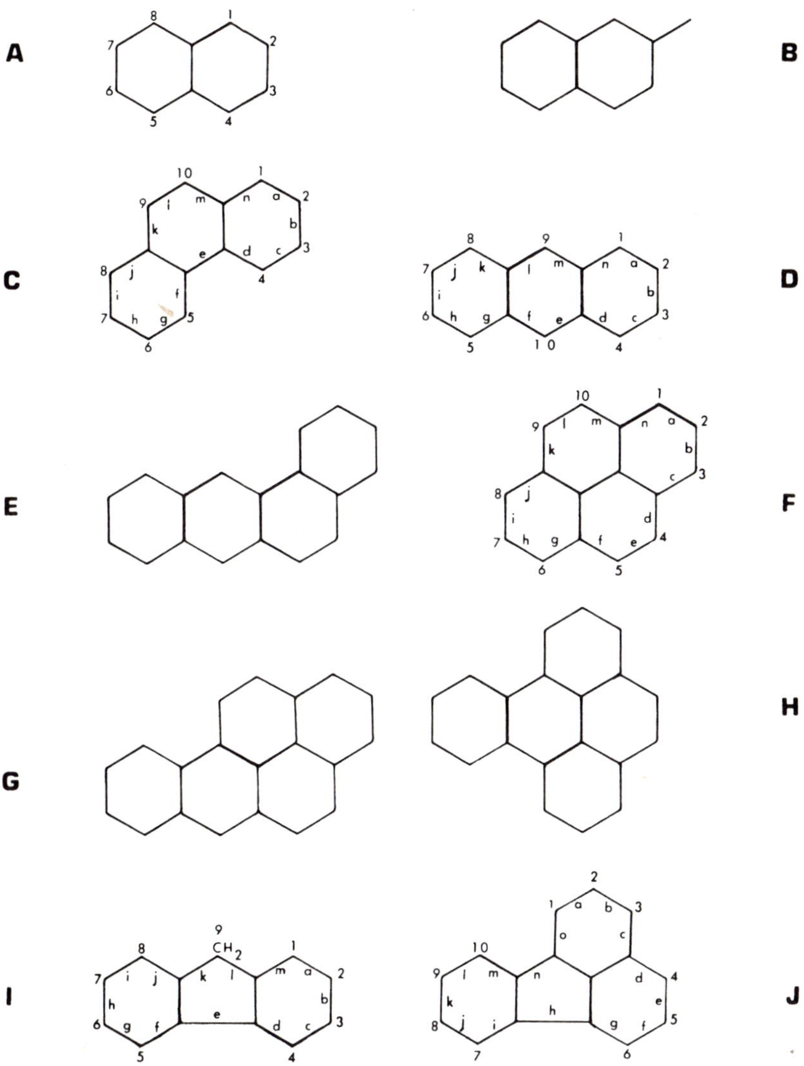


FIG. 1. Ring structures of representative polycyclic aromatic hydrocarbons. The numbering and lettering system for several PAH is also given. A, naphthalene; B, 2-methylnaphthalene; C, phenanthrene; D, anthracene; E, benz[*a*]anthracene; F, pyrene; G, benzo[*a*]pyrene; H, benzo[*e*]pyrene; I, fluorene; J, fluoranthene.

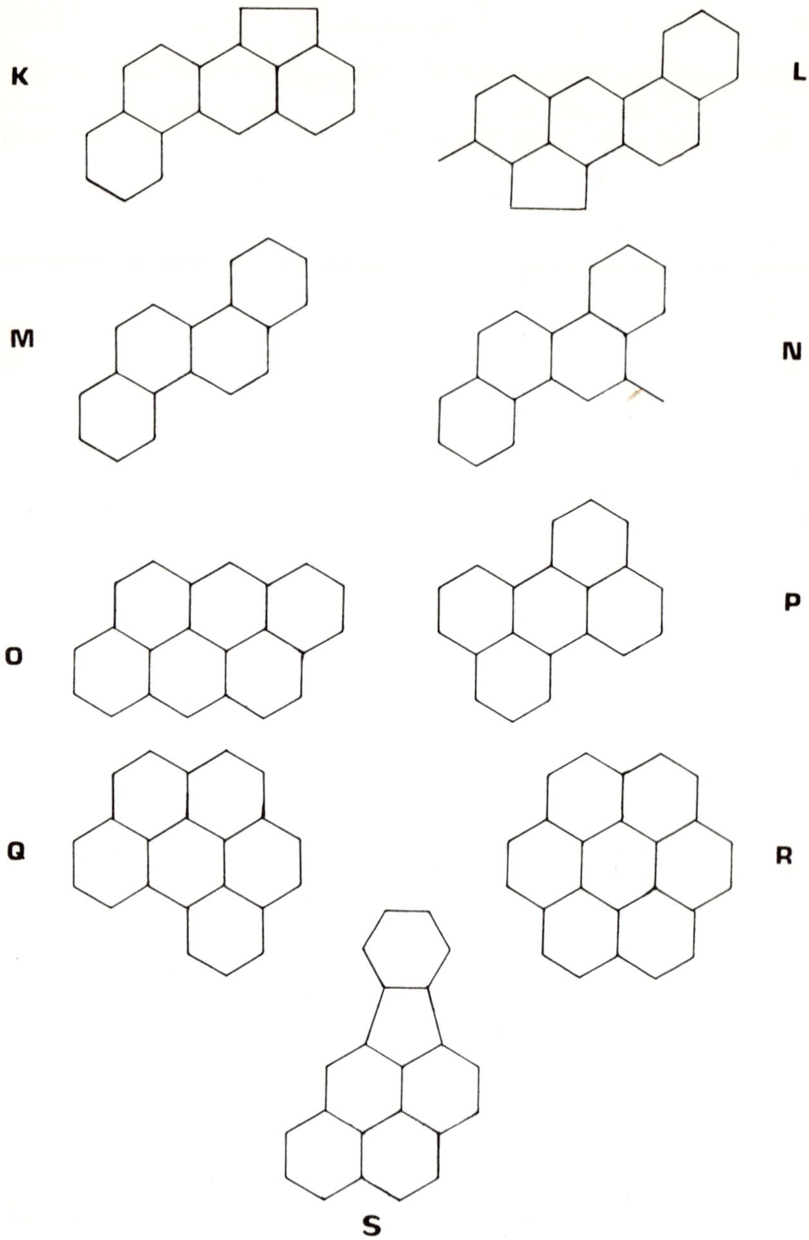


FIG. 1—*contd.* K, benz[*j*]aceanthrylene = cholanthrene; L, 3-methylcholanthrene; M, chrysene; N, 5-methylchrysene; O, dibenzo[*cd,jk*]pyrene = anthranthrene; P, perylene; Q, benzo[*ghi*]perylene; R, coronene; S, indeno[1,2,3-*cd*]pyrene.

molecule simultaneously, this is indicated by using the appropriate letters to denote the faces so involved.

6. The structural formulas used show aromatic rings as plain hexagons and a methylene group as $CH_2$.

Nomenclature for a few PAH, such as phenanthrene and anthracene, differs from these rules. Structural formulas of several PAH typical of those found in environmental samples are shown in Fig. 1. Examples of the numbering system for carbon atoms and lettering systems for ring faces are also included.

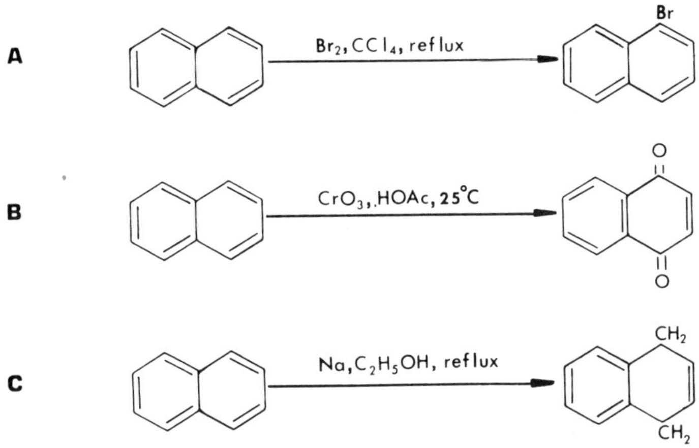


FIG. 2. Typical reactions of PAH. A, Electrophilic substitution; B, oxidation; C, reduction.

PAH undergo three types of chemical reactions characteristic of aromatic hydrocarbons—electrophilic substitution, oxidation, and reduction. Examples of these reactions are shown in Fig. 2. Oxidation and reduction reactions destroy the aromatic character of the affected benzene ring but electrophilic substitution does not.

The presence of PAH in water, aquatic sediments, and organisms has been recognized for more than 20 years. With the development of sensitive high resolution techniques for the analysis of PAH in environmental samples, it has become apparent that extremely complex mixtures of PAH, including the carcinogenic forms such as benzo[*a*]pyrene, are nearly ubiquitous trace contaminants of aquatic ecosystems. A considerable amount of research has been conducted in recent years on the sources, fates, and biological effects of these PAH mixtures in aquatic ecosystems. Much of the earlier literature has been reviewed by Andelman and Suess (1970) and Andelman and Snodgrass (1972). Special aspects of the problem relating to drinking water were reviewed by Harrison *et al.* (1975). Much of the recent literature up to May 1975 on the environmental fate of PAH was reviewed by Radding *et al.* (1976).

The data presented in the above reviews and in this review on the sources and distribution of PAH in the environment were collected over a period of about 20 years. Sample collection, extraction, and analysis techniques have varied widely (Harrison *et al.*, 1975) and the analytical results obtained by different techniques are often not strictly comparable. Earlier investigators often were not fully aware of the extreme compositional complexity of environmental PAH assemblages. Thus, estimates of total PAH based on the quantitative analysis of a relatively few PAH in a sample (often only 6–12 individual PAH in earlier studies) gave low and variable results. In addition, the extreme difficulty in isolating a really pure PAH fraction from environmental samples, particularly those with a complex organic matrix such as aquatic sediments and biological tissues, has not been appreciated fully by some investigators. This could result in high and variable estimates of total or individual PAH. Giger and Blumer (1974) have reviewed some of the difficulties in characterizing the PAH composition and concentration of complex environmental samples. Much of the data discussed in the sections which follow on the sources and distribution of PAH in the environment should be viewed as semiquantitative estimates of PAH distributions in nature.

*CHAPTER 2*

# *Sources of PAH in the Aquatic Environment*

Until relatively recently it was generally considered that PAH were formed only during the high temperature (e.g., 700 °C) pyrolysis of organic materials. The discovery of complex mixtures of PAH spanning a wide molecular weight range in fossil fuels such as coal and crude petroleum has led to the conclusion that, given sufficient time, pyrolysis of organic materials at temperatures as low as 100–150 °C can lead to the production of PAH (Blumer, 1976). In addition, there has been considerable speculation in recent years, and some experimental evidence in its favor, that PAH are synthesized by bacteria and plants. Thus, PAH may be formed in three ways: high temperature pyrolysis of organic materials, low to moderate temperature diagenesis of sedimentary organic material to form fossil fuels, and direct biosynthesis by microbes and plants.

Although PAH are produced by natural processes, a wide variety of human activities increases the environmental load of these substances. Industrial activities resulting in the production of PAH include: preparation of acetylene from natural gas; pyrolysis of kerosene to form benzene, toluene and other organic solvents; pyrolysis of wood to form charcoal, wood tars and carbon blacks; manufacture of electrolytic aluminium using graphite electrodes; coke production; gas production from petroleum; coal gasification; production of synthetic alcohol; and oil refinery operations (Andelman and Snodgrass, 1972). Incineration of industrial and domestic wastes, forest and grass fires, power generation from fossil fuels, and the combustion of fuels in internal combustion engines also produce emissions rich in PAH. These anthropogenic PAH may reach the aquatic environment in industrial and domestic sewage effluents, surface runoff from land, deposition of airborne particulates, and spillage of petroleum and petroleum products into water bodies. Cycling of PAH in aquatic environments is very poorly understood.

In the sections that follow, the available scientific literature on the possible biosynthesis of PAH, PAH in fossil fuels and related materials, and high temperature pyrolytic sources of PAH will be briefly reviewed.

## 2.1 BIOSYNTHESIS OF PAH

The extensive studies of Professor Borneff and associates in the German Federal Republic (GFR) and Professor Mallet and associates in France during the 1960s led to the conclusion that PAH were nearly ubiquitous microcomponents of freshwater, marine, and terrestrial environments, even in regions far removed from industrial activities. PAH, including the carcinogen benzo[*a*]pyrene, were detected in the sediments, water, and marine biota from the west coast of Greenland (Mallet *et al.*, 1963) and the Clipperton Lagoon in the eastern equatorial Pacific Ocean (Niaussat and Auger, 1970). In the past 15 years, a large number of publications have appeared providing circumstantial or direct evidence both for and against the biosynthesis of PAH.

It is well known that a wide variety of organic molecules containing fused-ring polyaromatic systems are synthesized by organisms, particularly bacteria, fungi, and higher plants (Gerarde and Gerarde, 1962). Many of these compounds are not true PAH since they contain oxygen, nitrogen, or sulfur substituents. Examples of these biogenic polycyclic compounds are shown in Fig. 3. Numerous reports, especially in older chemical literature, concern the isolation of PAH from plant material. However, many of these findings are questionable in that isolation of the PAH-containing oils and tars was accomplished by methods that might lead to the pyrolytic formation of PAH from suitable precursors (Guddal, 1959). For instance, Jonsson (1968) isolated and identified a group of alkylphenanthrenes from pine wood tar. The tar contained about 20 per cent retene, 1-methyl-7-isopropylphenanthrene. Small amounts of 1-methylphenanthrene, 2-ethylphenanthrene, 1,7-dimethylphenanthrene, 1-methyl-7-ethylphenanthrene, 9,10-dihydroretene and possibly 1,3-dimethyl-7-ethylphenanthrene, and 1,3-dimethyl-7-isopropylphenanthrene were isolated from the crude retene. Although Jonsson considered this phenanthrene assemblage natural, wood tars are produced by destructive distillation of wood under conditions that would favor PAH pyrosynthesis (Youngblood and Blumer, 1975). The high concentration of this unusual assemblage of PAH suggests instead a pyrosynthesis from natural products related to phenanthrene which are common in plant materials (Thomson, 1965).

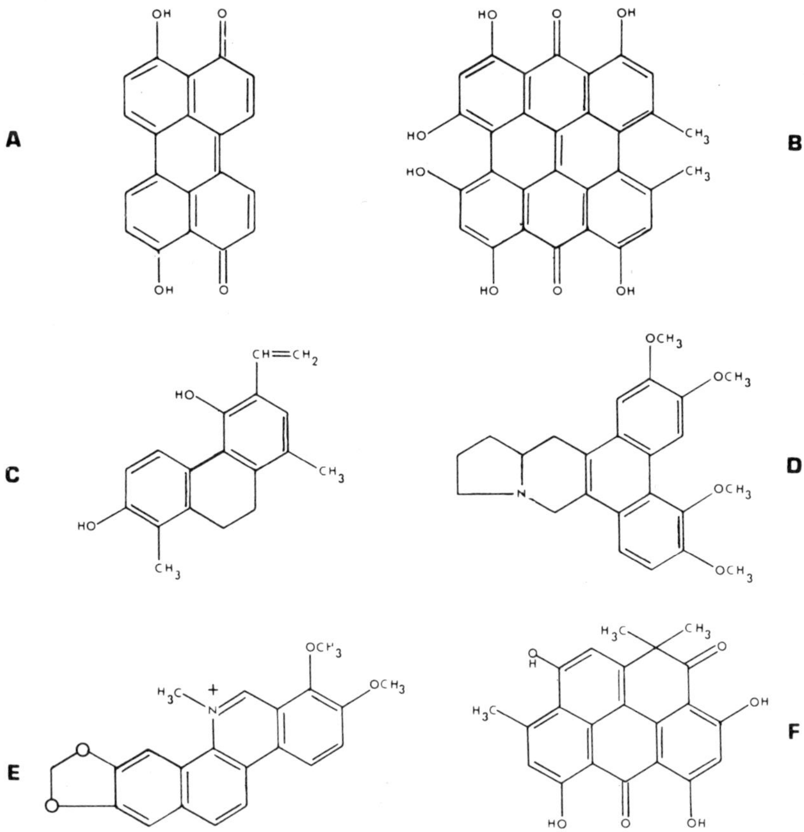


FIG. 3. Examples of natural products with a PAH-like structure. A, 4,9-dihydroxyperylene-3,10-quinone from the fungus *Daldinia concentrica.* Closely related pigments are found in the insect family Aphididae. B, hypercin from St. John's wort *Hypercium perforatum;* C, juncusol from the marine marsh grass *Juncus roemerianus;* D, tylophorine, the major alkaloid from *Tylophora asthamatica;* E, cherylthrine from roots of *Chelidonium majus;* F, resistomycin from *Streptomyces resistomycificus.*

### 2.1.1 Indirect Biosynthesis

Bacteria, fungi, higher plants, and some animals synthesize a wide variety of polycyclic quinone pigments (Thomson, 1971). Naphthoquinones are common in flowering plants. One example of a naphthoquinone is vitamin $K_2$. The largest group of naturally occurring quinones is the anthraquinones. Naphthacenequinones and phenanthraquinones are less

TABLE 1

PERYLENE CONCENTRATIONS IN MARINE SEDIMENTS (FROM ORR AND GRADY, 1967 AND AIZENSHTAT, 1973, WITH PERMISSION OF PERGAMON PRESS)

| *Location of sample* | *Depth below sea floor (m)* | *Sea floor depth (m)* | *Oxidation state of sediment*[a] | *% Organic carbon* | *Perylene µg/kg (ppb) dry sediment* |
|---|---|---|---|---|---|
| *Santa Barbara Basin* | | | | | |
| 34°13·5′N 120°01·9′W | 0–0·5 | 600 | R | 5·8 | 78 |
| 34°13·5′N 120°02·2′W | 0·5–0·8 | 600 | R | 5·8 | 68 |
| 34°11·0′N 120°03·0′W | 0·8–1·5 | 580 | R | 5·8 | 244 |
| 34°13·5′N 120°01·6′W | 1·2–1·4 | 600 | R | 5·8 | 260 |
| 34°08·9′N 120°01·6′W | 0–0·5 | 493 | O | 5·8 | <0·2 |
| *Tanner Basin* | | | | | |
| 32°57′N 119°44′W (5 samples) | 0–3·3 | 1 472–1 510 | R | 11·1–11·7 | <1·0 |
| *San Clemente Basin* | | | | | |
| 32°37·7′N 118°07·5′W | surface | 2 060 | R | 5·7 | <0·2 |
| *San Pedro Martin Basin, Gulf of California* | | | | | |
| 28°20′N 112°23′W | 0·15–0·25 | 883 | R | 4·75 | <1·0 |

| | | | | | |
|---|---|---|---|---|---|
| *Saanich Inlet (B.C., Canada)* | | | | | |
| 48°30·40′N 123°30·06′W | 0·0–0·1 | 200 | R | 3·87 | 343·2 |
| 48°30·40′N 123°30·06′W | 2–2·1 | 200 | R | — | 741·0 |
| 48°30·40′N 123°30·06′W | 17·1–17·4 | 200 | R | 2·82 | 1 675·8 |
| 48°30·40′N 123°30·06′W | 34·5–34·8 | 200 | R | 2·53 | 2 370·0 |
| *Gulf of Mexico* | | | | | |
| 23°01·8′N 92°02·6′W | 34 | 3 747 | R ⇌ O | 1·11 | 12·0 |
| 23°01·8′N 92°02·6′W | 209 | 3 747 | O | 0·82 | <0·1 |
| 23°01·8′N 92°02·6′W | 324 | 3 747 | O | 0·47 | <0·1 |
| 23°01·8′N 92°02·6′W | 534 | 3 747 | O | 0·47 | <0·1 |
| *Mid-Atlantic Ocean, Vema fracture zone* | | | | | |
| 10°53·55′N 44°02·57′W | 100 | 5 168 | R | 0·87 | 184·6 |
| 10°53·55′N 44°02·57′W | 230 | 5 168 | R | 1·00 | 331·3 |
| 10°53·55′N 44°02·57′W | 478 | 5 168 | O | 0·51 | 160·7 |
| *Bandaras Bay, Mexico* | | | | | |
| 20°39·9′N 105°16·1′W | surface | 94 | R | 1·64 | 254 |
| *Glastonbury peat, Avalon, England* | | | | | |
| 51°09′N 2°45′W | — | — | R | 53·3 | 3 010·0 |

[a] R = reducing; O = oxidizing.

abundant. Perylenequinones are biosynthesized by fungi by oxidative dimerization of naphthalene precursors. A group of perylenequinone pigments, the aphins, has been isolated from 20 species of aphids or plant lice (Insecta: Family Aphididae). Eight-ring quinones are elaborated by several fungi and some higher plants (e.g., hypericin from St John's wort, *Hypericum perforatum*) through dimerization of anthraquinones. Such aromatic quinones are readily reduced to the corresponding hydroquinones and to the parent PAH with further reduction. Conditions suitable for the reductive synthesis of PAH from biogenic extended quinones may occur frequently in highly anaerobic basin sediments and waterlogged soils. Orr and Grady (1967) and Aizenshtat (1973) analyzed the hydrocarbons in several marine sediments (Table 1). In basin floor sediments off southern California, USA, where water depth varied from 600 to 2060 m, perylene was found at concentrations as high as 260 $\mu$g/kg (parts per billion) dry sediment (Santa Barbara Basin) and represented 0·04 to 0·20 per cent of the total hydrocarbons isolated from sediments. Perylene concentrations increased with depth in the sediment, precluding a direct depositional origin. No perylene was detected in sediments from the wall of the basin, in tar from nearby oil seeps or in basin floor sediments from the Tanner or San Clemente Basins. Of the California basins, the Santa Barbara Basin had the highest sedimentation rate, shortest water column depth and lowest oxygen content in bottom waters. The anaerobic zone, negative redox potential ($-Eh$) extended to the surface of the sediment. Even higher concentrations of perylene were found in sediments from Saanich Inlet and the Vema fracture zone. High concentrations of perylene have also been detected in anaerobic freshwater sediments (Rose, 1977; Wakeham, 1977). It was hypothesized by these authors that perylene is formed in anaerobic sediments through reduction of perylenequinones and related extended quinones of biological origin which are sensitive to oxidative destruction during sedimentation. Benzo[*ghi*]perylene and coronene, which are also sometimes found at high concentrations in recent anaerobic sediments, may be formed by processes similar to those described (Meinschein, 1959). Sedimentary PAH assemblages formed in this way are typically quite simple in composition.

### 2.1.2 Complete Biosynthesis by Bacteria

Considerably more attention has been paid to the possibility of complete biosynthesis of PAH by microorganisms and plants. Knorr and Schenk (1968) grew several species of bacteria on highly purified nutrient agar and glycerin–grape-sugar agar and determined the concentrations of benzo[*a*]pyrene (B*a*P) that accumulated in the bacterial cells. Eight strains of saprophytic *Mycobacterium smegmatis* contained 2·0–4·8 μg B*a*P/100 g dry weight. Similar cultures of *Escherichia coli*, *Proteus vulgaris*, *Pseudomonas fluorescens*, *Serratia marcescens*, and *Myxobacteria* Z6 contained 2·0–6·0 μg B*a*P/100 g dry weight. The authors concluded that B*a*P and possibly other fluorescent PAH were synthesized by the bacteria. Brisou (1969) cultured the anaerobic bacteria *Welchia* sp. and *Clostridium putride* in a reducing thioglycollate medium with different lipid substrates for 15 to 20 days. No B*a*P or only traces were detected in cultures grown in the unsupplemented thioglycollate medium and in the media supplemented with pentadecanoic acid, heptadecanoic acid, laneicosane, cholesterol, and *p*-aminobenzoic acid. Low concentrations of B*a*P (0·0025–0·06 μg/100 g dry weight) were detected in the bacteria grown in the presence of nonadecanoic acid, mixed estrogens, and sodium oleate. Lima-Zanghi (1968) and Mallet *et al.* (1967) previously reported that *Clostridium putride* was able to synthesize B*a*P from a mixed lipid fraction of marine plankton (Table 2). In some experiments, the researchers measured B*a*P in the total culture while in others it was measured in harvested dried bacteria. Cultures containing *C. putride* plus the plankton extract contained more B*a*P than those containing extract alone after 15 days. Plankton extracts inoculated with the aerobic bacteria *Phytobacterium*, *Pseudomonas*, or *Flavobacterium* contained less B*a*P than uninoculated plankton extracts, indicating B*a*P degradation by aerobes. *Clostridium putride* accumulated up to 800 μg B*a*P/100 g dry weight during culture in the presence of a plankton lipid extract that initially contained 5·8 μg B*a*P/100 g. The authors considered these results to be evidence of B*a*P biosynthesis by the anaerobe, *C. putride*; they did not consider the possibility that the bacteria might accumulate B*a*P to high concentrations from very low or even undetectable levels in the culture medium. Mallet and Tissier (1969) cultured *Clostridium putride* and *Escherichia coli* (facultative anaerobe) in forest soil which had been dry sterilized at 170 °C. After six months the complete cultures contained 1·95–5·1 μg B*a*P/100 g. However, uninoculated control soils contained small amounts of B*a*P ranging from traces to 1·40 μg/100 g, suggesting that sterilization conditions were sufficient for the pyrosynthesis of a small amount of B*a*P. In a second experiment soil was sterilized in an autoclave at 120 °C and no B*a*P was detected in control soil samples. Autoclaved soil samples inoculated with *C. putride* contained 0·0032–0·0085 μg B*a*P/100 g after four months.

B*a*P biosynthesis by planktonic aerobic bacteria was investigated by Niaussat *et al.* (1969) and Niaussat (1970). They cultured the marine

TABLE 2

BIOSYNTHESIS OF BENZO[*a*]PYRENE (B*a*P) FROM MARINE PLANKTON LIPIDS BY AEROBIC AND ANAEROBIC BACTERIA (FROM MALLET *et al.*, 1967, WITH PERMISSION OF THE PUBLISHER)

| *Lipid extract* | *BaP concentration*[a,b] |
|---|---|
| Plankton 42 | 12·7[a] |
| Plankton 42 + *Flavobacterium* (aerobic) | 0·4[a] |
| Plankton 148 | ND |
| Plankton 148 + *Clostridium* (anaerobic) | 0·2[a] |
| Plankton 186 | traces[a] |
| Plankton 186 + *Phytobacterium* (aerobic) | ND |
| Plankton 49 | 4·95[a] |
| Plankton 49 + *Pseudomonas* (aerobic) | ND |
| Plankton 213 | ND |
| Plankton 213 + *Clostridium* (anaerobic) | 100[b] |
| Plankton 42 | 12·7[b] |
| Plankton 42 + *Clostridium* (anaerobic) | 107[b] |
| Plankton 211 | traces[b] |
| Plankton 211 + *Clostridium* (anaerobic) | 16[b] |
| Plankton 43 | 5·8[b] |
| Plankton 43 + *Clostridium* (anaerobic) | 800[b] |

[a] Expressed in μg/total culture.
[b] Expressed in μg/100 g dry bacteria.
ND, not detected.

planktonic algae *Coccolithus* sp., *Isochrysis* sp. and *Platymonas* sp. in an Erd–Schreiber medium and measured concentrations of B*a*P in the cells and culture medium at different times for a period of ten months. No B*a*P was detected in the medium or algal cells upon initial sampling or after one or two months. After four months 0–0·115 μg B*a*P/100 g was detected in the culture medium and 0·07–0·116 μg B*a*P/100 g was detected in the cells. B*a*P concentrations were highest in *Platymonas* cultures and lowest in *Isochrysis* cultures and generally dropped to lower levels after ten months. The four- and ten-month cultures contained bacterial contamination. To test the hypothesis that the appearance of B*a*P in the algal cultures was due to bacterial biosynthesis, the researchers cultivated an isolated bacterium, *Alcaligenes*, in the absence of algae. No B*a*P was detected in the culture after 24 h, but after one month the B*a*P concentration rose to 0·40 μg B*a*P/100 g. In addition, the alga *Platymonas pagilis* was cultured in the presence and absence of penicillin. After one month cultures without penicillin contained 0·18–0·20 μg B*a*P/100 g while no B*a*P was detected in cultures containing penicillin. The authors concluded that the accumulation of B*a*P in marine algal cultures was due to biosynthetic activity of associated aerobic bacteria.

Niaussat *et al.* (1970) studied the effect of different organic substrates on biosynthesis of B*a*P and perylene by the gram-positive aerobic-facultative anaerobic bacterium *Bacillus badius*. The experimental substrates lycopene

TABLE 3

CONCENTRATIONS OF PERYLENE AND BENZO[*a*]PYRENE IN 186-h CULTURES OF *Bacillus badius* IN PEPTONE BROTH WITH ADDED ORGANIC SUBSTRATES. CONCENTRATIONS ARE GIVEN IN μg/liter OF CULTURE. VALUES ARE MEAN CONCENTRATIONS ±SD, $n = 4$ (FROM NIAUSSAT *et al.*, 1970, WITH PERMISSION OF THE PUBLISHER)

| *Compound* | *Control (PB)*[a] | *PB + lycopene* | *PB + β-carotene* | *PB + sodium acetate* | *PB + naphthalene-acetic acid* | *PB + vitamin* $K_1$ |
|---|---|---|---|---|---|---|
| Benzo[*a*]pyrene | 0·084 | ±0·124 | ±0·098 | ±0·059 | ±0·135 | 0·119 |
| | ±0·001 3 | ±0·028 | ±0·023 | ±0·001 1 | ±0·005 4 | ±0·027 |
| Perylene | 0·30 | 0·47 | 0·31 | 0·25 | 0·415 | 0·365 |
| | ±0·060 5 | ±0·064 | ±0·075 | ±0·015 | ±0·088 | ±0·063 |

[a] PB, Peptone broth.

(a plant carotene), β-carotene, sodium acetate, or water-soluble vitamin $K_1$ were added to the culture medium at a concentration of 10 mg/liter. Naphthalene-acetic acid was added at 0·1 mg/liter. B*a*P was undetectable in the peptone broth culture medium before addition of the *B. badius* culture and rose to 0·015 μg B*a*P/liter after 18-h culture (Table 3). After seven days perylene was present in all cultures at a concentration 3–4 times higher than that of B*a*P. Addition of lycopene, β-carotene, naphthalene-acetic acid, or vitamin $K_1$ to the peptone culture medium resulted in a small increase in the concentration of both perylene and B*a*P in the cultures. Sodium acetate decreased apparent B*a*P and perylene biosynthesis.

Mallet *et al.* (1972a) and Brisou (1972) reviewed the data discussed above and concluded that (1) biosynthesis of B*a*P and certain other PAH by some anaerobic and possibly also aerobic bacteria does occur, (2) only certain strains of bacteria can synthesize PAH, and (3) starting substrates for bacterial PAH biosynthesis can include fatty acids, sterols, plant pigments, and aliphatic terpenes.

However, Hase and Hites (1976) recently questioned the biosynthesis of PAH by marine anaerobic bacteria. Mixed anaerobic bacterial cultures obtained from Charles River (Boston, Massachusetts, USA) sediment were grown in a chemically defined medium supplemented with dextrose as the carbon source. Previous chemical analyses of this anoxic sediment revealed the presence of PAH with at least 11 different aromatic ring structures and abundant alkyl substituted PAH, in some cases containing up to 15 carbon atoms in the alkyl groups (Hites and Biemann, 1975). Modern gas chromatographic–mass spectrometric (GC–MS) techniques revealed the presence of several PAH in the extract of both the mixed culture of anaerobic bacteria and in the medium blank (consisting of the complete chemical culture medium without bacteria, extracted and analyzed in the same way as the bacterial culture). The abundances of these PAH were estimated to be 10–1000 nanograms (ng) per component and their total abundance in the bacteria was about the same as in the blank. High-resolution mass spectrometry revealed that composition and distribution of PAH and their alkyl derivatives were approximately the same in the bacteria as in the medium in which the bacteria were grown. It was concluded that the bacteria did not synthesize the complex PAH assemblage but bioaccumulated it by a factor of about 500 from the culture medium. The investigators suggested that PAH in the medium blank could originate from tap water, test chemicals, or from autoclaving of the dextrose solution at 120 °C.

In comparing their negative results with previous positive findings, Hase and Hites (1976) reasoned that insufficient care was taken in earlier work in determining PAH concentration of the blanks. Failure to consider the possibility of bacterial bioaccumulation of PAH from very dilute solution in the medium could result in a serious misinterpretation of results. Therefore, the question of bacterial biosynthesis of PAH remains unresolved. It could be, as some of the French workers have suggested, that those bacteria able to do so at all require specific organic substrates such as fatty acids to carry out PAH biosynthesis.

### 2.1.3 Complete Biosynthesis by Plants

Several investigators have also attempted to demonstrate the biosynthesis of PAH by lower and higher plants. Borneff *et al.* (1968a,b) studied PAH biosynthesis by the freshwater alga *Chlorella vulgaris*. Algae were cultured for 19 days in the presence of $^{14}C$-acetate which served as the sole organic carbon source. The PAH fraction of the algae was extracted with benzene, purified by $Al_2O_3$ chromatography, separated by thin-layer chromatography (TLC), and the PAH spots identified and quantified by fluorescence spectrometry. Fluorescent spots were eluted from the TLC plate and counted by liquid scintillation. Seven PAH were identified in labeled and nonlabeled algae at concentrations ranging from 0·07 μg (benzo[*a*]pyrene) to 6·20 μg (fluoranthene) in 113·5 g dry algae (Table 4). All PAH eluates from the $^{14}C$-acetate labeled algae contained radioactive counts at least ten times higher than did the corresponding PAH eluates from algae grown in the presence of nonlabeled acetate. Exchange reactions between $^{14}C$-acetate and preformed PAH were negligible. These results very strongly

TABLE 4

PAH CONTENT AND RADIOACTIVITY DETECTED IN THE ALGA *Chlorella vulgaris* GROWN IN THE PRESENCE OF NONRADIOACTIVE ACETATE OR $^{14}C$-ACETATE AS A SOLE ORGANIC CARBON SOURCE (FROM BORNEFF *et al.*, 1968a, WITH PERMISSION OF ACADEMIC PRESS)

| *PAH compound* | *$^{14}C$-labeled* | | | *Nonlabeled* | | |
|---|---|---|---|---|---|---|
| | *PAH* (μg) | *cpm*[a] (*corrected*) | *cpm/μg C* | *PAH* (μg) | *cpm*[a] (*corrected*) | *cpm/μg C* |
| Fluoranthene | 6·20 | 1 973 | 334 | 5·80 | 126 | 23 |
| Benz[*a*]anthracene | 0·65 | 408 | 663 | 0·78 | 30 | 41 |
| Benzo[*b*]fluoranthene | 0·42 | 133 | 333 | 0·39 | 12 | 32 |
| Benzo[*a*]pyrene | 0·08 | 118 | 1 549 | 0·07 | 11 | 165 |
| Benzo[*ghi*]perylene | 0·23 | 934 | 4 247 | 0·22 | 57 | 271 |
| Benzo[*k*]fluoranthene | 0·15 | 158 | 1 106 | 0·14 | 22 | 165 |
| Indeno[1,2,3-*cd*]pyrene | 0·17 | 252 | 1 550 | 0·18 | 19 | 110 |

[a] cpm, counts per minute, corrected for background and external standardization.

suggest that $^{14}C$-acetate served as a metabolic substrate for the *de novo* synthesis of several PAH by the algae. However, contrary to the statement of the authors, the work does not provide unequivocal evidence of PAH biosynthesis by plants. Several of the eluted PAH including benzo[*a*]pyrene, benzo[*ghi*]perylene, benzo[*k*]fluoranthene, and indeno-[1,2,3-*cd*]pyrene had a higher $^{14}C$ specific activity than expected, based on calculations of the amounts of $^{14}C$-acetate originally added, counting efficiency, etc. (485 cpm/μg C expected). Borneff and co-workers suggested that this was due to the presence of labeled nonfluorescent PAH on the chromatogram in the same positions as the identified fluorescent PAH. Another explanation is that the benzene extraction–alumina-column clean-up procedure did not yield a pure PAH fraction. The fraction could contain a wide variety of nonpolar compounds of biological origin (e.g., alkanes, olefins, carotenoids). These compounds would not interfere with the quantitative determination of PAH by the method used; if they migrated with PAH on the TLC plates they could contribute significantly to the radioactivity measured in the eluted PAH spots. Labeling experiments of this type are perhaps the best approach to demonstrating the biosynthesis of PAH, but must be coupled with a method such as high-resolution gas chromatography–mass spectrometry (GC–MS) for unequivocally purifying the PAH fraction and identifying its composition.

TABLE 5

CHANGES IN THE CONCENTRATIONS OF BENZO[*a*]PYRENE DURING GERMINATION OF PLANT SEEDS GROWN HYDROPONICALLY IN A PAH-FREE MEDIUM (FROM GRAF AND DIEHL, 1966, WITH PERMISSION OF THE PUBLISHER)

| *Concentration before growth (μg/100 g dry material)* | | *Concentration after growth (μg/100 g dry material)* | | |
|---|---|---|---|---|
| *Plant* | *BaP* | *Plant* | *BaP (with light)* | *BaP (without light)* |
| Rye seeds | 0·2 | Rye seedlings | 1·8 | 2·6 |
| Rye seeds | 0·05 | Rye seedlings | 3·8 | — |
| Lentil seeds | 0·01 | Lentil seedlings | 2·4 | — |
| Wheat seeds | 0·05 | Wheat seedlings | 0·8 | 1·0 |

Numerous reports of the nearly ubiquitous presence of small amounts of PAH in a wide variety of higher plants led Graf and Diehl (1966) to attempt to demonstrate PAH biosynthesis in several species. Leaves of several species of trees in European areas remote from industrial and human activity were analyzed and the results demonstrated the consistent presence of eight PAH including the carcinogens benzo[*a*]pyrene, benzo[*b*]fluoranthrene, and benzo[*j*]fluoranthrene. The concentration of B*a*P ranged from 10 to 20 μg/kg in dried green leaves and increased three- to five-fold during wilting and yellowing. Seeds, fruits, and bulbs rich in starch contained only 1–10 per cent of the concentrations of B*a*P found in the green parts of the plants. Seeds of rye, wheat, and lentils germinated in a hydroponic (liquid) medium 'free' of PAH showed a significant increase in the concentration of B*a*P during germination even in the absence of light (Table 5). The authors concluded that there is a natural background of 10 to 20 μg B*a*P/kg dry plant material originating from endogenous biosynthesis. Circumstantial evidence supporting the conclusions of Graf and Diel was provided recently by Blum and Swarbrick (1975). In their studies, green beans, cantaloupes, and cotton were grown hydroponically in nutrient-containing aqueous solutions saturated with $^{14}$C-labeled B*a*P. The researchers could not demonstrate a significant uptake of $^{14}$C-B*a*P by the plants and concluded that contamination by B*a*P must be by another route.

Hancock *et al.* (1970) studied the seasonal presence of PAH in the leaves of post oak *Quercus stellata* and little bluestem *Schizachyrium scoparium* at an uncontaminated control station and a station adjacent to a railroad right-of-way in the vicinity of Bryan, Texas, USA. The experimental station was located in an area where it appeared the sole source of PAH contamination would be from diesel locomotive exhaust emissions. The investigators felt the control station was located sufficiently far from the railroad right-of-way, 3·2 km, to preclude PAH contamination from this source. It was located, however, 0·8 km away from a small airport and a construction site where diesel earth-moving equipment operated. Five PAH were identified and quantified in leaves sampled. Individual PAH concentrations ranged from 5 to 110 $\mu$g/kg in dry plant material. At most sampling times, specimens from the control area had higher PAH concentrations than those from the railroad right-of-way. Individual and total PAH concentrations were highest in dormant plants (winter) and lowest during periods of active plant growth (spring). Based on evidence obtained the authors concluded that most of the PAH found in leaf material were products of plant biosynthesis and not of locomotive exhaust.

Grimmer and Duvel (1970) provided convincing testimony that much of the apparent plant biosynthesis of PAH reported by Graf and Diehl (1966) and Hancock *et al.* (1970) could be explained as surface contamination of plant material by airborne particulate PAH. Lettuce, soybeans, rye, and tobacco plants were raised in an open field, a greenhouse, and a special environmental room in which the air supply was carefully filtered of all airborne particulate matter. Plants grown in open air and the greenhouse contained similar concentrations of seven PAH (Table 6) while those grown in the environmental chamber did not contain detectable levels of any PAH. Specimens reared under plastic sheets, a method used in some experiments by Graf and Diehl (1966), were contaminated by airborne particles. Despite the elaborate air filtration procedures and other precautions taken in the environmental chamber it was impossible to completely eliminate traces of phenanthrene, anthracene, fluoranthene, and pyrene from chamber air. The above results provide rather convincing evidence that most, if not all, of the 'natural background' of PAH found in plants is due to contamination by airborne particulate PAH. In further

TABLE 6

AMOUNTS OF PAH IN TEST PLANTS AND SEEDS GROWN IN DIFFERENT ENVIRONMENTS FOR 6–40 WEEKS. PAH CONCENTRATIONS IN μg/kg (FROM GRIMMER AND DUVEL, 1970, WITH PERMISSION OF THE PUBLISHER)

| | *Lettuce seeds* | *Lettuce* | *Lettuce* | *Lettuce* | *Lettuce* | *Tobacco* | *Tobacco* |
|---|---|---|---|---|---|---|---|
| *Conditions* | — | *Environmental chamber* | *Environmental chamber* | *Greenhouse* | *Open air* | *Environmental chamber* | *Greenhouse* |
| *Filter* | — | 624*ST* | *A*105*ST* | — | — | *A*105*ST* | — |
| *Weight* (*g*) | 1 000 | 900 | 500 | 300 | 1 000 | 320 | 400 |
| *PAH* | | | | | | | |
| Benzo[*e*]pyrene | 0 | 0 | 0 | 4·2 | 4·3 | 0 | 2·5 |
| Benzo[*a*]pyrene | 0 | 0 | 0 | 3·7 | 4·2 | 0 | 1·8 |
| Perylene | 0 | 0 | 0 | 0·4 | 0·4 | 0 | 0·25 |
| Anthanthrene | 0 | 0 | 0 | 0·2 | 0·2 | 0 | 0 |
| Benzo[*ghi*]perylene | 0 | 0 | 0 | 2·5 | 2·2 | 0 | 1·1 |
| Dibenz[*ah*]anthracene | 0 | 0 | 0 | 0·4 | 0·6 | 0 | Trace |
| Coronene | 0 | 0 | 0 | 0·6 | 0·7 | 0 | 0·1 |

TABLE 6—*contd.*

| | *Rye* | *Rye* | *Rye seeds* | *Soya* | *Soya* | *Soya* |
|---|---|---|---|---|---|---|
| *Conditions* | *Environmental chamber* | *Greenhouse* | *Open air* | *Environmental chamber* | *Environmental chamber* | *Greenhouse* |
| *Filter* | 624*ST* | — | — | *A*105*ST* | *Special* | — |
| *Weight (g)* | 800 | 320 | 1 000 | 150 | 250 | 220 |
| Benzo[*e*]pyrene | 0 | 3·4 | 1·2 | 0 | 0 | 4·3 |
| Benzo[*a*]pyrene | 0 | 1·6 | 0·64 | 0 | 0 | 3·1 |
| Perylene | 0 | 0·9 | 0·10 | 0 | 0 | 0·3 |
| Anthanthrene | 0 | 0·1 | 0·06 | 0 | 0 | Trace |
| Benzo[*ghi*]perylene | 0 | 0·9 | 0·34 | 0 | 0 | 1·5 |
| Dibenz[*ah*]anthracene | 0 | 0·1 | 0·10 | 0 | 0 | 0·1 |
| Coronene | 0 | 0·1 | 0·10 | 0·10 | 0 | 0·4 |

support of this conclusion, Payer *et al.* (1975) reported that the green alga *Scenedesmus acutus* contained lower concentrations of six PAH when cultured extensively in open unpolluted air in Thailand than when cultured in polluted air in Dortmund (GFR). Lognormal distributions of individual carcinogenic PAH concentrations in replicate samples from both sites indicated that the endogenous formation of PAH occurred only to such a limited extent, if at all, that it was completely masked by accumulation of the same substances from the environment.

Two possible modes of PAH biosynthesis have been discussed above. The literature reviewed provides substantial circumstantial evidence for the indirect biosynthesis of certain PAH, notably perylene, through chemical reduction in anoxic sediments of extended quinones and related PAH precursors synthesized by fungi, plants, and animals. If this process occurred the resulting PAH would be buried in anoxic sediments and not be readily available for uptake and translocation by biota. Therefore indirect biosynthesis, while it may be important locally, probably does not contribute significantly to the total environmental burden of PAH except possibly on a geological scale (e.g., the formation of fossil fuels). Support for direct biosynthesis of PAH by bacteria and plants is contradictory. The rates of PAH biosynthesis reported are all very low, generally in the range of a few micrograms PAH or less per kilogram dried biomass. It is likely that much of this apparent PAH biosynthesis is due to contamination from exogenous PAH sources since such contamination was not adequately excluded in most cases. It must be concluded that until convincing evidence to the contrary is provided, biosynthesis contributes little to the global PAH burden, but may represent a significant localized source of PAH under certain conditions (e.g., anoxic sediments rich in plankton lipids). Youngblood and Blumer (1975) predict that PAH assemblages formed by direct or indirect biosynthesis would have a relatively simple composition; this is illustrated when comparing the saturated hydrocarbon fraction in organisms to that of petroleum. Modern analytical methods have revealed that the vast majority of PAH assemblages are far more complex compositionally than could be explained by a biosynthetic origin.

## 2.2 PAH IN FOSSIL FUELS

### 2.2.1 PAH in Coal

The organic matter in coal consists largely of a macromolecular material of complex and variable composition. A large number of complex PAH and heteroaromatic compounds are formed when coal is pyrolyzed. Because of

TABLE 7

AROMATIC ACIDS FOUND IN THE AQUEOUS $Na_2Cr_2O_7$ OXIDATION OF COAL, AND IDENTIFIED AS METHYL ESTERS (FROM HAYATSU *et al.*, 1975, WITH PERMISSION OF MACMILLAN JOURNALS LTD)

| *Nucleus* | *Number of* $-COOCH_3$ | *Relative abundance* (+15–20%) *relative to benzene tricarboxylic acid methyl ester* |
|---|---|---|
| Benzene | 2 | 38 |
| | 3 | 100 |
| | 4 | 64 |
| | 5 | 11 |
| $OCH_3$-Benzene | 1 | 17 |
| | 2 | 17 |
| Naphthalene | 1 | 1 |
| | 2 | 9 |
| | 3 | 27 |
| | 4 | 8 |
| $CH_3$-Naphthalene | 1 | 1 |
| | 2 | 3 |
| | 3 | 5 |
| Phenanthrene | 2 | 20 |
| | 3 | 17 |
| | 4 | 8 |
| $OCH_3$-Phenanthrene | 2 | 8 |
| Fluoranthene | 3 | 4 |
| | 4 | 3 |
| $OCH_3$-Fluorenone | 1 | 1 |
| Anthraquinone | 2 | 8 |
| Dibenzofuran and/or naphthofuran | 1 | 18 |
| | 2 | 35 |
| | 3 | 28 |
| | 4 | 8 |
| Benzonaphthofuran | 2 | 5 |
| | 3 | 7 |

Methoxy compounds ($OCH_3$) may result from the reaction of phenols with diazomethane. Methoxyphenanthrene may be derived from the reaction of fluorene with diazomethane. Anthracene is oxidized to anthraquinone. The pyrene ring is extensively attacked by this procedure yielding fluoranthene and other compounds. Fluorene, methoxyfluorenone and anthraquinone may be oxidized further to benzene carboxylic acids in this procedure.

the complexity of pyrolytic reactions, it is difficult to relate the products formed in this way to the original structure of coal. Most workers in the field consider coal to be predominantly an aromatic material. Wiser (1973) suggested that 70 to 75 per cent of the carbon present in bituminous coal is in the aromatic form. Six-membered rings are dominant but five-membered rings are also present. Hydroaromatic structures contain 15 to 25 per cent of the carbon in coal. Hayatsu *et al.* (1975) identified 35 aromatic carboxylic acids resulting from the oxidation of an Illinois bituminous coal with aqueous $Na_2Cr_2O_7$. This procedure was chosen as it will oxidize substituted PAH in good yields to their corresponding PAH carboxylic acids with minimum ring degredation. The resulting acids weighed 53 per cent of the original weight of the coal. Relative abundances of different aromatic carboxylic acid methyl esters isolated and identified are given in Table 7. Sixteen additional aromatics were isolated but not identified. Recently Woo *et al.* (1978) used X-ray excited optical luminescence (XEOL) spectrometry to identify several PAH in samples of Iowa and Illinois coal. The PAH identified were benz[*a*]anthracene, benzo[*a*]pyrene, benzo[*e*]-pyrene, dibenzo[*cd,m*]pyrene, perylene and phenanthrene.

### 2.2.2 PAH in Petroleum

Crude petroleum and most refined petroleum products are extremely complex mixtures of many thousands of organic compounds. Hydrocarbons are most abundant, usually representing more than 75 per cent of the weight of the oil. The remainder is made up primarily of various sulfur, oxygen and nitrogen containing organic compounds (Speers and Whitehead, 1969). Petroleums from different sources vary tremendously in the relative concentrations of the different hydrocarbon types present. Gilchrist *et al.* (1972) analyzed the PAH distribution in 11 samples of mineral oil. Total aromatic content varied from 7·4 to 34·3 weight per cent of the whole oil (Table 8). Total tri- through to hexacyclic PAH concentrations varied from 0·2 to 7·4 per cent. Jones *et al.* (1977) reported that shale oil and Char-Oil-Energy Development (COED) pyrolysis process syncrude oil contain 5 per cent PAH. Woodward *et al.* (1976) reported concentrations of 7·7 to 14·6 per cent PAH for three samples of synthetic crude oil produced by experimental coal liquification processes. Eight refined oil products contained 0·34 to 3·7 per cent PAH and 12·2 to 37·7 per cent di- and tri-aromatic hydrocarbons (Jewell *et al.*, 1972). In addition, several 6-, 7-, and 8-ring PAH have been identified in high boiling petroleum distillates (McKay and Latham, 1973; Coleman *et al.*, 1973). Thus, crude and refined oils contain significant quantities of PAH.

TABLE 8

DISTRIBUTION OF AROMATIC HYDROCARBON TYPES IN 11 OIL SAMPLES, CORRECTED FOR SULFUR COMPOUNDS NOT DETECTED BY MASS SPECTROMETRY. VALUES IN PER CENT, WT./WT. (FROM GILCHRIST *et al.*, 1972, WITH PERMISSION OF THE SOCIETY FOR ANALYTICAL CHEMISTRY)

| *Compound type* | *Oil sample* | | | | | | | | | | |
|---|---|---|---|---|---|---|---|---|---|---|---|
| | A | B | C | D | E | F | G | H | I | J | K |
| Monocyclics | 20·4 | 11·5 | 17·3 | 10·7 | 6·1 | 6·8 | 9·4 | 5·5 | 9·2 | 7·9 | 6·7 |
| Dicyclics | 11·7 | 14·1 | 14·0 | 9·1 | 5·5 | 7·2 | 7·4 | 1·7 | 7·9 | 4·0 | 4·6 |
| Tricyclics | 1·0 | 3·5 | 2·2 | 4·1 | 1·2 | 3·0 | 1·3 | 0·1 | 2·2 | 0·7 | 0·8 |
| Tetracyclics | 0·2 | 1·7 | 0·7 | 2·5 | 0·9 | 0·2 | 0·7 | <0·1 | 1·8 | 0·3 | 0·2 |
| Pentacyclics | <0·1 | 0·2 | <0·1 | 0·5 | 0·1 | ND | 0·1 | ND | 0·1 | <0·1 | ND |
| Hexacyclics | ND | ND | ND | 0·3 | ND | ND | ND | ND | ND | ND | ND |
| Total aromatics | 33·3 | 31·0 | 34·3 | 27·2 | 13·8 | 17·2 | 18·9 | 7·4 | 21·2 | 13·0 | 12·3 |
| Total PAH (tri- through to hexacyclics) | 1·2 | 5·4 | 3·0 | 7·4 | 2·2 | 3·2 | 2·1 | 0·2 | 4·1 | 1·1 | 1·0 |

ND, not detected.

TABLE 9

CONCENTRATIONS OF PAH IN TWO CRUDE OILS AND TWO DISTILLATE FUEL OILS. THESE OILS ARE THE API REFERENCE OILS (NAPHTHALENE–FLUORENES FROM ANDERSON *et al.*, 1974c; PHENANTHRENE–BENZO[*ghi*]PERYLENE FROM PANCIROV AND BROWN, 1975, WITH PERMISSION OF THE PUBLISHERS)

| *Compound* | *Concentration in mg/kg (ppm)* | | | |
|---|---|---|---|---|
| | *South Louisiana crude* | *Kuwait crude* | *No. 2 fuel oil* | *Bunker C residual oil* |
| Naphthalene | 400 | 400 | 4 000 | 1 000 |
| 1-Methylnaphthalene | 800 | 500 | 8 200 | 2 800 |
| 2-Methylnaphthalene | 900 | 700 | 18 900 | 4 700 |
| Dimethylnaphthalenes | 3 600 | 2 000 | 31 100 | 12 300 |
| Trimethylnaphthalenes | 2 400 | 1 900 | 18 400 | 8 800 |
| Fluorenes | 200 | <100 | 3 600 | 2 400 |
| Phenanthrene | 70 | 26 | 429 | 482 |
| 1-Methylphenanthrene | 111 | — | 173 | 43 |
| 2-Methylphenanthrene | 144 | 89 | 7 677 | 828 |
| Fluoranthene | 5·0 | 2·9 | 37 | 240 |
| Pyrene | 3·5 | 4·5 | 41 | 23 |
| Benz[*a*]anthracene | 1·7 | 2·3 | 1·2 | 90 |
| Chrysene | 17·56 | 6·9 | 2·2 | 196 |
| Triphenylene | 10 | 2·8 | 1·4 | 31 |
| Benzo[*ghi*]fluoranthene | 1 | <1 | | |
| Benzo[*b*]fluoranthene | <0·5 | <1 | | |
| Benzo[*j*]fluoranthene | <0·9 | <1 | | |
| Benzo[*k*]fluoranthene | <1·3 | <1 | | |
| Benzo[*a*]pyrene | 0·75 | 2·8 | 0·6 | 44 |
| Benzo[*e*]pyrene | 2·5 | 0·5 | 0·1 | 10 |
| Perylene | 34·8 | <0·1 | — | 22 |
| Benzo[*ghi*]perylene | 1·6 | <1 | | |

The chemical complexity of petroleum has hampered attempts to identify and quantify specific PAH in oil samples. The four American Petroleum Institute (API) reference oils designated for biological studies have been extensively used in biological effects studies and considerable information is available about their chemical composition (Anderson *et al.*, 1974c; Pancirov, 1974; Pancirov and Brown, 1975; Youngblood and Blumer, 1975). Concentrations of several PAH in these oils are listed in Table 9. These four oils differ substantially in PAH concentrations. Naphthalene and alkyl naphthalenes are usually present at higher concentrations than are the higher molecular weight PAH. South Louisiana crude oil has unusually high concentrations of chrysene

TABLE 10

CONCENTRATIONS OF BENZO[*a*]PYRENE IN CRUDE OILS AND REFINED OIL PRODUCTS

| *Material* | *BaP concentration (mg/kg (ppm))* | *Reference* |
|---|---|---|
| Libyan Brega crude | 1·47[a] | Graf and Winter, 1968 |
| Venezuelan Tia Juana crude | 1·85[a] | Graf and Winter, 1968 |
| Persian Gulf Safanaya crude | 0·44[a] | Graf and Winter, 1968 |
| South Louisiana crude | 0·75 | Pancirov and Brown, 1975 |
| Kuwait crude | 2·8 | Pancirov and Brown, 1975 |
| (Mean B*a*P crude oils) | 1·46 | |
| Diesel oil | 0·029[a] | Graf and Winter, 1968 |
| Fuel oil | 0·029[a] | Graf and Winter, 1968 |
| No. 2 fuel oil | 0·6 | Pancirov and Brown, 1975 |
| Regular gasoline | 0·212 | Sullivan, 1974 |
| Premium gasoline | 0·480 | Sullivan, 1974 |
| Unused motor oil | 0·28 | Gross, 1974 |
| Unused motor oil | 0·029[a] | Graf and Winter, 1968 |
| Motor oil, used for 220 km | 6·4[a] | Graf and Winter, 1968 |
| Bunker C residual oil | 44 | Pancirov and Brown, 1975 |
| Petroleum bitumen | 3–5 | Schamp and van Wassenhove, 1972 |
| Asphalts | ND–27[b] | Wallcave *et al.*, 1971 |

[a] Reported as μg/100 ml. Converted to mg/kg based on assumed specific gravity of 0·9.
[b] Range for 8 asphalts.
ND, not detected.

(17·56 ppm) and perylene (34·8 ppm). No. 2 fuel oil contains a high concentration of 2-methylphenanthrene (7677 ppm) while the other refinery distillate product, Bunder C residual oil contains high concentrations of 2-methylphenanthrene (828 ppm), fluoranthene (240 ppm), chrysene (196 ppm), benzo[*a*]pyrene (44 ppm), and perylene (22 ppm). Typically, the alkyl homologs are present at higher concentrations in crude and refined oils than are the parent PAH (Pancirov and Brown, 1975; Youngblood and Blumer, 1975; Blumer, 1976). This is reflected in the relative concentrations of naphthalene and $C_1$–$C_3$ naphthalenes in the four API oils. Youngblood and Blumer (1975) determined the homolog distribution for the tri-, tetra-, and pentacyclic aromatic fractions of three of the API reference oils, a No. 2 fuel oil similar to the corresponding API reference No. 2 fuel oil, and three shale oils (Fig. 4). The alkyl homolog distribution is different for each PAH series and also for each oil. PAH in the crude oils and shale oils are generally more highly alkylated than those

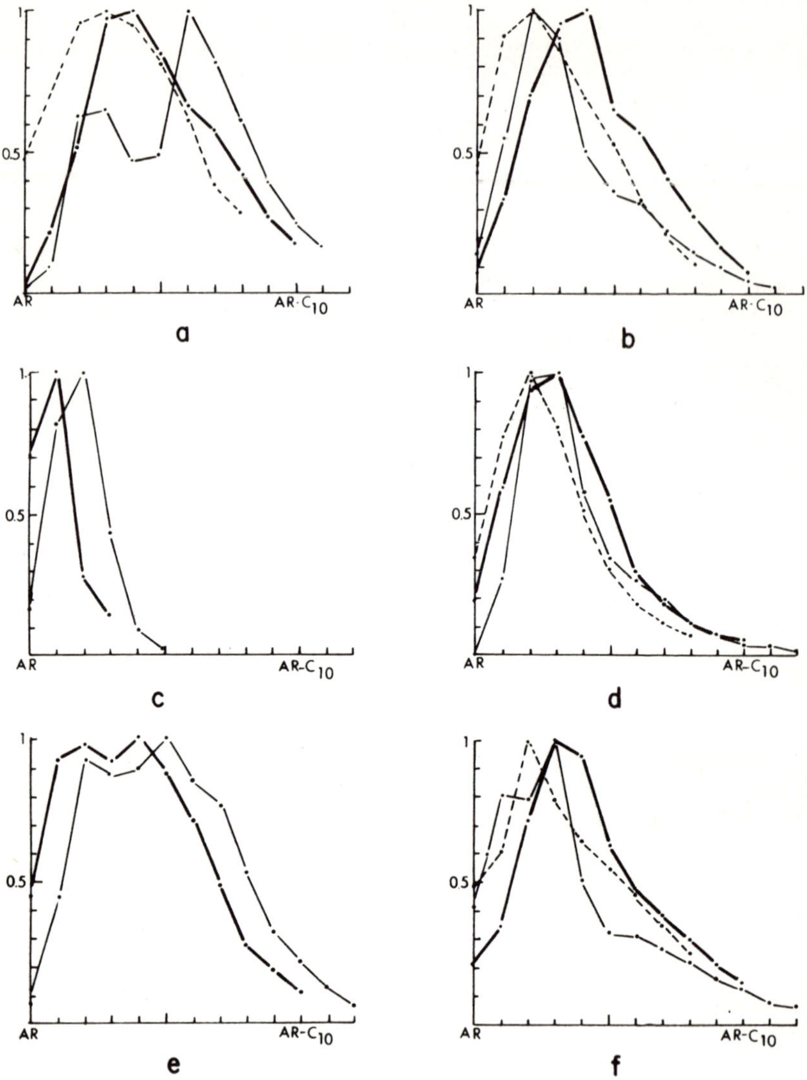


FIG. 4. Molecular weight distribution in the nominal tricyclic (●——●), tetracyclic (●——●), and pentacyclic (●-----●) PAH series in fossil fuels. Curves are normalized to the most abundant species in each series. AR, unsubstituted PAH. a, Kuwait crude oil; b, South Louisiana crude oil; c, No. 2 fuel oil; d, Bunker C residual oil; e, Serpiano shale oil; f, Leo shale oil. (From Youngblood and Blumer, 1975, with permission of Pergamon Press.)

in the furnace heating oils. Two- to four-carbon homologs are generally most abundant and there is a decreasing frequency of PAH with seven or more alkyl carbons.

A similar predominance of alkyl PAH is seen in petroleum asphalt. Wallcave *et al.* (1971) analyzed several PAH and their alkyl homologs in eight samples of asphalt. B*a*P concentrations ranged from 0·1 to 27 ppm while the concentration of total alkyl B*a*P ranged from 1·0 to 69 ppm. The ratio of alkyl B*a*P to B*a*P varied from 2·6 to 10·0.

Because so much more is known about the environmental distribution and fate of B*a*P than about other PAH, it would be informative to compare the B*a*P concentrations of different crude and distillate oil products (Table 10). The mean concentration of B*a*P in five crude oils is 1·46 ppm. B*a*P concentrations in refined petroleum products are quite variable. The lower boiling petroleum fractions, even though formed by high temperature catalytic cracking, contain smaller concentrations of B*a*P than the crude oils. Only the heavy petroleum fractions, residual oil, bitumen, and asphalt contain significantly higher B*a*P concentrations than the crude oils.

## 2.3 FORMATION OF PAH BY PYROLYSIS OF ORGANIC MATTER

It is generally agreed that the majority of PAH in the environment is derived during incomplete combustion of organic matter at high temperatures (Suess, 1976). Several mechanisms have been proposed for the formation of PAH by pyrolysis and pyrosynthesis (Badger, 1962; Schmeltz and Hoffmann, 1976). In pyrolysis, complex organic molecules are partially cracked to lower molecular weight free radicals. Pyrosynthesis of PAH then proceeds by the rapid combination of free radicals containing one, two, or many carbons (Badger *et al.*, 1960). Pyrolysis of styrene-like ($\phi$-CH═CH—R) compounds similar to caffeic acid found in tobacco smoke yields PAH which could be formed from a biphenyl cyclobutane intermediate, which is a dimerization product of styrene (Schmeltz and Hoffmann, 1976). Cyclic compounds, both saturated and unsaturated, are excellent PAH precursors. Schmeltz and Hoffman (1976) proposed a scheme for the pyrosynthesis of naphthalene and higher molecular weight PAH from benzene. At high temperatures in the presence of air, benzene is converted to phenol which upon loss of water would be converted to 'benzyne'. The latter, through a 1,4-addition across benzene, gives rise to

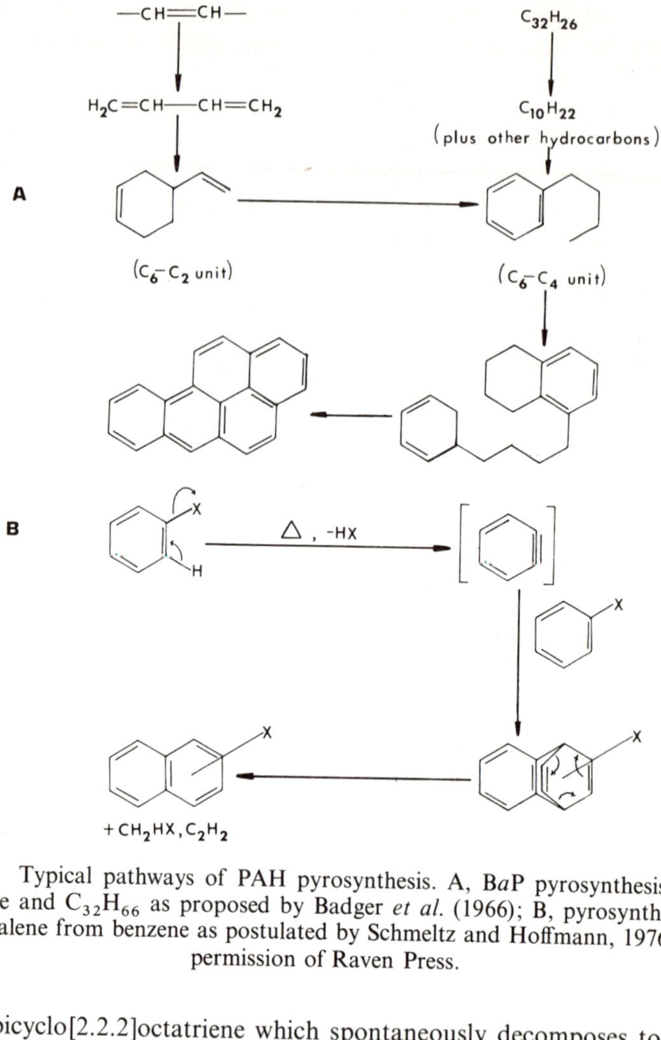


FIG. 5. Typical pathways of PAH pyrosynthesis. A, B*a*P pyrosynthesis from isoprene and $C_{32}H_{66}$ as proposed by Badger *et al.* (1966); B, pyrosynthesis of naphthalene from benzene as postulated by Schmeltz and Hoffmann, 1976, with permission of Raven Press.

benzobicyclo[2.2.2]octatriene which spontaneously decomposes to acetylene and naphthalene. Similarly, naphthalene could give rise to 'naphthalynes' and ultimately to higher PAH (Fig. 5). Thus, pyrolytic degredation to simple $C_1$ and $C_2$ precursors is not necessary for PAH pyrosynthesis.

Nearly all the airborne PAH produced by flame pyrolysis are associated with the particulate fraction produced during combustion (soots or carbon blacks). Soot can be thought of as an agglomerate of pericondensed PAH

(Thomas *et al.*, 1968). The fundamental chemical unit of a soot particle is thought to be a $C_{150}H_{30}$ molecule of 61 fused rings or a $C_{216}H_{36}$ molecule of 91 fused rings. These highly condensed aromatic systems in the soot particle are essentially inert. PAH of environmental concern are primarily adsorbed on the surface of the soot through hydrogen bonding. As indicated above, pyrolysis of organic matter produces a variety of polybenzenoid radicals. Continued polymerization of these radicals results in soot. Early termination of this polymerization by hydrogen atoms results

TABLE 11

PRODUCTION OF BENZO[*a*]PYRENE DURING PYROLYSIS OF DIFFERENT BIOORGANIC MATERIALS AT 840 °C IN $N_2$ (FROM SCHMELTZ AND HOFFMANN, 1976, WITH PERMISSION OF RAVEN PRESS)

| *Substance pyrolyzed* | *μg BaP/g pyrolyzed* |
|---|---|
| Glucose | 47·5 |
| Fructose | 98·4 |
| Cellulose | 288·8 |
| Stearic acid | 1 200 |
| Dotriacontane | 3 130 |
| β-Sitosterol | 3 750 |

in enhanced PAH production. Thus, the presence of a chemically reducing atmosphere favors PAH pyrosynthesis. Commins (1969) showed that as the ratio of fuel to air increases in the combustion mixture, the mixture gets richer, the concentrations of pyrene, fluoranthene, benzo[*a*]pyrene, and benzo[*e*]pyrene tend to increase while the concentrations of the higher molecular weight benzo[*ghi*]perylene, anthanthrene, and coronene decrease.

Several other factors affect the yield and molecular weight distribution of PAH formed by pyrosynthesis. These include the chemical composition of the fuel, pyrolysis temperature, duration of exposure to elevated temperature, and presence in the pyrolysis mixture of C,H-radical scavengers. The amounts of B*a*P produced by pyrolysis of several bioorganic compounds are listed in Table 11. Carbohydrates are relatively poor PAH precursors compared to more highly reduced compounds such as lipids. Straight-chain saturated compounds such as stearic acid and dotriacontane are even better PAH precursors. Chain length has little effect on the amount of PAH produced. Pyrolysis of substances displaying chain branching or unsaturation results in a substantial increase in the production of PAH, including alkyl PAH. Phytosterols such as $\beta$-sitosterol and branched-chain isoprenoid alcohols such as solanesol give high yields of PAH. Sterols are excellent precursors for naphthalenes, anthracene, phenanthrene, chrysene, and B*a*P (Schmeltz and Hoffmann, 1976). Similar variations are seen in the yield of PAH from the combustion of petroleum hydrocarbons (Table 12). Aliphatic hydrocarbons are relatively poor PAH precursors compared to gasoline (about 30 per cent aromatic hydrocarbons) and more so in comparison to aromatic hydrocarbons (Brunnemann and Hoffmann, 1976).

TABLE 12

BENZO[*a*]PYRENE AND BENZ[*a*]ANTHRACENE EMISSION RATES DURING COMBUSTION OF HYDROCARBON FUELS (1-MIN RUN) (FROM BRUNNEMANN AND HOFFMANN, 1976, WITH PERMISSION OF RAVEN PRESS)

| *Fuel* | *μg BaP* | *μg BaA* |
|---|---|---|
| Gasoline | 9·6 | 17·3 |
| 2,2,4-Trimethylpentane | 2·6 | 0·8 |
| 2,4,4-Trimethyl-1-pentane | 0·7 | 0·4 |
| 50% *o*-xylene–50% benzene | 25·8 | 56·3 |
| 25% 2,2,4-Trimethylpentane + 25% 2-methylbutane + 50% *o*-xylene | 9·0 | 32·3 |

Pyrolysis temperature plays an important role in PAH generation. The temperature optima depend on the nature of the fuel and the type of PAH formed. The optimum temperature for pyrosynthesis of B*a*P from n-butylbenzene is 700 °C and for benzo[*b*]fluoranthene production, 750 °C (Badger *et al.*, 1964). Above and below the temperature optima, yields of these PAH drop sharply. Pyrosynthesis of pyrene from tobacco terpenoids occurs at an optimum greater than 900 °C (Schmeltz and Hoffmann, 1976). Commins (1969) studied the effects of temperatures between 550 and 1000 °C on the pyrosynthesis of 12 PAH from styrene. Maximum yields of all PAH ranging in molecular weight from naphthalene to coronene were highest at 780 °C, decreasing at higher and lower temperatures.

In most cases PAH ratios do not vary substantially with pyrolysis temperature. Blumer (1976) and Youngblood and Blumer (1975) have suggested that the relative abundance and distribution of alkyl-substituted PAH is highly dependent on the temperature at which the PAH mixture was formed. At very high temperatures, near 2000 °C, for example in coal coking, the pyrolytic products consist of a relatively simple mixture of unsubstituted PAH. At lower temperatures, as in open flames (400–800 °C), significant quantities of alkyl-substituted PAH are formed. Long alkyl chains are not favored and unsubstituted PAH outnumber substituted ones. As pyrolysis temperature decreases, the ratio of alkyl-substituted to unsubstituted PAH increases. Crude oils, which are usually formed at temperatures below 150 °C, have a predominance of substituted over unsubstituted PAH. If Blumer's hypothesis is correct, then the source of an environmental PAH assemblage can be inferred from analysis of its alkyl homolog distribution.

Duration of exposure of organic precursors to pyrolysis conditions may also have a significant effect on the yield and composition of the PAH mixture produced. However, this aspect of PAH pyrosynthesis has not been adequately studied (Radding *et al.*, 1976). Pyrolytic reactions carried out at high temperatures are essentially instantaneous (Thomas *et al.*, 1968) while the molecular rearrangements that occur during the formation of petroleum require millions of years. Extended residence time at high temperature may favor dealkylation of alkyl PAH and destruction of less stable linear PAH (Blumer, 1975).

### 2.3.1 Sources of Pyrolytic PAH

A great many domestic and industrial activities result in pyrosynthesis of PAH. Resulting PAH may be released to the environment in airborne particulates or in the solid or aqueous byproducts of the pyrolytic process. The vast majority of research to date has concentrated on airborne particulate PAH emissions because of the obvious hygienic problem of inhalation carcinogenesis. Relatively little is known about the concentration or composition of PAH in solid wastes and liquid effluents from various industrial processes.

PAH of pyrolytic origin are ubiquitous in the human environment. Charcoal-broiled steak contains 11–50 $\mu$g/kg (ppb) B*a*P (Lijinsky, 1967). Panalaks (1976) analyzed 70 samples of commercially available smoked food products and seven samples of charcoal-broiled meats. Seventy per cent of the smoked food products contained PAH. Total PAH ranged from 0–141 ppb and individual PAH from 0–38 ppb. In the charcoal-broiled meats, total PAH and individual PAH ranged from 0–164 ppb and 0–60 ppb, respectively.

Conditions are ideal for PAH pyrosynthesis within a cigarette flame (Schmeltz and Hoffmann, 1976). Temperatures in the flame zone range from ambient to 900 °C and a reducing atmosphere exists in hot regions directly behind the oxidative zone in the burning cone of a cigarette. PAH are concentrated in smoke particulates which represent less than 8 per cent

TABLE 13

CONCENTRATIONS OF SELECTED PAH IN CIGARETTE SMOKE CONDENSATE FROM KENTUCKY REFERENCE 1R1 CIGARETTES (FROM SEVERSON *et al.*, 1976, WITH PERMISSION OF RAVEN PRESS)

| *Compound* | *Concentration* (*μg/100 cigarettes*) |
|---|---|
| Phenanthrene | 36·2 |
| Anthracene | 17·3 |
| Methylphenanthrene and methylanthracene | 149·4 |
| Fluoranthene | 12·3 |
| Pyrene | 16·2 |
| Methylpyrene | 20·3 |
| Benz[*a*]anthracene, chrysene and triphenylene | 7·6 |
| Benzo[*a*]pyrene and benzo[*e*]pyrene | 2·4 |
| Perylene | 0·3 |
| Indeno[1,2,3-*cd*]pyrene | 0·6 |
| Benzo[*ghi*]perylene | 0·5 |
| Anthanthrene | 0·3 |
| Methylbenzo[*ghi*]perylene | 0·5 |
| Coronene | 0·1 |

TABLE 14

ESTIMATED BENZO[*a*]PYRENE EMISSIONS FROM INTERNAL COMBUSTION ENGINES OF VEHICLES IN THE USA BEFORE THE INSTITUTION OF EMISSION CONTROLS AND DECREASE OF EMISSIONS IN EMISSION-CONTROLLED VEHICLES (REPRODUCED FROM *Particulate Polycyclic Organic Matter* (1972), p. 16, WITH PERMISSION OF THE NATIONAL ACADEMY OF SCIENCES, WASHINGTON, D.C.)

| *Vehicle type* | *Fuel consumed* (*gal/year*) | *BaP emission factor* (*μg/gal*) | *BaP emission* (*tons/year*) |
|---|---|---|---|
| Gasoline-powered | | | |
| automobiles | $56{\cdot}4 \times 10^9$ | 170 | 10 |
| trucks | $24{\cdot}2 \times 10^9$ | ~500 | ~12 |
| Diesel-fuel-powered | | | |
| trucks and buses | $5{\cdot}8 \times 10^9$ | 62 | 0·4 |
| Total | | | ~22 |
| 1968 emission-controlled vehicle | | 20–30 | |
| advanced systems | | ≤10 | |

of the total smoke. Considerable research has been conducted on the PAH composition of cigarette smoke condensate. An example of a typical analysis is given in Table 13. Low molecular weight PAH are present at higher concentrations than high molecular weight forms. Alkyl homologs seem to be present at higher concentrations than the parent compounds. Schmeltz and Hoffmann (1976) have shown that the hexane-extractable lipoidal fraction which represents 6 per cent of the dry tobacco leaf weight accounts for 60 per cent of B*a*P produced by pyrolysis of tobacco.

Burning of fossil fuels is an important source of PAH in the environment. The particulate fraction of vehicular exhaust may contain significant quantities of B*a*P and other PAH (Table 14). Before the adoption of exhaust emission strategies in the USA, annual emission of B*a*P to the environment from gasoline- and diesel-powered vehicles was estimated to be about 22 tons per year (NAS, 1972). Gasoline engines have a higher B*a*P emission factor than diesel engines. A large number of PAH, in addition to B*a*P, are present in automobile exhaust tars (Table 15). Particularly characteristic of gasoline exhaust emissions is the approximately 4:1 ratio of B*e*P/B*a*P and the high concentrations of pyrene and coronene. B*a*P emission factors have been reduced by more than 90 per cent through modern emission-control strategies.

Many heat and electrical generating facilities burn fossil fuels and produce as byproducts liquid, solid, and gaseous wastes that may be rich in PAH. Hangebrauck *et al.* (1964), Sawicki *et al.* (1965), and others have measured airborne PAH emissions from heat generation sources that utilize different fossil fuels and firing methods (Table 16). Coal-fired heating plants, especially those using old fashioned hand-stoked burners, generally emit larger amounts of PAH than do oil- and gas-fired units. However, there is considerable variation in PAH emissions even from units using the same fuel and firing methods.

Catalytic cracking of crude petroleum to produce hydrocarbon fuels and other refined petroleum products results in the production of PAH. As indicated earlier, much of the PAH produced becomes concentrated in the high boiling residual oil (Bunker C or No. 6 fuel oil) and asphalts. Significant quantities of PAH may also be released in the catalyst regeneration flue gas (Table 17). If discharged uncleaned, these flue gases would represent a substantial route of PAH input to the airborne particulate PAH load. However, as denoted in Table 17, an incinerator-type air-cleaning system removes 99·96 per cent of the PAH from the flue gas.

Coke is an important high energy fuel extensively used in the iron and

TABLE 15

CONCENTRATIONS OF SEVERAL PAH IN THE EXHAUST OF GASOLINE ENGINES BEFORE ADOPTION OF HYDROCARBON EMISSION STANDARDS FOR VEHICLES (FROM HOFFMANN AND WYNDER, 1963, WITH PERMISSION OF THE *Journal of the Air Pollution Control Association*)

| *PAH* | *Exhaust tar concentration (mg/kg)* | *Concentration (μg/1-min run)* |
|---|---|---|
| Benz[*a*]anthracene | 61·7 | 4·2 |
| Alkylbenz[*a*]anthracene | 2·0 | 0·14 |
| 11-H-Benzo[*b*]fluorene | 51·0 | 3·5 |
| Fluoranthene | 972 | 60·0 |
| Alkylfluoranthene | 261 | 17·9 |
| Benzo[*b*]fluoranthene | 64 | 4·4 |
| Benzo[*j*]fluoranthene | 17 | 1·2 |
| Benzo[*k*]fluoranthene | 54 | 3·7 |
| Benzo[*ghi*]fluoranthene | 150 | 10·2 |
| Chrysene | 175 | 12·0 |
| Alkylchrysene | 67 | 4·6 |
| Pyrene | 1 770 | 122·0 |
| Alkylpyrene | 8·0 | 0·06 |
| Triphenylene | 44·0 | 3·0 |
| Dibenz[*ah*]anthracene | 2·5 | 0·16 |
| Benzo[*a*]pyrene | 31·5 | 2·2 |
| Benzo[*e*]pyrene | 422·0 | 29·0 |
| Alkylbenzo[*a*]pyrene | 1·0 | 0·07 |
| Indeno[1,2,3-*cd*]pyrene | 82·0 | 5·7 |
| Indeno[1,2,3-*cd*]fluoranthene | 3·0 | 0·18 |
| Perylene | trace | trace |
| Anthanthrene | 44·0 | 3·1 |
| Benzo[*ghi*]perylene | 51·0 | 3·5 |
| Coronene | 338 | 23·2 |
| Total PAH measured | 4 671·7 | 314·01 |

steel industry. Approximately 87 × 10⁶ tons of bituminous coal were processed to produce 60 × 10⁶ tons of coke in 1972 (Guerin, 1977). The production of coke involves exposing hard coal to high temperatures (~1400 °C) in a reducing atmosphere, conditions ideal for pyrosynthesis of PAH. Therefore, it is not surprising that airborne particulate emissions from the coking process contain high concentrations of PAH. Lao *et al.* (1975) performed a detailed computerized gas chromatographic–mass spectrometric (GC–MS) analysis of PAH in four air-filter samples of gaseous coke oven emissions. Sixty-five PAH were identitied and quantified

TABLE 16

AIRBORNE PAH EMISSIONS FROM HEAT GENERATION SOURCES. CONCENTRATION IN μg/million BTU HEAT INPUT (FROM HANGEBRAUCK *et al.*, 1964, WITH PERMISSION OF THE *Journal of the Air Pollution Control Association*)

| *Source No.* | *Fuel used* | *Firing method* | *Benzo[a]pyrene*[a] | *Pyrene* | *Benzo[e]pyrene* | *Perylene* | *Benzo[ghi]perylene* | *Anthanthrene* | *Coronene* | *Anthracene* | *Phenanthrene* | *Fluoranthene* | *Benz[a]anthracene* |
|---|---|---|---|---|---|---|---|---|---|---|---|---|---|
| 1 | Coal | Pulverized | 19 | 150 | | | | | | | | 180 | 19 |
| 2 | | | 32 | 240 | 92 | | | | | 370 | | 550 | |
| 3 | | Chain grate stoker | 37 | 390 | 130 | | | | | | | 680 | |
| 4 | | Spreader stoker | 26 | 590 | 350 | | | | 26 | | | 360 | |
| 5 | | Underfeed stoker | 10 000 | 16 000 | 7 900 | 1 600 | 4 500 | 290 | 330 | 850 | 10 000 | 38 000 | 3 900 |
| 6 | | | 120 | 1 700 | 230 | | | | | | 1 000 | 3 200 | |
| 7 | | | 3 800 | 7 700 | 5 400 | | 580 | | 1 200 | | 29 000 | 47 000 | 560 |
| 8 | | Hand stoked | 400 000 | 600 000 | 100 000 | 60 000 | 300 000 | 90 000 | 30 000 | 400 000 | 1 000 000 | 1 000 000 | |
| 9 | Oil | Steam atomized | <20 | 49 | | | | | | | | 56 | |
| 10 | | | 47 | 300 | | | | | | | 1 800 | 270 | 27 |
| 11 | | Low pressure air atomized | 900 | 6 100 | | | 300 | | 2 100 | 3 900 | 3 500 | 1 900 | |
| 12 | | Centrifugal atomized | <40 | 1 800 | | | | | | | 8 900 | 5 000 | |
| 13 | | | <60 | 15 | | | | | | | | 76 | |
| 14 | | Vaporized | <100 | 1 200 | | | | | | | | 15 000 | |
| 15 | Gas | Premix burner | <20 | 160 | | | | | 14 | | | 100 | |
| 16 | | | 200 | 18 000 | 490 | | 1 800 | 200 | 5 300 | | | 2 900 | |
| 17 | | | <20 | 170 | | | | | | | | 320 | |
| 18 | | | <20 | 120 | 18 | | | | | | 77 | 110 | |

[a] 'Less than' values for benzo[*a*]pyrene were calculated for those samples having concentrations below the limit of quantitative determination (approximately 0·6 μg/sample). Similar calculations were not included for the other PAH (indicated by blanks in the table).

TABLE 17

CONCENTRATIONS OF PAH IN A CATALYST REGENERATION FLUE GAS OF A GAS–OIL STOCK FROM AN OIL REFINERY. SAMPLES III AND IV WERE TAKEN FROM THE SAME INDUSTRIAL PROCESS UNIT BEFORE AND AFTER AN INCINERATION-TYPE AIR-CLEANING SYSTEM. SAMPLE II IS FROM A SIMILAR UNIT OF DIFFERENT DESIGN WITH NO AIR-POLLUTION CONTROL EQUIPMENT (FROM SAWICKI *et al.*, 1965, WITH PERMISSION OF PERGAMON PRESS)

| *Compound* | *Concentration* (*mg*/1 000 $m^3$ *gas*) | | |
|---|---|---|---|
| | *Sample II* | *Sample III* | *Sample IV* |
| Anthracene | 250 | 17 | 0·11 |
| Phenanthrene | 810 | 310 | 1·10 |
| Benz[*a*]anthracene | | ~270 | ~0·10 |
| Chrysene | | ~1 600 | ~1·0 |
| Fluoranthene | | ~120 | ~0·30 |
| Pyrene | 1 400 | 1 700 | 0·50 |
| Benzo[*a*]pyrene | 1 200 | 2 700 | 0·60 |
| Benzo[*e*]pyrene | 1 200 | 4 300 | 1·30 |
| Perylene | 100 | 430 | 0·07 |
| Benzo[*ghi*]perylene | 740 | 4 200 | 1·70 |
| Anthanthrene | 45 | 200 | 0·04 |
| Coronene | | 200 | 0·11 |
| Total weight | 5 745 | ~16 047 | ~6·93 |

(Table 18). Total and individual PAH concentrations varied considerably from one sample to another and PAH concentrations were very high. In Sample No. 1, total PAH represented about 5 per cent (0·05 g/g) of the weight of the particulate emissions. Surprisingly, despite the high formation temperature, alkyl and naphtheno PAH were present in significant concentrations in the effluent, although their concentrations were usually lower than those of the unalkylated parent PAH. For example, in Sample No. 1, benzo[*a*]pyrene, methylbenzo[*a*]pyrene, and dimethylbenzo[*a*]pyrene were present at concentrations of 2630·92, 330·85, and 82·68 μg/g, respectively. This is somewhat contrary to the contention of Blumer (1976) that PAH assemblages produced at such high temperatures should be devoid of alkyl PAH.

Coal tars are also produced by the high temperature treatment of coal. Thus, coal tar contains PAH that could be derived either from PAH

TABLE 18

PAH CONCENTRATIONS IN FOUR AIR-FILTER SAMPLES OF COKE OVEN EMISSIONS (FROM LAO *et al.*, 1975, WITH PERMISSION OF ELSEVIER SCIENTIFIC PUBLISHING CO.)

<table>
<tr><th rowspan="2">Compound</th><th colspan="4">Concentration (μg/sample)</th></tr>
<tr><th>Sample No. 1[a]</th><th>Sample No. 2[a]</th><th>Sample No. 3[b]</th><th>Sample No. 4[b]</th></tr>
<tr><td>Fluorene</td><td>271·52</td><td>20·86</td><td>19·29</td><td>15·37</td></tr>
<tr><td>Dihydrophenanthrene</td><td>586·98</td><td>79·55</td><td>316·49</td><td>325·83</td></tr>
<tr><td>Dihydroanthracene</td><td>168·88</td><td>21·92</td><td>101·68</td><td>232·54</td></tr>
<tr><td>2-Methylfluorene</td><td>98·71</td><td>9·06</td><td>8·87</td><td rowspan="2">40·28</td></tr>
<tr><td>1-Methylfluorene</td><td>73·46</td><td>24·35</td><td>44·01</td></tr>
<tr><td>9-Methylfluorene</td><td>44·32</td><td>11·81</td><td>6·71</td><td>102·83</td></tr>
<tr><td>Methylfluorene</td><td>87·84</td><td>10·77</td><td>—</td><td>10·38</td></tr>
<tr><td>Phenanthrene</td><td>2 828·54</td><td>163·53</td><td>458·80</td><td>636·98</td></tr>
<tr><td>Anthracene</td><td>942·85</td><td>46·44</td><td>305·89</td><td>500·36</td></tr>
<tr><td>Methylphenanthrene</td><td>1 023·41</td><td>44·67</td><td>130·22</td><td>283·92</td></tr>
<tr><td>Methylanthracene</td><td>1 692·26</td><td>85·30</td><td>258·93</td><td>573·25</td></tr>
<tr><td>Ethylphenanthrene</td><td>1 578·60</td><td>58·04</td><td>202·97</td><td>197·39</td></tr>
<tr><td>Ethylanthracene</td><td>1 096·71</td><td>49·19</td><td>249·86</td><td>1 473·68</td></tr>
<tr><td>Octahydrofluoranthene and octahydropyrene</td><td>280·42</td><td>11·06</td><td>64·36</td><td>44·53</td></tr>
<tr><td>Dihydrofluoranthene</td><td>115·07</td><td>73·57</td><td>64·87</td><td>31·86</td></tr>
<tr><td>Dihydropyrene</td><td>575·06</td><td>36·65</td><td>52·29</td><td>404·25</td></tr>
<tr><td>Fluoranthene</td><td>5 979·74</td><td>269·74</td><td>451·95</td><td>1 097·82</td></tr>
<tr><td>Dihydrobenzo[a]fluorene</td><td>791·41</td><td>24·03</td><td>—</td><td>—</td></tr>
<tr><td>Dihydrobenzo[b]fluorene and dihydrobenzo[c]fluorene</td><td>213·53</td><td>28·77</td><td>99·07</td><td>46·38</td></tr>
<tr><td>Pyrene</td><td>4 627·33</td><td>206·35</td><td>472·09</td><td>1 446·64</td></tr>
<tr><td>Benzo[a]fluorene</td><td>971·18</td><td>87·42</td><td rowspan="2">62·97</td><td>37·21</td></tr>
<tr><td>Benzo[b]fluorene</td><td>109·45</td><td>16·70</td><td>22·63</td></tr>
<tr><td>Benzo[c]fluorene</td><td>627·02</td><td>38·96</td><td>290·03</td><td>98·47</td></tr>
<tr><td>Methylfluoranthenes</td><td>2 208·31</td><td>146·60</td><td>305·07</td><td>136·39</td></tr>
<tr><td>Methylpyrenes</td><td>1 873·67</td><td>137·46</td><td>331·17</td><td>296·08</td></tr>
<tr><td>Benzo[c]phenanthrene</td><td>220·45</td><td>82·70</td><td>1 510·34</td><td>2 156·14</td></tr>
<tr><td>Benzo[ghi]fluoranthene</td><td>677·35</td><td>164·25</td><td>201·74</td><td>151·88</td></tr>
<tr><td>Dihydrobenz[a]anthracene, dihydrochrysene, and dihydrotriphenylene</td><td>383·03</td><td>54·75</td><td>101·78</td><td>271·38</td></tr>
<tr><td>Benz[a]anthracene</td><td>2 740·45</td><td>105·15</td><td rowspan="2">5 509·43</td><td rowspan="2">2 673·65</td></tr>
<tr><td>Chrysene and triphenylene</td><td>4 202·02</td><td>119·04</td></tr>
<tr><td>Dihydromethylbenz[a]anthracene, dihydromethylchrysene, and dihydromethyltriphenylene</td><td>841·67</td><td>93·30</td><td>247·80</td><td>86·35</td></tr>
<tr><td>Methylbenz[a]anthracene</td><td>159·33</td><td>22·36</td><td>1 015·70</td><td>1 669·72</td></tr>
<tr><td>Methyltriphenylene</td><td>463·99</td><td>40·81</td><td>371·97</td><td>369·10</td></tr>
</table>

TABLE 18—*contd.*

| *Compound* | *Concentration (µg/g sample)* | | | |
|---|---|---|---|---|
| | *Sample No. 1*[a] | *Sample No. 2*[a] | *Sample No. 3*[b] | *Sample No. 4*[b] |
| Methylchrysene | 1 151·61 | 107·59 | 193·68 | 448·19 |
| Dihydromethylbenzo[*k* and *b*]-fluoranthene and dihydromethylbenzo[*a* and 1]-pyrenes | 434·38 | 8·63 | 185·08 | 137·26 |
| Dimethylbenz[*a*]anthracene, dimethyltriphenylene, and dimethylchrysene | 246·35 | 46·73 | 30·16 | 541·70 |
| Benzo[*j*]fluoranthene | 176·92 | 18·66 | 80·04 | 285·34 |
| Benzo[*k*]fluoranthene and benzo[*b*]fluoranthene | 3 930·34 | 155·03 | 2 170·92 | 2 556·98 |
| Methylbenzo[*k*]fluoranthene and methylbenzo[*b*]fluoranthene | 735·95 | 33·05 | 430·05 | 492·13 |
| Benzo[*e*]pyrene | 103·86 | — | — | — |
| Benzo[*a*]pyrene | 2 630·92 | 122·15 | 2 007·75 | 2 297·83 |
| Perylene | 702·12 | 22·07 | 616·85 | 698·44 |
| Methylbenzo[*a*]pyrene | 330·85 | 5·88 | 344·12 | 247·96 |
| Dimethylbenzo[*k*]fluoranthene and dimethylbenzo[*b*]fluoranthene | 116·74 | — | 73·13 | — |
| Dimethylbenzo[*a*]pyrene | 82·68 | — | 70·18 | — |
| Dibenzanthracene | 123·66 | — | 84·52 | — |
| *o*-Phenylenepyrene | 101·54 | — | — | — |
| Benzo[*ghi*]perylene and anthanthrene | 72·35 | — | — | — |
| Methyldibenzanthracene | 89·04 | — | — | — |
| Methylbenzo[*ghi*]perylene | 36·79 | — | — | — |
| Coronene | 864·55 | — | 833·30 | 766·58 |
| Dibenzopyrene | 693·21 | — | 587·05 | 493·27 |

[a] Sample Nos. 1 and 2, glass-fiber filter.
[b] Sample Nos. 3 and 4, 0·8 µm porosity silver-membrane filter.

indigenous to the coal or from pyrolysis of coal hydrocarbons. The gaseous emissions associated with coal tar production may also contain significant quantities of particulate PAH. Table 19 summarizes the data of Lao *et al.* (1975) regarding PAH composition of coal tar and gaseous emissions from coal tar production. Airborne particulate emissions from coal tar production are quite rich in PAH. The PAH measured represent approximately 87 per cent of the weight of the particulate sample. Coal tar

TABLE 19

CONCENTRATIONS OF PAH IN COAL TAR AND AIRBORNE COAL TAR EMISSIONS (FROM LAO *et al.*, 1975, WITH PERMISSION OF ELSEVIER SCIENTIFIC PUBLISHING CO.)

| *Compound* | *Coal tar air emissions* | | *Coal tar* |
|---|---|---|---|
| | *mg/g sample* | *μg/m³ air sample* | *mg/g sample* |
| Fluorene | 2·41 | 85 | 27·39 |
| Dihydrophenanthrene and dihydroanthracene | 0·68 | 24 | 1·99 |
| Methylfluorenes | 0·74 | 26 | 3·12 |
| Phenanthrene and anthracene | 142·90 | 5 020 | 81·07 |
| Methylphenanthrene and methylanthracene | 2·18 | 77 | 2·53 |
| Ethylphenanthrene and ethylanthracene | 3·88 | 136 | 4·36 |
| Octahydrofluoranthene | 0·15 | 5 | 0·23 |
| Octahydropyrene | 0·15 | 5 | 0·21 |
| Dihydrofluoranthene and dihydropyrene | 0·15 | 5 | 1·05 |
| Fluoranthene | 144·85 | 5 090 | 54·28 |
| Dihydrobenzofluorenes | 8·85 | 311 | 5·26 |
| Pyrene | 105·47 | 3 705 | 46·50 |
| Benzofluorenes | 24·53 | 862 | 20·13 |
| Methylfluoranthenes | 3·91 | 137 | 2·25 |
| Methylpyrenes | 3·88 | 136 | 2·06 |
| Trimethylfluoranthene and trimethylpyrene | 1·29 | 45 | 0·66 |
| Benzo[*ghi*]fluoranthene | 3·29 | 115 | 4·40 |
| Dihydrobenz[*a*]anthracene, dihydrochrysene, and dihydrophenylene | 2·65 | 93 | 3·87 |
| Benzo[*a*]anthracene, chrysene, and triphenylene | 45·74 | 1 607 | 42·66 |
| Dihydromethylbenz[*a*]anthracene, chrysene, and triphenylene | 0·5 | 18 | 2·12 |
| Methylbenz[*a*]anthracene, methylchrysene, and methyltriphenylene | 4·26 | 150 | 11·04 |
| Dimethylbenz[*a*]anthracene, chrysene, and triphenylene | 0·41 | 15 | 1·87 |
| Benzo[*j*]fluoranthene | 0·12 | 4 | 0·73 |
| Benzo[*k*]fluoranthene | 9·96 | 350 | 32·54 |
| Methylbenzo[*k*]fluoranthene and methylbenzo[*j*]fluoranthene | 2·47 | 88 | 7·43 |
| Benzo[*a*]pyrene and benzo[*e*]pyrene | 13·83 | 486 | 33·13 |
| Methylbenzo[*a*]pyrene and methylbenzo[*e*]pyrene | 2·15 | 76 | 8·76 |
| Dibenzanthracenes | 0·14 | 14 | 2·13 |
| Benzo[*b*]chrysene and *o*-phenylenepyrene | 3·88 | 137 | 22·79 |
| Benzo[*ghi*]perylene and anthanthrene | 2·62 | 92 | 17·94 |

TABLE 20

LEVELS AND DAILY EMISSIONS OF PAH FROM MUNICIPAL INCINERATORS (FROM DAVIES *et al.*, 1976, WITH PERMISSION OF THE AMERICAN CHEMICAL SOCIETY)

| Compound | Stack gases | | Residues | | Water | | |
|---|---|---|---|---|---|---|---|
| | *mg/1 000 m*³ | *mg/day*[a] | *µg/kg* | *mg/day*[b] | *Input (µg/liter)* | *Output (µg/liter)* | *Output (mg/day*[c]*)* |
| Fluoranthene | 0·58 | 274 | 58 | 1 360 | 0·08 | 0·62 | 15·5 |
| Pyrene | 1·58 | 745 | 49 | 1 150 | 0·08 | 0·54 | 13·5 |
| Benz[*a*]anthracene and chrysene | 0·72 | 340 | 171 | 4 010 | 0·03 | 0·64 | 16·0 |
| Benzo[*b*]fluoranthene, benzo[*k*]fluoranthene, and benzo[*j*]fluoranthene | 0·32 | 151 | 292 | 6 850 | 0·03 | 0·14 | 3·5 |
| Benzo[*a*]pyrene and benzo[*e*]pyrene | 0·02 | 9 | 147 | 3 450 | 0·03 | 0·14 | 3·5 |
| Perylene | 0·18 | 85 | 82 | 1 920 | 0·02 | 0·13 | 3·3 |
| Benzo[*ghi*]perylene | 0·42 | 198 | 47 | 1 100 | <0·01 | 0·03 | 0·8 |
| Indeno[1,2,3-*cd*]pyrene | 0·18 | 85 | <10 | <230 | <0·01 | <0·01 | <0·3 |
| Coronene | 0·04 | 19 | <20 | <470 | <0·01 | <0·01 | <0·3 |
| Total of PAH measured | | 1 906 | | 19 840 | | | 56·1 |

[a] Assuming 12-h daily operation; levels measured at flow 655 m³/min corrected to dry, 20 °C, 760 mmHg.
[b] Assuming 23 per cent average moisture content, and 30 tons daily output.
[c] Assuming a discharge of 25 000 liters/day from the quench trough.

itself is approximately 44·4 per cent PAH by weight. As might be expected, the gaseous emissions are enriched in lower molecular weight PAH in comparison to coal tar, while coal tar has relatively higher concentrations of higher molecular weight PAH than the emissions. Alkyl PAH are well represented in both coal tar and the emissions but their concentrations are generally much lower than those of the unalkylated parent PAH.

Other coal conversion processes include coal carbonization, liquification, and gasification. Several alternative methods are currently being investigated for carrying out these processes in an effort to find alternative sources of high-quality fuels and hydrocarbon feedstocks to replace the dwindling domestic supplies of crude oil. Several of these conversion processes are carried out at high temperatures and result in the production of large quantities of PAH (Huffstetler and Rickert, 1977; Dailey, 1977). If these coal conversion processes become economically feasible and are widely adopted, they may represent an important new source of PAH in the environment.

Davies *et al.* (1976) measured the concentrations of several PAH in the stack gas, solid residues, and wastewaters from a modern continuous feed incinerator rated at 9·14 tons of refuse per hour (Table 20). All PAH measured showed a similar distribution among the three types of discharge. Over ten times more PAH were released each day in the solid residues than in the stack gas and wastewaters combined. Several of the PAH present at high concentrations in the residue are known carcinogens; these include benz[*a*]anthracene, benzo[*b*]fluoranthene, benzo[*j*]fluoranthene, and benzo[*a*]pyrene. The solid residues (ash) are often placed in sanitary land fills. PAH from land-deposited incinerator residues may reach water bodies in surface run-off from land or by leaching into ground water.

There are, of course, many other natural and anthropogenic pyrolytic sources of PAH. Youngblood and Blumer (1975) have suggested that forest and prairie fires represent an important source of atmospheric PAH. Farrington and Meyers (1975) estimated that such fires inject 6 million metric tons/year of hydrocarbons into the atmosphere of which a significant portion may be PAH. Suess (1976) approximated that worldwide emissions of B*a*P into the atmosphere from forest and agricultural fires is 420 metric tons/year. Virtually any natural, industrial, or domestic process in which organic carbon is subjected to high temperatures will result in the production of some PAH. The tremendous diversity of sources of PAH and our extremely limited knowledge about them makes any estimate of the global flux of these organic molecules very difficult.

*CHAPTER 3*

# *Rates and Routes of PAH Entry into the Aquatic Environment*

## 3.1 PAH FROM BIOSYNTHESIS

PAH formed by the processes discussed in the previous section may reach the aquatic environment by a wide variety of routes. Andelman and Snodgrass (1972) suggested that most of the PAH in the marine environment are derived from endogenous sources (e.g., biosynthesis). However, as indicated earlier, direct PAH biosynthesis has not been conclusively demonstrated. Recent detailed gas chromatographic–mass spectrometric (GC–MS) analyses of shallow water marine sediments from the New England, USA, area have revealed the presence of PAH assemblages far more complex than would be expected of biosynthetic formation (Youngblood and Blumer, 1975; Hites, 1976; Hase and Hites, 1977). Similar conclusions have been drawn concerning PAH assemblages in terrestrial soils (Blumer *et al.*, 1977). In that case, the authors concluded that the main source of PAH in soils was vehicular exhaust condensate.

If we assume that some aquatic algae and microbes are, in fact, able to synthesize PAH, a rough estimate can be made of a biosynthetic contribution to the PAH burden in the aquatic environment. Borneff *et al.* (1968a, b) provide biosynthetic rates by the alga *Chlorella* of 67 $\mu$g total PAH/kg dry algal biomass and 0·617 $\mu$g B*a*P/kg dry algal biomass. The annual production of organic carbon in the world's oceans is estimated to be about $40 \times 10^9$ tons/year (Koons and Monaghan, 1976). Most of this production is due to phytoplankton. Based on these figures, the annual production of PAH by marine phytoplankton is $2{\cdot}7 \times 10^3$ metric tons PAH/year and that of B*a*P is 25 metric tons/year. Although these values are subject to many sources of error and uncertainty, they do give a rough estimate of the potential magnitude of hypothetical PAH biosynthesis in the aquatic environment.

Suess (1976) suggested that PAH formed by biosynthesis are quite static and tend to remain in the organisms or ecosystems in which they were formed. However, PAH synthesized by soil bacteria may be leached into ground water or into surface runoff from land. PAH from other sources may reach the aquatic environment by spillage of crude and refined oil, in industrial and domestic effluents, runoff water from land, and by dry or wet precipitation from the atmosphere.

## 3.2 PAH FROM FOSSIL FUELS

World petroleum production for 1971 has been estimated by the National Academy of Sciences at 2478·4 million tons (NAS, 1975). Crude oil production has risen rapidly since that time and will continue to do so for several years to come (Flower, 1978). Assuming that the mean concentration of total PAH in crude oil is 2·8 per cent (Table 8) and that the mean concentration of B*a*P in crude oil is 1·46 ppm (Table 10), then petroleum production results in the mobilization of 69·4 million tons PAH/year and 3618·5 tons B*a*P/year. Most of this crude petroleum is refined to produce various fuels, gasoline, and hydrocarbon feedstocks; the majority of the resulting petroleum distillate products are eventually combusted. Additional PAH may be formed during the refining and combustion of petroleum products and some of these may reach the aquatic environment.

The best estimate of the amounts of crude oil and distillate products entering the aquatic environment each year from natural and anthropogenic sources was reported by the NAS (1975) to be approximately 6·1 million metric tons per year. This represents approximately 0·25 per cent of the annual crude oil production. Marine transportation operations (2·13 million tons/year), river runoff (1·6 million tons/year), natural submarine oil seeps (0·6 million ton/year), and atmospheric fallout (0·6 million tons/year) are the major sources of marine petroleum pollution. Based on these figures Pancirov and Brown (1975) estimated that the annual input of B*a*P to the marine environment from petroleum spillage is 20–30 tons. Total PAH entry into the marine environment from this source would be about $17 \times 10^4$ tons/year.

The use of creosoted pilings for docks and other shoreline structures may represent an important localized source of PAH pollution in aquatic systems (Zitko, 1975; Dunn and Stich, 1976a) but it does not contribute significantly to the total input of PAH to the aquatic environment.

Creosote may contain as much as 93 g PAH/kg and 22 mg B*a*P/kg (Lijinsky *et al.*, 1963).

## 3.3 PAH FROM INDUSTRIAL AND DOMESTIC WASTEWATER

A large number of industries produce wastewaters containing significant quantities of PAH and most current wastewater treatment practices do not effectively remove all PAH impurities (Harrison *et al.*, 1975). Commonly cited industrial sources of PAH-contaminated wastewater effluents are oil refineries, industries utilizing solid and liquid hydrocarbon feedstocks for manufacture of chemical byproducts, the plastics and dyestuffs industries, high temperature furnaces (especially those using anthracite electrodes), the lime industry, and many others. The majority of detailed studies of PAH contamination of industrial wastewaters have been performed by Russian and British scientists and were summarized by Andelman and Suess (1970) and Harrison *et al.* (1975) (Table 21). B*a*P concentrations in the wastewaters listed vary from undetectable to 1000 $\mu$g/liter. Several other PAH, some of them also carcinogens, have been detected also and many more would probably be identified by application of modern high-resolution gas chromatographic techniques. It should be emphasized that a great many industries probably producing significant aqueous PAH emissions are not included in Table 21. For instance, metal smelting industries, including some aluminum smelters, use Söderberg electrodes. These electrodes are made from anthracite, coketar pitch, and anthracene oil. Palmork *et al.* (1973) showed that sediments in a Norwegian fjord into which effluents from an aluminum smelter were discharged contained up to 1000 mg PAH/kg. Production of 2 tons of aluminum require 1 ton of electrode. Estimated production of 3 million tons of aluminum by this method in Europe would result in the release of 150 tons of benzopyrenes, 180 tons of fluoranthene and anthracene, and a similar quantity of phenanthrene. The use of specially precooked electrodes instead of unbaked electrodes may substantially reduce these PAH emissions (Johnston, 1976).

Domestic wastewaters including raw sewage and storm sewer runoff may contain significant quantities of PAH. In most cases, PAH concentration increases as the industrial contribution to the sewage increases (Borneff and Kunte, 1965; 1967) (Table 22). PAH concentrations in sewage may also increase substantially following heavy rains,

TABLE 21

CONCENTRATIONS OF BENZO[*a*]PYRENE IN INDUSTRIAL WASTEWATER EFFLUENTS (FROM ANDELMAN AND SUESS, 1970; HARRISON *et al.*, 1975, WITH PERMISSION OF THE PUBLISHERS)

| *Industry* | *Wastewater source* | *BaP concentration (μg/liter)* | *Other PAH identified*[a] |
|---|---|---|---|
| Shale oil | | | |
| | After dephenolization | 2–230 | |
| Coke byproducts | | | |
| | After biochemical treatment | 12–16 | |
| | After oil separation | 6·5–290 | |
| | Spent gas liquor | very small quantity | AT, B*a*A, CH, FL, PR, PY |
| Coke or oil–gas works | | | |
| | Before discharge to sewer | not indicated | AC, AN, BPR, FR, PA |
| | Before discharge to sewer (two plants) | 340 and 1000 | |
| | After filtration through coke beds | 20 | |
| Oil–gas works | | | |
| | After oil separation | 3–30 | |
| Oil refinery | | | |
| | Coking of residues after direct distillation of oil | 0·48–50 | |
| | Catalytic cracking of kerosene gas oil fraction at 450 °C | 0·11–0·29 | |
| | Catalytic cracking | 0·07–0·11 | |
| | Thermal cracking | 0·09–0·23 | |
| | Pyrolysis of ethane–ethylene fractions at 700–800 °C | 3·6 | |
| | After settling ponds of various refineries | up to 0·22 | |
| Acetylene | | | |
| | Not indicated | 0·015–0·100 | |
| Ammonium sulfate | | | |
| | After cooling and settling | ~10 | AP, AT, B*a*A, B*e*P, CH, FL, PR, PY, TP |

[a] AC, Anthracene; AN, acenaphthylene; AP, alkylpyrene; AT, anthanthrene; B*a*A, benz[*a*]anthracene; B*e*P, benzo[*e*]pyrene; BPR, benzo[*ghi*]perylene; CH, chrysene; FL, fluoranthene; FR, fluorene; PA, phenanthrene; PR, perylene; PY, pyrene; TP, triphenylene.

TABLE 22

CONCENTRATIONS OF PAH IN MUNICIPAL WASTEWATER EFFLUENTS IN THE GFR. ALL VALUES ARE IN μG/LITER (FROM BORNEFF AND KUNTE, 1965, WITH PERMISSION OF THE PUBLISHER)

| *Compound* | *Effluents to Rotach River* | | *Mixed sewage, high % industry* | |
|---|---|---|---|---|
| | *Domestic* | *Industrial* | *Stockach* | *Rodolfzell* |
| Fluoranthene | 0·273 | 2·198 | 3·420 | 2·660 |
| Pyrene | — | 1·957 | 3·120 | 2·560 |
| Benz[*a*]anthracene | 0·191 | 0·167 | 1·360 | 0·343 |
| Benzo[*b*]fluoranthene | 0·036 | 0·114 | 0·870 | 0·525 |
| Benzo[*j*]fluoranthene | 0·037 | 0·045 | 1·740 | 1·100 |
| Benzo[*k*]fluoranthene | 0·031 | 0·032 | 0·460 | 0·336 |
| Benzo[*a*]pyrene | 0·038 | 0·100 | 0·100 | 0·368 |
| Benzo[*ghi*]perylene | 0·040 | 0·073 | 0·480 | 0·120 |
| Indeno[1,2,3-*cd*]pyrene | 0·022 | 0·057 | 0·930 | 0·476 |
| Total identified PAH | 0·668 | 4·743 | 12·480 | 8·488 |
| Total unidentified PAH | 0·130 | 0·390 | 2·500 | 0·070 |
| Total PAH | 0·798 | 5·133 | 14·980 | 8·558 |

TABLE 23

CONCENTRATIONS OF PAH IN DOMESTIC SEWAGE FROM HEGNE (GFR) DURING DRY AND WET WEATHER. ALL VALUES ARE IN μG/LITER (FROM BORNEFF AND KUNTE, 1965, WITH PERMISSION OF THE PUBLISHER)

| *Compound* | *Dry weather* | *During heavy rain* |
|---|---|---|
| Fluoranthene | 0·352 | 16·350 |
| Pyrene | 0·254 | 16·050 |
| Benz[*a*]anthracene | 0·025 | 10·360 |
| Benzo[*b*]fluoranthene | 0·039 | 10·790 |
| Benzo[*j*]fluoranthene | 0·057 | 9·910 |
| Benzo[*k*]fluoranthene | 0·022 | 1·840 |
| Benzo[*a*]pyrene | 0·001 | 3·840 |
| Benzo[*ghi*]perylene | 0·004 | 4·180 |
| Indeno[1,2,3-*cd*]pyrene | 0·017 | 4·980 |
| Total identified PAH | 0·771 | 78·300 |
| Total unidentified PAH | 0·075 | 9·200 |
| Total PAH | 0·846 | 87·500 |

presumably owing to increased storm sewer runoff from roadways (Table 23). The PAH concentration (3·1 μg/liter total for 11 PAH, including 0·57 μg B*a*P/liter) was much higher in a sample of runoff water from the M4 motorway, London, England, than in water samples from the River Thames, London (Acheson *et al.*, 1976). PAH in road runoff probably arise from wear and leaching of asphalt road surfaces, wear of vehicle tires which contain carbon black, and condensation from vehicular exhaust. Household sewage also contains PAH derived from cooking and washing activities and from urine and feces (Andelman and Suess, 1970). Human urine was shown to contain 1·3 μg/liter B*a*P (Mallet and Heros, 1960). A sample of sewage sludge from the German Federal Republic (GFR) contained 1·7 mg/kg B*a*P and 26·8 mg/kg total PAH (Grimmer *et al.*, 1978).

It is extremely difficult to estimate the quantity of PAH entering fresh and marine waters from industrial and domestic wastewater effluents. Information on concentrations of PAH in different effluents, especially in those after sewage treatment, is very incomplete and even less reliable information is available on the total volume of wastewater produced from different sources each year. Much of the PAH burden entering fresh and marine waters in this form is from non-point-sources, making quantification difficult. However, the amounts of PAH entering fresh and marine waters by these routes are undoubtedly very large (Borneff and Kunte, 1965; Andelman and Suess, 1970; Harrison *et al.*, 1975). It is estimated that more than $60 \times 10^{12}$ liters of water are released each year in the USA in the form of domestic sewage and liquid waste effluents from the types of industries described above (USDI, 1968). If it is assumed that the ratio of US wastewater production to that of the rest of the world is 1:3, then worldwide wastewater production would be $24 \times 10^{13}$ liters/year. The mean concentration of B*a*P in 11 samples of wastewater analyzed by Borneff and Kunte (1965) was 0·277 μg/liter. Excluding the single, unusually high value of 1·84 μg B*a*P/liter, the mean value is then 0·120 μg B*a*P/liter. The mean concentration of total PAH in the same wastewater samples was 18·18 μg/liter. Hellmann (1974) reported values ranging from 0·02–>0·4 μg B*a*P/liter and 1·0–>20 μg total PAH/liter for wastewaters of the GFR. He considers wastewaters to be the main source of PAH in the water and sediments of the Rhine River at Koblenz. Using the average values reported above, the annual contribution of domestic and industrial wastewaters to the PAH burden of the aquatic environment is 29 tons B*a*P/year and $4 \cdot 4 \times 10^3$ tons total PAH/year. These estimates are admittedly quite conservative.

## 3.4 PAH FROM RUNOFF FROM LAND

Runoff from land may also contribute substantial quantities of PAH to fresh and marine waters. As mentioned before, runoff from road surfaces may contain particularly high concentrations of PAH (Acheson *et al.*, 1976). The total annual runoff from the North American continent is estimated at $3 \times 10^{15}$ liters/year. Of course this includes some of the $60 \times 10^{12}$ liters/year of domestic and industrial waste effluents. Hellmann (1974) reported that the mean concentration of PAH in clean brook water in the GFR was about 10 ng B*a*P/liter and 250 ng total PAH/liter. These values are in the lower part of the range reported by Harrison *et al.* (1975) for various fresh surface waters. Using Hellmann's (1974) values and a runoff rate of $2{\cdot}94 \times 10^{15}$ liters/year, the estimated input of PAH to water bodies from this source would be 29·4 tons B*a*P/year and 735 tons total PAH/year. Using a ratio of 1:3 to convert from North American to worldwide values, total worldwide inputs of PAH to the aquatic environment from surface land runoff would be 117·6 tons B*a*P/year and 2940 tons total PAH/year.

## 3.5 ATMOSPHERIC FALLOUT OF PAH

It is generally assumed that virtually all the PAH released into the atmosphere are associated with aerosols and airborne particulate matter (Thomas *et al.*, 1968; NAS, 1972; Suess, 1976). Lower molecular weight PAH (naphthalene–pyrene) have sufficiently high vapor pressures that they can exist at significant concentrations in the vapor phase. The vapor pressures and thus equilibrium vapor concentrations of PAH decrease as molecular weight increases and as ambient temperature decreases (Tables 24 and 25). Air samples for PAH analysis are usually collected by pumping the air sample through a particle filter. The vapor phase PAH may not be adequately retained, giving erroneously low results. Pupp *et al.* (1974) approximated the amount of adsorption of vapor phase PAH by the filter necessary to make filter sampling of air efficient. It is apparent from Table 25, that because of the effect of temperature on the equilibrium vapor concentration of PAH, sampling of airborne PAH by this method will be more efficient in the winter than in the summer. It is now considered likely that most of the high molecular weight chlorinated hydrocarbons (pesticides and polychlorinated biphenyls, etc.) in the atmosphere are present in the vapor phase and not adsorbed to particulates (Risebrough *et*

TABLE 24

VAPOR PRESSURE AND SATURATION VAPOR CONCENTRATION OF SEVERAL PAH AT 25 °C (DATA FROM RADDING *et al.*, 1976). VAPOR PRESSURE AND SATURATION VAPOR CONCENTRATION OF THE CHLORINATED HYDROCARBONS DDT AND AROCLOR 1260 (AN INDUSTRIAL POLYCHLORINATED BIPHENYL MIXTURE) ARE INCLUDED FOR COMPARISON (DATA FROM SPENCER AND CLIATH, 1972, AND MACKAY AND WOLKOFF, 1973)

| *Compound* | *Molecular weight* | *Vapor pressure (mmHg)* | *Saturation vapor concentration* ($\mu g/1000\,m^3$) |
|---|---|---|---|
| Anthracene | 178 | $1{\cdot}95 \times 10^{-4}$ | $1{\cdot}8 \times 10^{7}$ |
| Phenanthrene | 178 | $6{\cdot}80 \times 10^{-4}$ | $6{\cdot}51 \times 10^{7}$ |
| Benz[*a*]anthracene | 228 | $1{\cdot}10 \times 10^{-7}$ | 1333 |
| Pyrene | 202 | $6{\cdot}85 \times 10^{-7}$ | 74 400 |
| Benzo[*a*]pyrene | 252 | $5{\cdot}49 \times 10^{-9}$ | 75 |
| Benzo[*e*]pyrene | 252 | $5{\cdot}54 \times 10^{-9}$ | 75 |
| Benzo[*k*]fluoranthene | 252 | $9{\cdot}59 \times 10^{-11}$ | 13 |
| Benzo[*ghi*]perylene | 276 | $1{\cdot}01 \times 10^{-10}$ | 1·5 |
| Coronene | 300 | $1{\cdot}47 \times 10^{-12}$ | 0·02 |
| *p,p′*-DDT | 354·5 | $4{\cdot}90 \times 10^{-7}$ | 13 000 |
| Aroclor 1260 | 361 | $4{\cdot}00 \times 10^{-5}$ | $7{\cdot}3 \times 10^{5}$ |

*al.*, 1976; Bidleman *et al.*, 1976). Since the lower molecular weight PAH have vapor pressures and equilibrium vapor concentrations similar to those of the chlorinated hydrocarbons, they can be expected to behave in the atmosphere in a manner similar to that of the chlorinated hydrocarbons.

Atmospheric processes that result in deposition of airborne PAH on land and water include rain, dry fallout, and vapor phase deposition onto surfaces. The first two processes are the primary mechanisms for the deposition of airborne particulate PAH. Residence time of particulate PAH in the atmosphere and their dispersal by wind currents are governed by particle and atmospheric physics and meteorological conditions, all of which have been extensively reviewed (NAS, 1972; Radding *et al.*, 1976). The size of particulates to which PAH are adsorbed is a primary determinant of particulate PAH residence time in the atmosphere and thus dispersal distance. Broddin *et al.* (1977) showed that more than 90 per cent of the PAH from a coke oven emission source was associated with particles smaller than 3 μm. The largest fraction of PAH was associated with particles 0·85 to 1·8 μm in diameter. Distribution of all 14 PAH investigated (anthracene to benzo[*ghi*]perylene) was similar. An analogous pattern was seen in airborne particulate PAH collected at several stations in and around Toronto, Ontario, Canada (Pierce and Katz, 1975; Katz and Pierce, 1976);

TABLE 25

EQUILIBRIUM VAPOR CONCENTRATIONS (EVC) OF SELECTED PAH AT A COMMON WINTER TEMPERATURE (−10 °C) AND A COMMON SUMMER TEMPERATURE (30 °C) AND THE REDUCTION FACTORS OWING TO ADSORPTION NECESSARY FOR PARTICULATE SAMPLE COLLECTION OF AIRBORNE PAH TO BE EFFICIENT (FROM PUPP *et al.*, 1974, WITH PERMISSION OF PERGAMON PRESS)

| *Compound* | *Winter EVC* ($\mu g/1000\,m^3$) | *Winter reduction factor* | *Summer EVC* ($\mu g/1000\,m^3$) | *Summer reduction factor* |
|---|---|---|---|---|
| Pyrene | 580 | 16–450 | 140 000 | 4 000–108 000 |
| Benz[*a*]anthracene | 3·4 | 0–34 | 2 800 | 127–28 000 |
| Benzo[*a*]pyrene | 0·15 | 0 | 160 | 4–800 |
| Benzo[*e*]pyrene | 0·15 | 0 | 160 | 6–160 |
| Benzo[*k*]fluoranthene | 0·013 | 0 | 30 | 1·5–30 |
| Benzo[*ghi*]perylene | $1{\cdot}8 \times 10^{-3}$ | 0 | 3·4 | 0–3 |
| Coronene | $1{\cdot}8 \times 10^{-5}$ | 0 | 0·058 | 0 |

size distribution of PAH-containing particulates followed an approximately lognormal relationship for urban and suburban sampling sites. The majority of PAH were associated with particles possessing mass median diameters of 1·2 to 2·6 μm. During the winter season, however, the mass median diameters of PAH-containing particulates were found to be in the submicrometer range, 0·6 to 0·8 μm. The authors concluded that the submicrometer PAH-containing particulates were derived primarily from automobile exhasut.

It has been estimated that without rainfall submicrometer particles will remain in the lower atmosphere from a few days to 4–6 weeks, whereas particles in the 1–10 μm range will be deposited in a few days or less (Suess, 1976). Rainfall can be expected to reduce the residence time of airborne particulates. Thus, PAH-containing particulates may remain in the atmosphere for a sufficiently long time to allow dispersal over hundreds or even thousands of miles.

### 3.5.1 Photooxidation of Airborne Particulate PAH

The degree to which airborne particulate PAH will contribute to inputs of PAH to the aquatic environment will depend not only on the residence time of the particles in the atmosphere but also on the sensitivity of the adsorbed PAH to various chemical and photooxidative reactions.

There is substantial evidence that PAH in the atmosphere or in surface waters are readily degraded by photooxidation, reaction with atmospheric oxidants, or by reaction with sulfur oxides. These reactions produce oxygenated compounds including quinones which may be carcinogenic. In experimental studies, ultraviolet (UV) irradiation of B*a*P in benzene or carbon disulfide solvents resulted in photooxidation. Photooxidation products included 5-phenoxybenzo[*a*]pyrene (Masuda and Kuratsune, 1966; Inomata and Nagata, 1972). Different PAH vary significantly in relative sensitivity to photooxidation. Thomas *et al.* (1968) reported a 60 per cent decrease in the concentration of B*a*P on soot particles following exposure to light for 40 min. Several other PAH, when subjected to similar conditions, showed little or no evidence of photooxidation. Pierce and Katz (1975) exposed equimolar concentrations of pure B*a*P and benz[*a*]anthracene to ozone and UV irradiation in a simulated atmosphere. More than half the original B*a*P was degraded after 30 min compared to only one fifth of the original amount of benz[*a*]anthracene. Estimates of the half-lives of selected PAH in reactions with major oxidizers in air and

TABLE 26

RELATIVE HALF-LIVES IN HOURS FOR DEGRADATION OF PAH BY SOME ENVIRONMENTAL OXIDIZERS (FROM RADDING *et al.*, 1976, WITH PERMISSION OF THE ENVIRONMENTAL PROTECTION AGENCY)

| *Compound* | *Half-life* ($t_{1/2}$) *degraded by:* | | | | | |
|---|---|---|---|---|---|---|
| | $RO_2\cdot$ | *Singlet oxygen* | $O_3$*(water)*[a] | $O_3$*(air)*[b] | $Cl_2$ | $HO\cdot$ |
| Anthracene | 38 000 | 5 | | | | |
| Dimethylanthracene | | 0·05 | | | | |
| Diphenylanthracene | | 0·6 | | | $t_{1/2} < 0{\cdot}5$ h for all PAH | $t_{1/2} \sim 10$ h for all PAH |
| Phenanthrene | $2 \times 10^8$ | | | | | |
| Pyrene | $2{\cdot}4 \times 10^5$ | | 0·68 | 560 | | |
| Perylene | 38 000 | | | | | |
| Tetracene | 96 | <5 | | | | |
| Benzo[*a*]pyrene | $2{\cdot}4 \times 10^5$ | 5 | 1·05 | 870 | | |
| Benz[*a*]anthracene | | 10 | 0·45 | 370 | | |
| Dimethylbenz[*a*]anthracene | | <5 | | | | |
| Dibenzanthracene | | | 0·42 | 340 | | |
| Dimethyldibenzanthracene | | <5 | 0·17 | <140 | | |
| Rubrene | | 0·02 | | | | |

[a] $[O_3] = 10^{-4}$ M.
[b] $[O_3] = 2 \times 10^{-9}$ M.

water—$RO_2$ · radical, singlet oxygen, ozone, chlorine, and HO · radical—were summarized by Radding *et al.* (1976) (Table 26).

The two major oxidizing species in urban air are HO · radical and ozone. The HO · radical is formed through a cycle involving photolysis of $NO_2$, $H_2O$, CO, and simple organic compounds, and ozone is formed from the oxygen atom and molecular oxygen (Radding *et al.*, 1976). Singlet oxygen is thought to be a less important oxidant of airborne organics owing to its very low concentration in the atmosphere and the variety of processes that rapidly quench it. Photooxidation (photooxygenation) by singlet oxygen appears to be the dominant chemical process for degradation of PAH and other organics in water (Zafiriou, 1977). In the presence of ozone and light, half-lives of a few minutes to a few hours may be expected for most PAH. Alkyl PAH seem to be more sensitive to photooxidation than the parent PAH (Radding *et al.*, 1976).

It is difficult to use such data to predict the half-life of complex PAH assemblages adsorbed to airborne particulates. For instance, McGinnes and Snoeyink (1974) observed an apparent autoinhibition of B*a*P photooxidation of B*a*P suspended in water adsorbed to 1·5 $\mu$m diameter microspheres. B*a*P degradation was rapid until approximately 60 per cent of the B*a*P was oxidised, then degradation proceeded very slowly and only 20 per cent more of the B*a*P decomposed in the following 4 days at the highest light flux. The mechanism of inhibition is not clearly understood. In environmental mixtures of PAH, inhibition by benzopyrene or other PAH quinones could slow degradation of other PAH which would, by themselves, undergo more rapid and complete oxidation. Furthermore, air, water, and soils containing PAH may also contain other compounds that affect the kinetics of PAH photooxidation. There is some evidence that phenols inhibit PAH photodegradation (Gubergrits *et al.*, 1972). Some metal–ion complexes quench singlet oxygen, reducing the rate of photooxidation by this route (Guillory and Cook, 1973). Little is known about the rates of PAH photooxidation in the vapor phase. It may proceed more slowly than photooxidation of particle-adsorbed PAH (see Section 5.1).

Circumstantial evidence that photooxidation of B*a*P and other airborne PAH is not as rapid as would be predicted from the results of model laboratory studies is provided by the observation that detectable levels of PAH are nearly always found in urban and even remote rural air samples. The Environmental Protection Agency (EPA) measured annual average ambient B*a*P concentrations in air samples from 139 urban and 21 nonurban stations from 1966 to 1970 (EPA, 1975). B*a*P concentrations in

TABLE 27

SEASONAL FLUCTUATIONS IN THE GEOMETRIC MEAN PAH CONCENTRATIONS IN AIR SAMPLES COLLECTED AT 13 STATIONS IN THE LOS ANGELES, CALIFORNIA, AREA (FROM GORDON, 1976, WITH PERMISSION OF THE AMERICAN CHEMICAL SOCIETY)

| *Compound* | *Quarter concentration* (*ng*/*m*$^3$) | | | |
|---|---|---|---|---|
| | 1 | 2 | 3 | 4 |
| Pyrene | 0·58 | 0·23 | 0·25 | 1·24 |
| Fluoranthene | 0·38 | 0·15 | 0·24 | 0·68 |
| Benz[*a*]anthracene | 0·30 | 0·06 | 0·10 | 0·59 |
| Chrysene | 0·70 | 0·26 | 0·44 | 1·57 |
| Benzo[*e*]pyrene | 1·30 | 0·42 | 0·62 | 1·96 |
| Benzo[*a*]pyrene | 0·77 | 0·17 | 0·26 | 1·27 |
| Benzo[*b*]fluoranthene | | 0·24 | 0·33 | 1·30 |
| Benzo[*j*]fluoranthene | 0·26 | 0·06 | 0·12 | 0·43 |
| Benzo[*k*]fluoranthene | 0·27 | 0·07 | 0·15 | 0·52 |
| Perylene | 0·22 | | 0·06 | 0·22 |
| Anthanthrene | 0·33 | 0·08 | 0·14 | 0·79 |
| Benzo[*ghi*]perylene | 3·80 | 1·35 | 2·71 | 8·25 |
| Indeno[1,2,3-*cd*]pyrene | 1·79 | 0·68 | 1·00 | 2·64 |
| Coronene | 2·49 | 1·13 | 1·66 | 4·44 |
| Total PAH measured | 13·19 | 4·90 | 8·08 | 25·90 |

urban air ranged from 0·2 to 29·5 ng/m$^3$. Mean B*a*P concentration at the urban stations was 2·96 ng/m$^3$ in 1966 and 2·62 ng/m$^3$ in 1970. B*a*P concentrations at nonurban stations ranged from 0·1 to 2·1 ng/m$^3$ with mean values of 0·44 ng/m$^3$ in 1966 and 0·23 ng/m$^3$ in 1970. Olsen and Haynes (1969) observed that B*a*P concentrations in urban air samples were higher in the winter (mean, 21·26 ng/m$^3$) than in the summer (mean, 2·05 ng/m$^3$). Similar seasonal variations in the concentrations of other PAH in airborne particulates have been reported (NAS, 1972; Pierce and Katz, 1975; Gordon, 1976), and may represent enhanced degradation of PAH at higher temperatures and light intensities during the summer or increased consumption of fuels for heating and other purposes during the winter. A third contributing factor may be the lower efficiency of airborne PAH sampling methods at high temperatures (Pupp *et al.*, 1974). However, high molecular weight PAH show the same seasonal concentration trends as do the volatile low molecular weight PAH (Table 27).

Pierce and Katz (1975) measured the PAH quinone concentrations in airborne particulates collected at urban and nonurban stations around Toronto, Ontario. Total PAH quinone concentrations were lower than those of PAH and did not show seasonal variations in the concentration that PAH did. Ratios of B*a*P quinone to B*a*P varied from 0·039 in February–March to 0·092 in August, while those of benz[*a*]anthracene quinone to benz[*a*]anthracene varied from $1{\cdot}2 \times 10^{-3}$ in February–March to $9{\cdot}2 \times 10^{-3}$ in September.

Further evidence that airborne particulate PAH are more persistant in the atmosphere than would be predicted from laboratory photooxidation studies is presented in a recent paper by Lunde and Bjorseth (1977). Lunde and Bjorseth analyzed PAH in four air samples collected in southern Norway. The origin of the air samples was inferred from meteorological data on air mass trajectories. Samples originating in England/France and northern England/Scotland contained a total of 32·099 and 28·252 ng/m$^3$ PAH, respectively (22 PAH analyzed), approximately 20 times higher than PAH concentrations in air samples from northern Norway or from stationary air from southern Norway. These preliminary data indicate that PAH can persist at relatively high concentrations in long-range transported aerosols.

The concentrations, in ng/mg particulates of PAH, in emission sources are substantially higher than those measured in ambient air. Mean B*a*P concentration in a large number of air samples collected in and around Los Angeles, California, was 5·24 ng/mg (Gordon, 1976), while the corresponding mean value for the Toronto, Ontario, area was 17·4 ng/mg (Pierce and Katz, 1975). Total PAH concentrations (11 PAH measured) in the California samples ranged from 41 to 62 ng/mg with a mean of 49·9 ng/mg (Gordon, 1976). It seems likely from the considerations discussed above that these ambient air PAH concentrations represent an approximate equilibrium between PAH emission rates and photodegradation rates.

Suess (1976) estimated that the annual worldwide emission of B*a*P to the atmosphere amounted to 5044 tons based on data for 1966–1969 (Table 28). The largest contribution was from heating and power generation, particularly where coal served as a fuel. Vehicular exhaust contributed less than 10 per cent of the total. However, Gordon (1976) and Pierce and Katz (1975) suggested that vehicular exhaust emissions were the primary source of airborne particulate PAH in Los Angeles and Toronto, respectively.

Owing to large gaps in our knowledge about the temporal and spatial distribution of PAH in air, airborne PAH residence times, and dispersal distances under natural conditions, any quantitative estimate of the contribution of airborne PAH to the total PAH burden of fresh and marine waters is extremely difficult. Giger and Schaffner (1978) reported that B*a*P represented approximately 1 per cent of the total PAH fraction extracted

TABLE 28

ESTIMATED ANNUAL RATE OF BENZO[*a*]PYRENE EMISSIONS TO THE ATMOSPHERE BASED ON DATA FOR THE PERIOD 1966–1969 (FROM SUESS, 1976, WITH PERMISSION OF ELSEVIER SCIENTIFIC PUBLISHING CO.)

| *Source of formation* | | *BaP emission (tons/year)* *USA* | *Worldwide (excl. USA)* | *Total worldwide* |
|---|---|---|---|---|
| *Heating and power generation using:* | | | | |
| coal | | 431 | 1 945 | 2 376 |
| oil | | 2 | 3 | 5 |
| gas | | 2 | 1 | 3 |
| wood | | 40 | 180 | 220 |
| | subtotal | 475 | 2 129 | 2 604 |
| *Industrial processes* | | | | |
| coke production | | 192 | 841 | 1 033 |
| catalytic cracking | | 6 | 6 | 12 |
| | subtotal | 198 | 847 | 1 045 |
| *Refuse and open burning* | | | | |
| Enclosed incineration | | | | |
| commercial and industrial | | 23 | 46 | 69 |
| other | | 11 | 22 | 33 |
| Open burning | | | | |
| coal refuse fires | | 340 | 340 | 680 |
| forest and agriculture | | 140 | 280 | 420 |
| other | | 74 | 74 | 148 |
| | subtotal | 588 | 762 | 1 350 |
| *Vehicles* | | | | |
| trucks and buses | | 12 | 17 | 29 |
| automobiles | | 10 | 6 | 16 |
| | subtotal | 22 | 23 | 45 |
| Grand totals | | 1 283 | 3 761 | 5 044 |

from airborne particulate samples from Zurich, Switzerland. Using the value of Gordon (1976) of 5·24 ng/mg particulates as a conservative estimate of the equilibrium airborne particulate B*a*P concentration and 524 ng/mg particulates as the estimated concentration of total airborne particulate PAH, a rough estimate of the rate of B*a*P and total PAH input to the oceans from this source can be made. Robinson (1976) estimated the annual input of nonsilicate anthropogenic particles from the air to oceans to be $100 \times 10^6$ tons. Assuming a B*a*P concentration of 5 ng/mg, B*a*P input to the oceans from this source would be 500 tons/year or 10 per cent of the B*a*P emitted to the atmosphere. If total PAH are present at a concentration of approximately 500 ng/mg, total PAH input to the oceans from this source would be 50 000 tons/year. Inputs to freshwaters would be much lower owing to the smaller total surface area of freshwater bodies. Deposition of PAH onto water from the vapor phase is not considered since there is a complete lack of information about concentrations of PAH in the vapor phase, their residence time here, and their dispersal properties. It should be

TABLE 29

ESTIMATED INPUTS OF BENZO[*a*]PYRENE AND TOTAL PAH TO THE AQUATIC ENVIRONMENT FROM VARIOUS SOURCES

| *Sources* | *Input in metric tons/year* | |
|---|---|---|
| | *BaP* | *total PAH* |
| Biosynthesis | 25 | 2 700 |
| Petroleum spillage | 20–30 | 170 000 |
| Domestic and industrial wastes | 29 | 4 400 |
| Surface runoff from land | 118 | 2 940 |
| Fallout and rainout from air | 500 | 50 000 |
| Total input | 697 | 230 040 |

kept in mind, however, that vapor-phase deposition onto water surfaces is considered to be the major mode of deposition of airborne DDTand PCBs in the oceans (Risebrough *et al.*, 1976; Bidleman *et al.*, 1976).

In summary, the estimated annual input of PAH to the aquatic environment from all sources is approximately 697 tons B*a*P/year and 230 040 tons PAH/year (Table 29). It is recognized that these estimates are based in most cases on very incomplete and inadequate data. Many extrapolations and assumptions, some of which may not be completely valid, had to be made to arrive at the estimates. However, it is believed that the order of magnitude of the estimates is correct and presents a reasonable picture of the relative contribution of different sources to the total B*a*P and PAH inputs. Surface runoff from land and fallout from the air appear to be the main sources of B*a*P in the aquatic environment, while petroleum spillage is the main source of total PAH. This reflects the relatively high PAH content of petroleum in comparison to other major PAH sources.

However, it should be noted that the composition of PAH assemblages from the different input sources varies substantially. The PAH assemblage from crude and refined petroleum is dominated by the tricyclic aromatic hydrocarbons. In the 11 crude oil samples analyzed by Gilchrist *et al.* (1972) (Table 8), tricyclics represented 50 to 93 per cent (mean, 67 per cent) of the total PAH assemblage (tri- and hexacyclics). PAH assemblages from other sources are usually not dominated to the same extent by low molecular weight PAH. For example, tricyclic aromatics represented 20·5 to 55 per cent (mean, 33·7 per cent) of the total PAH in four samples of airborne particulate PAH coke oven emissions analyzed by Lao *et al.* (1975) (Table 18). Since all the proven PAH carcinogens are in the four-to-six ring PAH groups (see Table 86), the contribution of potentially carcinogenic PAH to the aquatic environment from oil spillage is less than might be implied from the estimates in Table 29 of inputs of total PAH to the aquatic environment from different sources. A better index of the contribution of different input sources to the carcinogenic PAH burden of the aquatic environment is provided by the estimates of B*a*P inputs from different sources.

*CHAPTER 4*

# *Distribution of PAH in the Aquatic Environment*

The amounts of B*a*P and total PAH estimated to enter the aquatic environment from all sources are significantly lower than the estimated inputs to the atmosphere and to land. If these aquatic PAH were evenly distributed through the world's oceans and freshwater bodies, their concentrations would be so low as to be completely undetectable and inconsequential. However, they are not evenly distributed. Most of the environmental PAH burden remains relatively near the point-source of the PAH and concentrations of PAH can be expected to decrease approximately logarithmically with distance from the source. Thus, the majority of PAH entering the aquatic environment are localized in rivers, estuaries, and coastal marine waters.

PAH in the aqueous phase are probably less sensitive to photooxidation than airborne PAH. Therefore, PAH are more persistent in water than in air. When incorporated into anoxic sediments they may persist for long, even geological, times. The cycle of PAH in the aquatic environment appears to be relatively simple. PAH entering water from various sources quickly become adsorbed to organic and inorganic particulate matter. Much of the particulate PAH are deposited in bottom sediments. Leaching or biological activity in the sediment may return a small fraction of sediment PAH to the water column. PAH are readily accumulated by aquatic biota to levels higher than those in the ambient medium. Relative concentrations of PAH in aquatic ecosystems are generally highest in the sediments, intermediate in aquatic biota, and lowest in the water column. Routes of removal of PAH from the aquatic environment include volatilization of PAH from the water surface (mainly low molecular weight PAH), photooxidation, chemical oxidation, microbial metabolism, and metabolism by higher metazoans.

TABLE 30

PUBLISHED VALUES FOR THE SOLUBILITY OF PAH IN DISTILLED WATER AT 25–27 °C

| *Compound* | *Number of aromatic rings* | *Solubility (μg/liter)* |
|---|---|---|
| Naphthalene | 2 | 12 500,[b] 22 000,[d] 31 200,[h] 31 300,[c] 34 000[f] |
| 1-Methylnaphthalene | 2 | 25 800[c] |
| 1,5-Dimethylnaphthalene | 2 | 2 740[c] |
| 2-Ethylnaphthalene | 2 | 8 000[c] |
| Acenaphthalene | 2 | 3 470[c] |
| Fluorene | 2 | 1 685[k] |
| Anthracene | 3 | 30,[d] 44·6,[e] 75,[a] 75[b] |
| 2-Methylanthracene | 3 | 21·3[e] |
| Phenanthrene | 3 | 1 002,[e] 1 070,[c] 1 180,[h] 1 600,[a] 1 600[b] |
| 1-Methylphenanthrene | 3 | 269[e] |
| 4,5-Methylenephenanthrene | 3 | 1 100[a] |
| Fluoranthene | 3 | 206,[k] 236,[d] 240,[a] 265[b] |
| Pyrene | 4 | 132,[k] 165,[a] 171,[d] 175[b] |
| Naphthacene | 4 | 1·0,[a] 1·5[b] |
| Chrysene | 4 | 1·5,[a] 1·8,[k] 6[b] |
| Triphenylene | 4 | 38,[a] 43[b] |
| Benz[*a*]anthracene | 4 | 9·4,[k] 10[b] 11[a] |
| 10-Methylbenz[*a*]anthracene | 4 | 55[a] |
| 10-Ethylbenz[*a*]anthracene | 4 | 45[a] |
| 10-Butylbenz[*a*]anthracene | 4 | 8[a] |
| 10-Amylbenz[*a*]anthracene | 4 | 0·8[a] |
| 9,10-Dimethylbenz[*a*]anthracene | 4 | 43[a] |
| Dibenz[*ac*]anthracene | 5 | 0·5,[a] 0·6[b] |
| Benzo[*a*]pyrene | 5 | 0·172,[j] 0·5,[g] 4·0,[a] 6·05[i] |
| 5-Methylbenzo[*a*]pyrene | 5 | 0·8[a] |
| Benzo[*e*]pyrene | 5 | ~4[d] |
| Perylene | 5 | <0·5[a] |
| Picene | 5 | 2·5[a] |

[a] Davis *et al.*, 1942.
[b] Klevens, 1950.
[c] Eganhouse and Calder, 1976.
[d] Schwarz and Wasik, 1976.
[e] May *et al.*, 1978a.
[f] Bohon and Claussen, 1951.
[g] Eisenbrand, 1971.
[h] Wauchope and Getzen, 1972.
[i] Weil-Malherbe, 1946.
[j] Lu *et al.*, 1977.
[k] May *et al.*, 1978b.

## 4.1 PAH IN WATER

### 4.1.1 Solubility of PAH

One of the most important characteristics of PAH relative to their incidence in water is solubility. Although PAH have higher aqueous solubilities than alkanes of similar molecular weight (McAuliffe, 1966), their solubilities are quite low, reflecting their nonpolar hydrophobic nature. The published values for solubility of PAH in distilled water show considerable variation (Table 30), undoubtedly owing to differences in the methods of preparation and measurement of saturated solutions of PAH. Several of the problems in preparing saturated solutions of PAH have been discussed by May *et al.* (1978a). Nevertheless, some definite trends do emerge. As might be expected, solubility tends to decrease as the number of aromatic rings or molecular weight increases. Naphthalene has a solubility of about 30 ppm while five-ring PAH have solubilities in the 0·5–5·0 ppb range. Alkyl PAH usually have lower solubilities than unalkylated parent compounds and solubility decreases with increasing degrees of alkylation. There may be exceptions—benz[*a*]anthracene is less soluble than methyl- or ethylbenz[*a*]anthracene, and chrysene is less soluble than methyl- and dimethylchrysene (Davis *et al.*, 1942). Linear PAH are generally much less soluble than angular or periconsidered PAH isomers. Thus, anthracene is less soluble than phenanthrene and naphthacene is less soluble than chrysene, benz[*a*]anthracene, or triphenylene.

The limited data available indicate that PAH are slightly less soluble in seawater than in freshwater owing to salting out (Table 31). These

TABLE 31
SOLUBILITY OF SELECTED PAH IN SEAWATER

| *Compound* | *Solubility (mg/liter)* |
|---|---|
| Naphthalene | $20 \pm 2$,[a] $22{\cdot}0 \pm 0{\cdot}293$[b] |
| 2,6-Dimethylnaphthalene | $2{\cdot}4 \pm 0{\cdot}5$[a] |
| 2,3,6-Trimethylnaphthalene | $1{\cdot}7 \pm 0{\cdot}6$[a] |
| Fluorene | $0{\cdot}8 \pm 0{\cdot}2$[a] |
| Phenanthrene | $0{\cdot}6 \pm 0{\cdot}1$,[a] $0{\cdot}71 \pm 0{\cdot}025$[b] |
| 1-Methylphenanthrene | $0{\cdot}3 \pm 0{\cdot}1$[a] |
| Fluoranthene | $0{\cdot}1 \pm 0{\cdot}06$[a] |

[a] 32‰ salinity, 22 ± 2 °C (Rossi and Neff, 1978).
[b] 35‰ salinity, 25·0 ± 0·5 °C (Eganhouse and Calder, 1976).

TABLE 32

EFFECT OF TEMPERATURE ON THE SOLUBILITY OF PAH IN DISTILLED WATER (FROM MAY *et al.*, 1978a, WITH PERMISSION OF THE AMERICAN CHEMICAL SOCIETY)

| *Temperature (°C)* | *Solubility (µg/kg)* Phenanthrene | 1-*Methyl-phenanthrene* | *Anthracene* | 1-*Methyl-anthracene* |
|---|---|---|---|---|
| 5·2 | | | 12·7 ± 0·4 | |
| 6·3 | | | | 7·06 ± 0·18 |
| 6·6 | | 95·2 ± 0·2 | | |
| 8·5 | 423 ± 4 | | | |
| 8·9 | | 114 ± 4 | | |
| 9·1 | | | | 8·48 ± 0·09 |
| 10·0 | 468 ± 2 | | 17·5 ± 0·3 | |
| 10·8 | | | | 9·43 ± 0·37 |
| 12·5 | 512 ± 1 | | | |
| 13·9 | | | | 11·1 ± 0·3 |
| 14·0 | | 147 ± 1 | | |
| 14·1 | | | 22·2 ± 0·1 | |
| 15·0 | 601 ± 7 | | | |
| 18·3 | | | 29·1 ± 0·6 | 14·5 ± 0·1 |
| 19·2 | | 193 ± 1 | | |
| 21·0 | 816 ± 8 | | | |
| 22·4 | | | 37·2 ± 1·1 | |
| 23·1 | | | | 19·1 ± 0·6 |
| 24·1 | | 255 ± 5 | | |
| 24·3 | 995 ± 1 | | | |
| 24·6 | | | 43·4 ± 0·1 | |
| 26·9 | | 304 ± 1 | | |
| 27·0 | | | | 24·2 ± 0·1 |
| 28·7 | | | 55·7 ± 0·7 | |
| 29·9 | 1 277 ± 11 | 355 ± 2 | | |
| 31·1 | | | | 32·1 ± 0·3 |

differences are not large. Temperature has a much greater effect on the aqueous solubility of PAH (Table 32). Solubility of phenanthrene increased from 423 to 1277 ppb between 8·5 and 29·9 °C, while that of anthracene increased from 12·7 to 55·7 ppb between 5·2 and 28·7 °C (May *et al.*, 1978a). Eganhouse and Calder (1976) showed that, in some cases, the presence of one or two PAH in solution affected the solubility of an additional PAH in that solution. The solubility of naphthalene was unaffected by the presence of other PAH. Both naphthalene and phenanthrene enhanced the solubility of acenaphthalene, but biphenyl and phenanthrene together exhibited mutual solubility reduction.

### 4.1.2 Solubilization of PAH

Natural waters, and even more so wastewaters, may contain significant quantities of dissolved or colloidal organic matter which may act to solubilize PAH. PAH could be solubilized by incorporation into micelles. A micelle is composed of an aggregate of surface-active molecules (surfactants), each possessing a hydrophobic hydrocarbon chain and an ionizable hydrophilic group (Elsworthy *et al.*, 1968). The solubilization of a hydrocarbon is thought to occur by association of the hydrocarbon with the hydrophobic core of the micelle, with the polar surface of the micelle, or both. However, a critical micellar concentration must be reached before solubilization of PAH is effected. The concentration of detergent such as linear alkylbenzene sulfonate must be as high as 10–50 mg/liter before significant solubilization of B*a*P occurs. Detergent concentrations of 0·1–1·0 mg/liter have no effect on B*a*P solubility (Bohm-Gossl and Kruger, 1965; Il'nitskii and Ershova, 1970; Il'nitskii *et al.*, 1971b).

Another mechanism for solubilization of PAH in water is hydrotropy, the enhancement of the solubility of organic compounds as a result of the introduction of other organic compounds not associated with colloid or micelle formation. A wide variety of organic compounds increase the apparent solubility of PAH in water. These include butyric and lactic acids (Ekwall and Sjoblöm, 1952), purines (Weil-Malherbe, 1946), organic solvents such as ethanol, dioxane and acetone (Suess, 1972b), and nitrogen-containing organics (Eisenbrand, 1971). Eisenbrand concluded that the effectiveness of nitrogen-containing organics as solubilizers of PAH increases with an increase in the nitrogen content of the molecule, methylation of the nitrogen, and presence of double bonds. The increase in solubility of the PAH is either directly proportional to the concentration of the solubility mediator or to the square of its concentration.

Natural waters contain substances such as humic acids, fulvic acids, and other degradative products of biological materials that may act as PAH solubilizers. Boehm and Quinn (1973) studied the effects of removal of these natural solubilizers on the solubility of alkane, isoprenoid, and aromatic hydrocarbons in natural seawater. Removal of dissolved organic matter from seawater resulted in a 50–99 per cent decrease in the amounts of n-alkanes and isoprenoid hydrocarbons solubilized, but had no effect on the solubility of phenanthrene and anthracene.

### 4.1.3 Particulate PAH

Because of their low aqueous solubility and hydrophobic character, PAH readily adsorb to particulate materials and solid surfaces in water.

Adsorption and concentration occur in the presence of such substrates as activated carbon, calcareous material, silica, glass, soil particles, and organic particles (Andelman and Suess, 1970; Herbes, 1977; May *et al.*, 1978a). Herbes (1977) studied the adsorption of anthracene from solution by organic particles (autoclaved yeast cells). He found that a constant fraction (0·45 ± 0·01) of the total anthracene was adsorbed by 35 mg/liter yeast cells over equilibrium anthracene concentrations ranging from 0·02 to 31 μg/liter. The fraction of anthracene adsorbed increased with increased concentration of yeast cells to 72 per cent at 250 mg particles/liter. As temperature increased the fraction adsorbed decreased. He concluded that a significant fraction (0·15–0·65) of anthracene would be associated with detrital and living organic matter in natural waters containing moderate levels of suspended organic solids and suggested that the role of suspended mineral particulate material may be far less significant in adsorption of PAH in natural waters than is the role of suspended organic matter. In support of this conclusion, Meyers and Quinn (1973) reported that only 22 per cent of the anthracene (100 μg/liter) in saline solution was adsorbed by bentonite clay at 25 °C. Thus, bentonite clay was a less efficient adsorbent than yeast cells for anthracene.

Lewis (1975) analyzed five PAH (fluoranthene, B*a*P, indeno[1,2,3-*cd*]pyrene, benzo[*k*]fluoranthene, and benzo[*ghi*]perylene) in the particulate and soluble phases of a large number of samples from different stations along the Severn and Trent Rivers, UK. Concentrations of particulate PAH varied more than those of the soluble PAH. Elevated particulate PAH were generally accompanied by elevated soluble PAH concentrations in the same sample. The concentration of soluble PAH usually represented 2–16 per cent of the concentration of particulate PAH. As might be expected, fluoranthene, the most soluble PAH analyzed, was present in solution at higher concentrations than the other PAH in most samples. Similar results were obtained by Borneff and Kunte (1964) for water samples from the River Main at Seligenstadt, GFR. These results strongly suggest that in natural water systems an exchange equilibrium exists between adsorbed and soluble PAH. While this equilibrium favors the particulate form, a significant fraction of the total PAH is present in solution except in very heavily PAH-contaminated systems.

### 4.1.4 PAH Concentrations in Natural Waters

The concentrations of PAH in fresh waters are quite variable, reflecting the variety of sources of these waters and uses to which they have been put. River water often contains significant quantities of B*a*P and other PAH

TABLE 33

TYPICAL CONCENTRATION RANGES OF BENZO[*a*]PYRENE AND TOTAL PAH IN VARIOUS RIVER WATERS

| *Source* | *Concentration (ng/liter)* *BaP* | *Concentration (ng/liter)* *Total PAH* | *Reference* |
|---|---|---|---|
| River Rhine at Mainz, GFR | 50–110 | 730–1 500 | Borneff and Kunte, 1964 |
| River Rhine at Koblenz, GFR | 10–60 | 500–3 000 | Hellmann, 1974 |
| River Aach at Stockach, GFR | 4–43 | 1 440–3 100 | Borneff and Kunte, 1965 |
| Other GFR rivers | 0·6–10 | 200–1 000 | Borneff and Kunte, 1964, 1965 |
| Pskov region, USSR, remote from human activity | 0·01–0·1 | — | Il'nitskii *et al.*, 1971b |
| Sunzha River, USSR, below discharge of an oil refinery | 50–3 500 | — | Samoilovich and Red'kin, 1968 |
| Thames River, England | 170–280 | 800–2 350 | Acheson *et al.*, 1976 |
| Trent River, England | 5·3–504 | 25–3 790 | Lewis, 1975 |
| Severn River, England | 1·5–48 | 20–266 | Lewis, 1975 |
| Oyster River, Connecticut, USA. | 78–150 | — | Keegan, 1971 |
| Monongahela River at Pittsburgh, Pennsylvania, USA | 42–77 | 600–663 | Basu and Saxena, 1978 |
| Ohio River at Huntington, West Virginia, USA | 5·6 | 57·9 | Basu and Saxena, 1978 |
| Delaware River at Philadelphia, Pennsylvania, USA | 41·1 | 351·8 | Basu and Saxena, 1978 |

(Table 33). In most cases, there is a direct relationship between PAH concentrations in river water and the degree of industrialization and other human activity along the banks and adjacent flood plain. Rivers remote from human activity are relatively uncontaminated. Ground water and well water contain PAH concentrations which are lower than those in river water by a factor of ten or more (Table 34). PAH in ground water are thought to be derived from the leaching of surface water through soils containing PAH (Borneff and Kunte, 1969; Hellmann, 1974; Suess, 1976). Purified tap water and reservoir water contain a PAH concentration similar to, or slightly higher than, that of ground water. Rain water may contain significant concentrations of PAH and represents an important route to land and water contamination from airborne particulate PAH.

TABLE 34

TYPICAL CONCENTRATION RANGES OF BENZO[*a*]PYRENE AND TOTAL PAH IN VARIOUS FRESHWATER SOURCES AND DRINKING WATER

| *Source* | *Concentration (ng/liter)* *BaP* | *Total PAH* | *Reference* |
|---|---|---|---|
| Ground water (GFR) | 0·4–7·0 | 10·9–123·5 | Borneff and Kunte, 1969 |
| Ground water (GFR) | 2–4 | 100–200 | Hellmann, 1974 |
| Ground water (USA) | 0·2 | 8·3 | Basu and Saxena, 1978 |
| Well water (GFR) | 2–15 | 100–750 | Hellmann, 1974 |
| Well water (England) | 0·2–0·6 | 3·6–5·8 | Lewis, 1975 |
| Tap water (GFR) | 0·5–9·0 | 29·2–125·5 | Borneff and Kunte, 1969 |
| Tap water (9 US cities) | 0·2–1·6 | 0·9–14·9 | Basu and Saxena, 1978 |
| Reservoirs (Moscow) | 4–13 | — | Il'nitskii and Rozhnova, 1970 |
| Reservoirs (England) | 0·7–3·8 | 9·1–43·2 | Lewis, 1975 |
| Rain water (Koblenz, GFR) | 4–80 | 200–4 000 | Hellmann, 1974 |
| Lake Constance (GFR) | 0·2–11·5 | 25–234 | Borneff and Kunte, 1964 |
| Lake Erie at Buffalo, New York, USA | 0·3 | 4·7 | Basu and Saxena, 1978 |

Relatively little information is available concerning the concentrations of PAH in estuarine and oceanic waters. Apparently very few attempts have been made to precisely identify and quantify PAH concentrations in the oceans. Most of the data available are based on estimates of total aromatics by UV, infrared (IR), or fluorescence techniques, methods which may be subject to considerable interference from non-PAH material. Typical of this type of approach is the work of Zsolnay (1977) (Table 35). UV-absorbing 'aromatics' represented 1–10 per cent of the total hydrocarbons in the water samples. Aromatic hydrocarbon concentrations decreased with depth in most cases. Recently, Marty *et al.* (1978) measured the concentrations of aliphatic, alicyclic, and aromatic hydrocarbons in surface water, sea surface microlayer, and airborne particulates from the tropical Atlantic Ocean west of the Canary and Cape Verde Islands. Aromatic hydrocarbons represented 13 to 55 per cent of the total hydrocarbons in the samples (Table 36). PAH in the water samples included phenanthrene, alkylphenanthrene, perylene, fluoranthene, and pyrene.

TABLE 35

DISTRIBUTION OF HYDROCARBONS IN MARINE AND ESTUARINE WATERS. AROMATIC HYDROCARBON VALUES ARE RELATIVE AND ARE BASED UPON THE USE OF A PHENANTHRENE STANDARD AT 254 nm (FROM ZSOLNAY, 1977, WITH PERMISSION OF ELSEVIER SCIENTIFIC PUBLISHING CO.)

| *Area and date* | *Depth (m)* | *Total hydrocarbons, mean and SD (μg/liter)* | *Aromatic hydrocarbons (based on UV absorbance) (ng/liter)* |
|---|---|---|---|
| Baltic Sea (1973) | | | |
| | 1 | 13·9 ± 3·8 | 277 ± 121 |
| | 10–50 | 5·4 ± 1·8 | 52 ± 9 |
| | 1 m above sediment | 3·8 ± 1·4 | 47 ± 13 |
| Nova Scotia to Gulf Stream (1973) | | | |
| | 1 | 4·7 ± 1·3 | 30·8 ± 10 |
| | 10 | 3·5 ± 1·8 | 16·8 ± 4 |
| | 25 | 6·1 ± 1·8 | 8·6 ± 9 |
| Sargasso Sea off Bermuda (1974–76) | | | |
| | 1 | 0·32 ± 0·12 | 31 ± 10 |
| | 30 | 0·47 ± 0·25 | 1 ± 0 |
| | 300 | 0·26 ± 0·16 | 1 ± 1 |
| | 1 200 | 0·00 | 4 |
| | 2 000 | 0·00 | 0 |
| Mediterranean Sea (1975) | | | |
| | 1 | 16·0 | 148 ± 36 |

TABLE 36

MEAN CONCENTRATIONS OF HYDROCARBONS IN SURFACE SAMPLES OF SEAWATER, THE SEA SURFACE MICROLAYER, AND AEROSOLS COLLECTED 12 m ABOVE THE SEA SURFACE IN THE TROPICAL EASTERN ATLANTIC OCEAN (FROM MARTY *et al.*, 1978, WITH PERMISSION OF THE PUBLISHER)

| *Sample* | *Concentration of total hydrocarbons* | *Composition of the hydrocarbon fraction* | |
|---|---|---|---|
| | | *Aliphatic and alicyclic hydrocarbons* | *Aromatic hydrocarbons* |
| Seawater | 10 μg/liter | 67 %, 9 % n-alkanes | 33 % (3·3 μg/liter) |
| Surface microlayer | 39 μg/liter | 45 %, 6 % n-alkanes | 55 % (21·4 μg/liter) |
| Aerosols | 9 μg/m$^3$ | 87 %, 2 % n-alkanes | 13 % (1·2 μg/m$^3$) |

Traces of benzofluoranthenes and benzopyrenes were also detected. The dominant PAH in the surface microlayer and airborne aerosols was phenanthrene (40 per cent of the aromatic fraction). Mono-, di-, and tri-methylphenanthrenes were also abundant. No perylene, benzofluoranthenes, or benzopyrenes were detected. The authors hypothesized that the phenanthrenes were of biological origin and that the other PAH were from industrial smoke and petroleum spillage. It would appear, therefore, that the concentrations of aromatics in marine surface waters are similar to those in ground water and well water. Barbier *et al.* (1973) analyzed the hydrocarbons in surface water off the French coast at Brest. The concentration of total hydrocarbons was 137 μg/liter. Analysis of the hydrocarbon type distribution by UV spectrophotometry and mass spectrometry revealed that bicyclic aromatics represented 3·5 per cent and PAH represented 2·5 per cent of the total hydrocarbons present. Thus, the concentration of PAH in Brest seawater was approximately 3·4 μg/liter. Brown and Huffman (1976) analyzed a large number of water samples from the Atlantic, Mediterranean, and the Indian Ocean along the Persian Gulf oil-tanker route. The mean concentration of nonvolatile hydrocarbons in surface water from the Atlantic Ocean was 4 μg/liter with values from 1·3 to 13 μg/liter falling within one standard deviation of the mean. Mean relative aromatic concentration was 24 per cent of the total nonvolatile hydrocarbons. Included in the total aromatic fraction were 1–4 ring aromatic plus sulfur-containing aromatics (benzo- and dibenzo-thiophenes). PAH represented less than one-half of the total aromatic fraction, so PAH concentrations in Atlantic Ocean surface water were about 0·4 μg/liter with a range of 0·13 to 1·3 μg/liter.

There are only a few reports dealing with concentration of specific PAH in seawater. Seawater at Brest, France was reported to contain traces of B*a*P (Saliot, 1969, as cited by Barbier *et al.*, 1973). Niaussat and Auger (1970) examined the concentrations of B*a*P and perylene in the water and biota of the Clipperton Lagoon in the equatorial Pacific. The Lagoon seawater contained 1·6 μg/liter B*a*P and 3·05 μg/liter perylene. Since this area is remote from industrial activity, the PAH were thought to have originated from biosynthesis. Subsequently, Niaussat *et al.* (1975) reported B*a*P concentrations in lagoon waters of the Polynesian atolls of Moruroa and Hao ranging from nondetectable to 0·4 μg/liter. Armstrong *et al.* (1977) reported that a water sample taken approximately 15 m from an oil separator platform brine outfall in Trinity Bay, Texas, USA, contained 0·4, 0·2, and 0·6 μg/liter naphthalene, 1-methylnaphthalene, and 2-methyl-naphthalene, respectively (Table 37). No higher molecular weight PAH

TABLE 37

CONCENTRATIONS OF AROMATIC HYDROCARBONS IN THE BRINE EFFLUENT FROM AN OIL SEPARATOR PLATFORM IN TRINITY BAY, TEXAS, USA, AND IN THE WATER AND SEDIMENTS 15 m FROM THE BRINE OUTFALL. (FROM ARMSTRONG *et al.*, 1977, WITH PERMISSION OF THE AMERICAN PETROLEUM INSTITUTE)

| *Compound* | *Brine Effluent* ($\mu g/liter$) | *Water* ($\mu g/liter$) | *Sediment* ($\mu g/kg$ *wet weight*) |
|---|---|---|---|
| Benzene | 3 300 | 1·5 | — |
| Toluene | 3 500 | 3·2 | — |
| Xylenes | 2 400 | — | — |
| $C_2$-Benzenes | — | 3·1 | — |
| $C_3$-Benzenes | 650 | 0·80 | 600 |
| $C_4$-Benzenes | — | — | 400 |
| Naphthalene | 300 | 0·40 | 200 |
| 1-Methylnaphthalene | 370 | 0·20 | 200 |
| 2-Methylnaphthalene | 500 | 0·60 | 300 |
| Dimethylnaphthalenes | 190 | 0·70 | 8 000 |
| Trimethylnaphthalenes | 260 | — | 10 000 |
| $C_4$-Naphthalenes | — | — | 9 000 |
| Biphenyl | 15 | — | — |
| Methylbiphenyls | 11 | — | 200 |
| Dimethylbiphenyls | 30 | — | 800 |
| Fluorene | 14 | — | 100 |
| Methylfluorenes | 63 | — | 700 |
| Dimethylfluorenes | 42 | — | 900 |
| Trimethylfluorenes | — | — | 600 |
| Phenanthrene | 36 | — | 100 |
| Methylphenanthrenes | 63 | — | 600 |
| Dimethylphenanthrenes | 20 | — | 1 000 |
| Trimethylphenanthrenes | — | — | 500 |
| Total aromatics | 11 764 | 10·50 | 34 200 |

were detected in the water despite the fact that the undiluted brine effluent contained significant concentrations of PAH in the dimethylnaphthalene–dimethylphenanthrene range.

## 4.2 PAH IN SEDIMENTS

### 4.2.1 Sedimentation of PAH

Freshwater and marine sediments often contain concentrations of PAH of higher magnitudes than those in the overlying water. As indicated in the previous section, because of their low solubility and hydrophobic nature, PAH tend to be greatly enriched in the inorganic and organic suspended particulate fraction in the water column in comparison to PAH in solution. The particulates gradually settle out of the water column carrying adsorbed PAH with them. Flocculation of suspended clay-sized particulates, as occurs in the increasing salinity gradient of an estuary, can be expected to increase the rate of particle deposition. Once deposited in sediments, PAH are less subject to photochemical or biological oxidation, especially if the sediment is anoxic. Thus, sedimentary PAH tend to be quite persistent and may accumulate to high concentrations. Partitioning between seawater and sediments of aromatic hydrocarbons derived from a brine effluent produced by a separator platform in a shallow Texas bay was reported by Armstrong *et al.* (1977). The light aromatics, benzene through to methylnaphthalenes, were present at low concentrations in the water column 15 m from the brine outfall while the higher molecular weight aromatics, dimethylnaphthalenes through to trimethylphenanthrenes, were undetectable in the water column (limit of detection $\sim 0{\cdot}1\ \mu g$/liter) (Table 37). Nearly all the aromatics above $C_2$-benzene were present in the sediments 15 m from the outfall at concentrations equal to, or greater than, those in the undiluted brine, and at concentrations much higher than those in the overlying water column. Within an aromatic series, the enrichment of alkylated aromatics in the sediment was greater than that of the unalkylated parent compound. Thus, for all but the most soluble low molecular weight aromatics, partitioning between the aqueous and sedimentary phases greatly favors the sedimentary phase. Because of the favorable partition coefficients and greater persistence of sedimentary PAH than PAH in solution, we can expect that sediments will nearly always contain concentrations of PAH greater by a factor of 1000 or more than the overlying water. Therefore, sediment PAH assemblages can serve as useful indices of the rates of PAH inputs to the aquatic environment. Sediment samples have a substantial integrating effect on temporal patterns of PAH input and offer good geographical resolution, especially when current patterns, sediment origins, and settling rates are known (Dunn and Stich, 1976a). However, the composition of the PAH assemblage in the sediment may be different from that of the PAH source owing to differential partitioning of PAH between the sediment and aqueous phases.

Alternative sources of sedimentary PAH—indirect biosynthesis via reduction of extended quinones and related polycyclic materials of biological origin, and direct biosynthesis by sediment microbiota—were discussed earlier. These sources could be responsible for significant

TABLE 38

CONCENTRATION OF BENZO[*a*]PYRENE IN SELECTED FRESHWATER SEDIMENTS

| *Source* | *Depth in sediment (m)* | *BaP (μg/kg dry sediment)* | *Reference* |
|---|---|---|---|
| River Seine, about 40 km downstream from Paris | | | |
| core sample I | 5·9 | 3 000 | Mallet, 1965, 1966 |
| | 7·1 | 29 | |
| | 9 | 40 | |
| | 10 | 40 | |
| core sample II | 0·8 | 8 500 | |
| | 3 | 6 500 | |
| | 10 | 50 | |
| core sample III | 2 | 10 000 | |
| core sample IV | 5 | 390 | |
| core sample V | 9 | 100 | |
| River Seine mud | — | 15 000 | Depuis, 1960 |
| River sediment below a coke byproduct plant, USSR | — | 8 200–17 000 | Fedorenko, 1964 |
| Rublevskoye Reservoir, USSR | — | 44 | Il'nitskii *et al.*, 1972 |
| Khiminskoe Reservoir, USSR | — | 390–500 | |
| River Sunzha, USSR | | | |
| 3–4 km below oil-refinery discharge | — | 9·2–19·0 | Samoilovich and Red'kin, 1968 |
| 25 km below discharge | — | trace–3.1 | |
| Pskov region, USSR remote from exogenous sources | — | 1–2 | Ershova, 1968 |
| Grosser Ploner See, GFR | | | |
| 100 m from north shore | — | 1 610 | Grimmer and Bohnke, 1975 |
| 100 m from south shore | — | 260 | |
| Lake Constance, GFR | — | 443 | Müller *et al.*, 1977 |
| River Usk, UK | — | 11 000 | Eglinton *et al.*, 1975 |

localized contributions of relatively simple PAH assemblages to the total sediment PAH burden.

### 4.2.2 PAH Concentrations in Sediments

The concentration of B*a*P has been measured in a large number of freshwater and marine sediments, primarily from European waters. B*a*P concentrations in river and lake sediments varied widely (Table 38). Rivers flowing through heavily populated or industrial areas contained bottom

TABLE 39

CONCENTRATION OF BENZO[*a*]PYRENE (B*a*P) IN SELECTED MARINE SEDIMENTS

| *Source* | *Water depth (m)* | *BaP (μg/kg dry sediment)* | *Reference* |
|---|---|---|---|
| Port of Dunkerque | — | 400 | Mallet *et al.*, 1963 |
| Canal d'Ille-Jeanty | — | 1 760 | |
| Southern regions of the North Sea | — | traces–122·5 | |
| Godthaab, Greenland | 0·20 | 5 | |
| Rhône River delta | — | 0–906 | Lalou, 1963 |
| Villefranche Bay, France | 4–82 | traces–1 500 | |
| Mouth of the Var River | 0–102 | 0–34 | |
| Port-Vendres | 16 | 23 | |
| Banyuls | 8 | 150 | |
| Cerbere | 75 | 200 | |
| Bay of St. Malo, France | | | |
| Pissot | — | 160 | Mallet *et al.*, 1967 |
| Fours-a-Chaux | — | 1 320 | |
| Bassin | — | 170 | |
| Bay of Naples, Italy | 2–120 | 1·4–3 000 | Boucart and Mallet, 1965 |
| English Channel coast, France | — | 15 000 | Mallet *et al.*, 1963 |
| Southern Beaufort Sea | — | 0·4–20 | Wong *et al.*, 1976 |
| Near oil-refinery tanker jetty, USA | 10 | 80 | Gump *et al.*, 1977 |
| Severn estuary at Aust, UK. | — | 470 | Thompson and Eglinton, 1978 |
| Polynesian atolls | — | 0·3–6·5 | Niaussat *et al.*, 1975 |

sediments with high concentrations of B*a*P, sometimes reaching 10–15 ppm. Sediments from river beds in more remote areas contained concentrations of B*a*P in the low ppb range.

Marine sediments, most of them collected from near-shore coastal and estuarine locations in Europe by Mallet and co-workers, contained B*a*P concentrations varying to an even greater degree (Table 39). The highest sedimentary B*a*P levels always occurred adjacent to regions of high population or intense industrial activity. However, small amounts of B*a*P were detected even in sediments from relatively remote regions such as Godthaab, Greenland, the southern Beaufort Sea, and the Clipperton Lagoon in the tropical eastern Pacific Ocean. Sediment B*a*P concentrations may vary by a factor of 1000 or more within a single bay or estuary, possibly owing to different rates of sediment deposition or degrees of anoxia within the sediment in different regions of the bay. A rising gradient of sedimentary B*a*P can also be expected as one moves toward a major point source of PAH (Perdriau, 1964). Dunn and Stich (1976a) analyzed samples of bottom sediments in the vicinity of the Iona Island sewage treatment plant outfall in the Fraser River estuary, south of Vancouver, British Columbia, Canada.

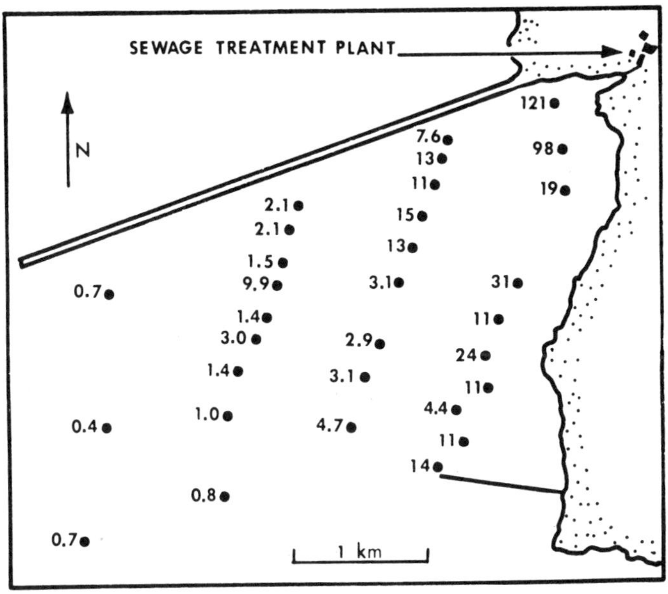


FIG. 6. Concentration of B*a*P in sediments in the vicinity of the Iona Island sewage treatment plant, Vancouver, British Columbia, Canada. Values are $\mu$g B*a*P/kg dry sediment (from Dunn and Stich, 1976a, reproduced by permission of the Minister of Supply and Services Canada).

A sharp gradient of contamination toward the outfall was observed (Fig. 6). The two samples taken closest to the outfall had B*a*P concentrations more than three times greater than the next most polluted sediment sample. Sediment B*a*P concentrations varied from 121 $\mu$g B*a*P/kg dry sediment near the outfall to 0·4–0·7 $\mu$g B*a*P/kg dry sediment 4–5 km downstream from the outfall. The authors concluded that the sewage treatment plant was the major contributor to the Fraser River estuary PAH burden.

Mallet *et al.* (1972b) reported that there was a seasonal cycle of B*a*P concentration in sediments from the Bay of St. Malo, France. B*a*P concentrations in the sediments were highest in January and December, 1967 (128 and 52·8 μg B*a*P/100 g dry sediment, respectively) and lowest in June, 1967 (15·4 μg B*a*P/100 g sediment). They concluded that this seasonal sedimentary B*a*P cycle resulted from a changing equilibrium between B*a*P biosynthesis by sediment anaerobes and B*a*P degradation by sediment aerobes. Possible seasonal variations in the rate of B*a*P input to the bay from wastewater and aerial fallout were not explored, although such variations would appear to be a reasonable explanation for a seasonal cycle of B*a*P in surface sediments.

The distribution of PAH within vertical cores of sediments from lake and coastal waters has been studied for clues to the sources of PAH in aquatic sediments. These studies are based on the reasonable assumptions that the rate of PAH deposition in sediments is proportional to their rate of release to the water body from nearby sources and to the sediment deposition rate, and that once deposited in the sediments PAH are relatively persistent and immobile. Thus, if either the rate of sedimentation or the age of sediments at different depths in the column can be determined, an accurate picture of the historical rate of PAH introduction to the water body can be inferred from an analysis of PAH at different depths within the sediment.

Lalou (1963) measured PAH concentrations at 15 depth intervals from the surface to 200 cm in a sediment core taken in the Bay of Villefranche, France. B*a*P concentrations increased from 1·8 ppm in the 0·3 cm sediment layer to 5 ppm at the 8–13 cm interval and then decreased gradually with depth to 0·026 ppm at 100 cm and 0·016 ppm at 200 cm. It was estimated that the sediments at 100 cm in the core were deposited 275 years ago and those at 200 cm were 550 years old. The initial increase in B*a*P concentrations with depth was not explained but may be attributed to sediment compaction or bioturbation in the surface layers of sediment. Sediments in the 8–13 cm depth interval were only 20–35 years old and could represent a historical period of increased B*a*P emissions to the Bay of Villefranche. The decrease in B*a*P concentration with greater depth in the sediment suggests a correlation of PAH input to industrial and domestic sources of PAH. Similar results were obtained in core samples from the River Seine about 40 km downstream from Paris (Mallet, 1965; 1966) (Table 38).

Grimmer and Bohnke (1975) investigated PAH in sediment layers dating back to 1915 at the maximum depth studied from the Grosser Ploner See (Schleswig-Holstein, GFR). A total of 64 PAH were identified and 25 of these were examined in detail at different levels in cores collected 100 m from the north and south lake shores. The southern shore of the lake is

FIG. 7. B*a*P concentrations in the annual layers of sediments from the (Nord) northern (built-up area) and (Süd) southern (forest) regions of the Grosser Ploner See (Schleswig-Holstein, GFR) from 1913 to 1972 (from Grimmer and Bohnke, 1975, with permission of Elsevier Scientific Publishing Co.).

woodland without settlements while the northern shore has been continuously settled and is crossed by a highway and main railway line. Sediments from the southern shore showed no significant variation in B*a*P concentration with depth but samples from the northern shore showed five times more PAH in the surface layers than in those deposited about 1913 (Fig. 7). Other PAH were distributed in the sediment cores in a similar fashion. Based on a comparison of the ratios of individual PAH in the sediment and those from various anthropogenic sources, the authors concluded that PAH in the northern shore sediments were derived primarily from airborne particulates produced by coal combustion. Automobile exhaust and petroleum spillage contributed only a minor fraction of the total PAH in the sediments. It was also concluded that the burden of carcinogenic PAH air pollutants has increased five-fold from 1915–70 along the northern lake shores.

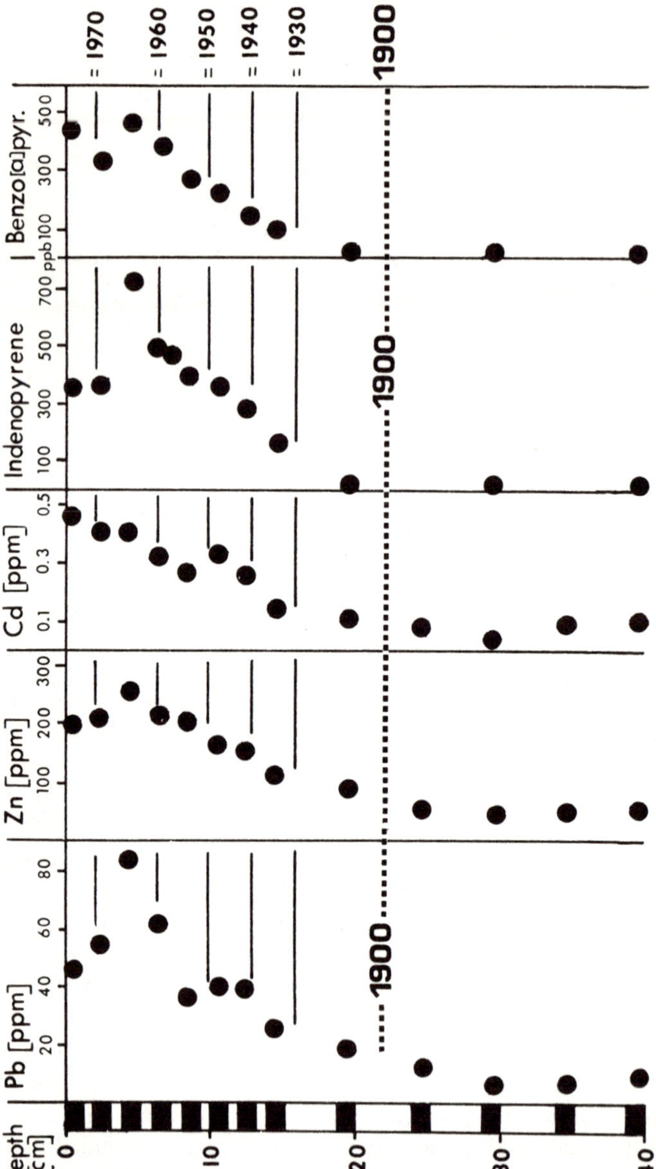


FIG. 8. Relationship between depth in dated sediments from Lake Constance, GFR, and concentrations of selected heavy metals and PAH (from Müller *et al.*, 1977, with permission of Springer-Verlag).

More recently Müller *et al.* (1977) performed a similar investigation with dated sediments from Lake Constance, GFR. Heavy metals and PAH were measured in sediment layers dating back to before 1900. Both showed a similar distribution. In most cases, the highest concentrations of individual PAH or heavy metals were observed in the top 5 cm of the sediment and decreased with depth below this level (Fig. 8). Vehicular exhaust and petroleum spills were excluded as major sources of PAH enrichment in surface layers of the sediment. Coal combustion was thought to be the main source of both PAH and heavy metals in the surface sedimentary layer. PAH were more enriched in surface layers in comparison to the 1900 stratum than were heavy metals. Enrichment factors—the ratio of the highest concentration of a PAH in the section to its concentration in the 19–20 cm stratum, which is closest to the 1900 time mark—varied between 15·5 and 56·9, indicating a substantial increase in PAH inputs to the lake since the turn of the century.

Hites *et al.* (1977) examined the distribution of PAH in three sections of dated sediment core corresponding to times of deposition of approximately 1970, 1900, and 1850, from Buzzards Bay, Massachusetts (Table 40). The investigators observed that PAH distribution has remained qualitatively constant over the past 125 years. Distribution of alkyl homologs of the PAH was unchanged with depth, within experimental error. While the absolute PAH concentrations remained essentially constant between 1970 and 1900, those for 1850 were substantially lower. The authors concluded that the nature of the PAH source has been constant since 1850 but that the intensity of this source increased considerably between 1850 and 1900 and has not changed much since. Analysis of historical information concerning the types and quantities of fossil fuels used in the area indicated that the probable source of the sedimentary PAH in this location is from the anthropogenic combustion of fossil fuels. In support of this conclusion, Farrington *et al.* (1977) reported that, in a sediment core from the same area, phenanthrene and $C_1$- and $C_2$-phenanthrene concentrations decreased with increasing depth in the sediment. The ratio of phenanthrene to $C_1$- and $C_2$-phenanthrene remained relatively constant with depth at 1·00:0·8:0·8. This ratio suggests a pyrolytic origin for the sedimentary PAH assemblage.

Wakeham (1977) used synchronous fluorescence spectroscopy to study the PAH composition of sediments from Lake Washington, Washington, USA. Multiple sources for the sedimentary PAH were inferred from PAH type distribution in the sediment strata and from analysis of PAH distribution in potential PAH sources. Naphthalene and three- and four-

TABLE 40

CONCENTRATIONS OF PAH (DRY WEIGHT BASIS) AND DISTRIBUTIONS OF ALKYL HOMOLOGS IN THREE SECTIONS OF A SEDIMENT CORE TAKEN FROM BUZZARDS BAY, MASSACHUSETTS, USA. THE ESTIMATED ERRORS IN THE PAH CONCENTRATIONS ARE ±20 PER CENT. THE RELATIVE ABUNDANCES OF PAH ARE EXPRESSED AS A PERCENTAGE OF THE TOTAL UNSUBSTITUTED PAH (VALUES IN PARENTHESES). THE ERROR IN DATES IS ±15 YEARS (FROM HITES *et al.*, 1977, WITH PERMISSION OF AAAS. COPYRIGHT, 1977, BY THE AMERICAN ASSOCIATION FOR THE ADVANCEMENT OF SCIENCE)

| *Year of deposition* | *Depth (cm)* | *PAH concentration (ppb) and relative abundance* | | | | | | *Homolog distribution* | |
|---|---|---|---|---|---|---|---|---|---|
| | | *Phenanthrene* | *Fluoranthene* | *Pyrene* | $C_{18}H_{12}$[a] | $C_{20}H_{12}$[b] | *Total*[c] | *Slope*[d] | *SD*[e] |
| 1970 | 0–4 | 53 (7) | 130 (16) | 120 (15) | 160 (20) | 340 (43) | 800 | −1500 | ±500 |
| 1900 | 20–24 | 42 (5) | 130 (15) | 120 (14) | 200 (23) | 380 (44) | 870 | −1400 | ±600 |
| 1850 | 38–42 | 8 (13) | 11 (17) | 7 (11) | 12 (19) | 26 (41) | 64 | −1800 | ±300 |

[a] Includes chrysene, benzanthracenes and triphenylene.
[b] Includes benzofluoroanthenes and benzopyrenes.
[c] Total unsubstituted PAH concentration, dry weight basis.
[d] Slope of a least squares line fitted to the average abundance (composited over the $C_{16}H_{10}$, $C_{18}H_{12}$ and $C_{20}H_{12}$ species) of the various alkyl homologs (normalized to $10^4$ for the unsubstituted species) versus the number of alkyl carbon atoms (see Fig. 9).
[e] Standard deviation of the slope of the homolog distribution.

ring components of the hydrocarbon spectra of surface sediments were quite similar to components observed in spectra of the aromatic fractions isolated from urban stormwater runoff. PAH containing five or more rings in the surface sediments were believed to have come primarily from atmospheric dustfall as these components were present in dustfall and absent in stormwater. They were probably of pyrolitic origin—combustion of fossil fuels. PAH assemblages isolated from the 14–16-cm sediment layer deposited about 1900 were qualitatively similar to those in the surface sediments. However, these deeper sediments contained higher concentrations of perylene than surface sediments. Perylene was the predominant PAH in the 40–42-cm sediment layer which was dated at about 1850. Perylene was undetected in surface sediments, present at approximately 1·0 ppm at

14–16 cm, and at 1·4 ppm at 40–42 cm. Perylene was absent from stormwater, river sediments, humus, and plankton. Wakeham (1977) suggested that perylene was formed in the sediment by the reduction of extended quinones of biological origin. Thus, a single sediment sample may contain PAH from a variety of sources.

### 4.2.3 PAH Composition in Sediments

Because freshwater and coastal marine sediments may receive PAH from various sources, the resulting sedimentary PAH assemblages are often extremely complex. These PAH assemblages may contain thousands of individual PAH and are so intricate that they cannot be fully resolved into individual components even by the most modern analytical techniques. Characterization is limited primarily to a description of the molecular weight distribution of the many overlapping homologous series. Giger and Blumer (1974) isolated the PAH fraction of sediments collected at two stations in Buzzards Bay, Massachusetts, by a combination of gel filtration, adsorption chromatography, and charge-transfer complexation. Aromatic ring-type concentrates were characterized by UV, visible, and mass spectrometry. Twelve major PAH types were identified and quantified, but summed weights of PAH derived from UV measurement accounted for only 6–9 per cent of the actual weight of the total PAH fraction or 29 per cent of the weight of combined fractions from the final PAH chromatogram (Table 41). It was concluded that the true PAH weight lay between the actual weight of chromatographic fractions and the weight that was reconstructed from individual UV contributions. The unidentified portion of the total PAH assemblage included a principal series of alkyl homologs of the identified parent PAH as well as unidentified PAH and their alkyl homologs, while a minor series of equal complexity included naphthenologs and thienologs of the basic PAH series.

PAH assemblages of these sediments as well as others of terrestial and marine origin have been characterized further (Blumer and Youngblood, 1975; Youngblood and Blumer, 1975). These investigations confirmed the presence in soils and recent marine sediments of a wide range of alkylated aromatic hydrocarbons. Homologs extending to at least seven alkyl carbons occurred in all samples and in all PAH molecular weight series. Within a given PAH series, the unsubstituted hydrocarbons were most abundant and the integrated intensities of isomeric homologs decreased nearly two-fold for each additional carbon atom. This is in marked contrast to the pattern seen in crude oil, shale oil, and refined petroleum products. As indicated earlier, in fossil fuels the initial members of each series are less

TABLE 41

CONCENTRATIONS OF DIFFERENT PAH TYPES IN SEDIMENTS FROM BUZZARDS BAY, MASSACHUSETTS. STATION A IS 2·1 km WEST OF WEST FALMOUTH, MASSACHUSETTS, IN 15 m WATER AND STATION B IS 2·9 km OFF THE ENTRANCE TO NEW BEDFORD HARBOR, MASSACHUSETTS, IN 7·5 m WATER (FROM GIGER AND BLUMER, 1974, WITH PERMISSION OF THE AMERICAN CHEMICAL SOCIETY)

| *Compound* | *Concentration (μg/kg dry sediment)* | |
|---|---|---|
| | *Station A* | *Station B* |
| Phenanthrene | 33 | ND[a] |
| Anthracene | 8 | 170 |
| Pyrene | 100 | 960 |
| Fluoranthene | 110 | 790 |
| Benz[*a*]anthracene | 41 | 330 |
| Chrysene | 40 | 240 |
| Perylene | 26 | 94 |
| Benzo[*a*]pyrene | 75 | 370 |
| Benzo[*e*]pyrene | 59 | 310 |
| Anthanthrene | 7 | 34 |
| Benzo[*ghi*]perylene | 66 | 280 |
| Coronene | 5 | 20 |
| Total from UV spectrometry | 570 | 3 600 |
| Total actual weight of PAH concentrate | 6 700 | 63 000 |
| Total actual weight of all fractions in final chromatogram | 1 630 | |

[a] Not determined. Fraction contains much material, but spectra unsuited for quantitative analysis because of excessive peak broadening.

abundant and the abundance curves peak at $ArC_2$–$ArC_4$. In addition, tricyclic aromatics predominate over four-ring aromatics to a much greater degree in fossil fuels than in recent soils and sediments. Sediment samples that were known to contain a contribution from fossil fuels exhibited within the tetracyclic and pentacyclic series the same distribution patterns as unpolluted recent sediments. However, the tricyclic series exhibited an alkyl distribution pattern intermediate between that of fossil fuels and unpolluted recent sediments. The authors considered several possible sources for the complex PAH assemblages in unpolluted recent soils and sediments and hypothesized that they were derived from deposition of airborne particulate PAH produced in forest and prairie fires.

Hites and co-workers have analyzed PAH assemblages in sediments from the Charles River, Boston, Massachusetts, and elsewhere by gas chromatography–mass spectrometry (GC–MS) and high-resolution mass spectrometry (Hites and Biemann, 1975; Hites, 1976; Hase *et al.*, 1976; Hase and Hites, 1977). The Charles River Basin sediments contained PAH with at least 11 different ring structures and with abundant alkyl-substituted homologs of these compounds (Table 42). Within most series of

TABLE 42

PAH FOUND IN SEDIMENT OF THE CHARLES RIVER BASIN, BOSTON, MASSACHUSETTS, USA, BY GAS CHROMATOGRAPHY–MASS SPECTROMETRY (FROM HITES AND BIEMANN, 1975, AND HASE *et al.*, 1976, WITH PERMISSION OF THE AMERICAN CHEMICAL SOCIETY)

| *Compound found* | *Unsubstituted compounds* | | *Maximum number of alkyl C atoms* |
|---|---|---|---|
| | *Elemental composition* | *Molecular weight* | |
| Anthracene or phenanthrene | $C_{14}H_{10}$ | 178 | 12 |
| 4,5-Methylenephenanthrene | $C_{15}H_{10}$ | 190 | 9 |
| Pyrene[a] | $C_{16}H_{10}$ | 202 | 15 |
| Fluoranthene[a] | $C_{16}H_{10}$ | 202 | 15 |
| Benzofluorenes[b] | $C_{17}H_{12}$ | 216 | — |
| Benz[*a*]anthracene | $C_{18}H_{12}$ | 228 | 10 |
| Chrysene or triphenylene | $C_{18}H_{12}$ | 228 | 10 |
| Benzo[*b* or *k*]fluoranthene | $C_{20}H_{12}$ | 252 | 5 |
| Benzo[*a*]pyrene | $C_{20}H_{12}$ | 252 | 5 |
| Benzo[*e*]pyrene | $C_{20}H_{12}$ | 252 | 5 |
| Perylene[c] | $C_{20}H_{12}$ | 252 | 5 |
| $C_{22}H_{12}$[b] | $C_{22}H_{12}$ | 276 | 3 |
| Dibenzopyrenes or dibenzofluoranthenes[b] | $C_{24}H_{14}$ | 302 | 3 |

[a] UV confirmation.
[b] Isomers unknown.
[c] Trace quantities.

homologs abundance monotonically decreased as alkyl substitution increased, such that a semilog plot of abundance versus number of alkyl carbons produced a straight line with a large negative slope (Fig. 9). The pyrene–fluoranthene series was most abundant and the benzofluoranthene–benzopyrene series was second most abundant. Among the six-ring PAH, dibenzopyrenes were more abundant than coronene. A higher molecular weight series with a nominal composition of $C_{26}H_{14}$ and including its methyl, ethyl, or dimethyl homologs was also present. The authors questioned the conclusion of Youngblood and Blumer (1975), that the sedimentary PAH assemblage originated from forest fire particulates,

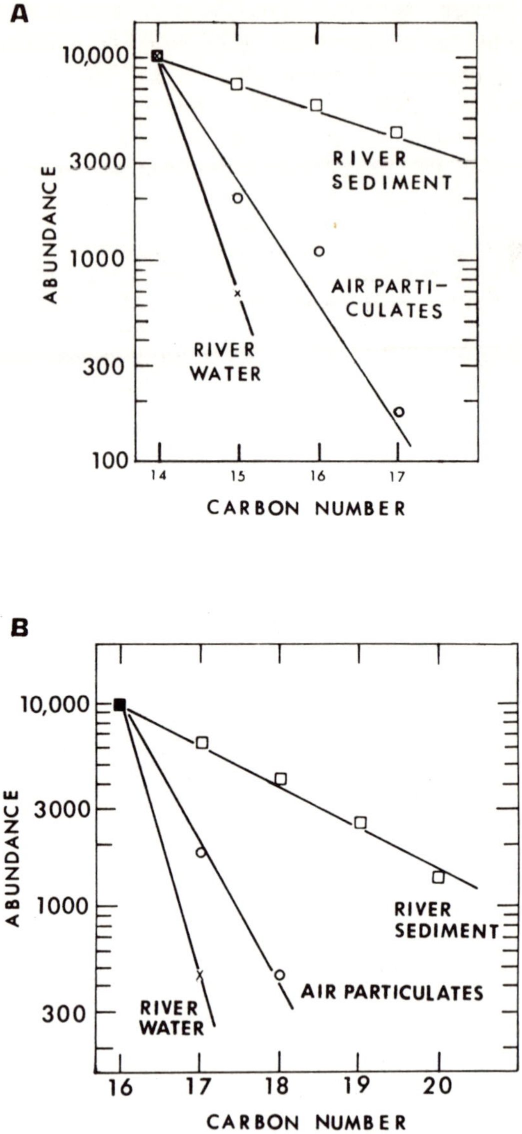


FIG. 9. Alkyl homolog distribution in the phenanthrene-type series (A) and pyrene-type series (B) in Charles River Boston, Massachusetts, sediment and water and in urban air samples. Data are normalized such that the unsubstituted species is 10 000 units in each case (from Hase and Hites, 1977, with permission of Ann Arbor Science Publishers).

for the following reason: PAH assemblages produced by pyrolysis are qualitatively similar nearly regardless of the fuel type and combustion conditions. Pyrolysates, including those produced in wood burning, contain PAH mixtures that have a very low abundance of alkyl homologs, resembling airborne particulate PAH assemblages more than recent sedimentary PAH assemblages. Instead, Hites and associates hypothesized a mechanism by which the steep slope of the alkyl homolog distribution observed in urban air particulates could be changed into a more gradual slope such as that observed for unpolluted sediments. This involves the differential aqueous solubility of the parent PAH and their alkyl homologs. Within a PAH series, in most cases, solubility in water decreases as the number of alkyl carbons increases. Thus, after airborne particulate deposition on soil or in water the lowest homologs (including unsubstituted species) continuously fractionate into the water phase to an extent proportional to their carbon number. The remaining species, which accumulate on particulate matter and in sediments, are therefore deficient in the lowest homologs, thus increasing the relative abundance of higher homologs.

To test this hypothesis, Hase and Hites (1977) examined the PAH in water and sediment samples from the Charles River Basin and compared them to urban airborne particulate PAH collected in Boston, Massachusetts (Fig. 9). The slope of the air particulate homolog plot was intermediate between that found in water and sediments for both the anthracene–phenanthrene series and the fluoranthene–pyrene series. Therefore, the water was enriched in the lower homologs while sediments were enriched in higher homologs, relative to the air particulates. The authors concluded that the main source of PAH in the aqueous environment is anthropogenic, airborne, combustion-produced PAH. It should be stressed that PAH-containing wastewater effluents such as those from industrial and domestic sewage and stormwater runoff would undergo similar fractionation in the aquatic environment as evidenced by the data of Lewis (1975) and Armstrong *et al.* (1977), discussed earlier. The hypothesis of Hase and Hites (1977) is also supported by estimates made earlier in this review of the relative importance of anthropogenic airborne particulates as a source of PAH in the aquatic environment.

## 4.3 PAH IN AQUATIC ORGANISMS

Zechmeister and Koe (1952) were the first to isolate and identify PAH from the tissues of a marine animal. Several highly fluorescent crystalline fractions were prepared from the lipid extract of thatched barnacles *Tetraclita squamosa rubescens* collected from creosoted wood pilings near the Kerckhoff Marine Laboratory in Corona del Mar, California. A number of PAH were identified including anthracene, phenanthrene, chrysene, fluoranthene, benzo[*ghi*]perylene, and benzo[*a*]pyrene. The crystalline fraction containing B*a*P was tested for carcinogenic activity by Shimkin *et al.* (1951). A single subcutaneous injection into mice of 0·5 mg of the fraction caused tumors which were identified as spindle-cell sarcomas. These tumors were indistinguishable from those induced by authentic B*a*P. Another sample of the same species collected at a different but unspecified site did not contain detectable quantities of PAH. Subsequently, Koe and Zechmeister (1952) isolated 11 chromatographically homogeneous and strongly fluorescent compounds in crystalline form from the fleshy peduncles of goose barnacles *Mitella polymerus* collected from pier pilings at the Scripps Institution of Oceanography, La Jolla, California. Six of the hydrocarbons were identified by their UV spectra; these were chrysene, fluoranthene, benzo[*ghi*]perylene, coronene, benzo[*e*]pyrene, and benzo[*a*]pyrene.

### 4.3.1 Benzo[*a*]pyrene

Since that time, there has been a large number of reports published on the presence of PAH, and in particular B*a*P, in tissues of aquatic organisms. Available data on the concentrations of B*a*P in assorted aquatic organisms are summarized in Table 43. The majority of analyses have been performed on marine organisms, especially ones obtained from the Atlantic and Mediterranean coasts of France and from the Pacific coast of the USA. Bivalve molluscs, particularly the commercially important mussel *Mytilus edulis* and oyster genera *Ostrea*, *Crassostrea*, and *Gryphea*, have received far more attention than other aquatic invertebrates or aquatic plants and fish. They are considered by many to be excellent subjects for monitoring pollutants in the aquatic environment. Because pelycopods are sedentary and tend to bioaccumulate from low concentrations in the ambient medium, concentrations of pollutants in their tissues represent a time-integrated index of pollutant inputs to their immediate surroundings.

B*a*P concentrations in the specimens analyzed ranged from undetectable (usually $<0{\cdot}1$ $\mu$g/kg dry weight) to as high as 5000 $\mu$g/kg. There are relatively few obvious trends in the B*a*P data. Organisms collected from coastal regions near major industrial or domestic point sources of PAH, such as the Seine estuary (France), the Bay of Naples (Italy), and Norfolk

TABLE 43

CONCENTRATIONS OF BENZO[*a*]PYRENE IN THE TISSUES OF AQUATIC ORGANISMS (MOSTLY MARINE). WET WEIGHT VALUES WERE CONVERTED TO DRY WEIGHT BY A FACTOR OF 4

| *Organism* | *Location* | *BaP*[a] (*μg/kg dry wt*) | *Reference* |
|---|---|---|---|
| 1. *Aquatic plants* | | | |
| Marine algae | Holsteinborg, Greenland | 60 | Mallet *et al.*, 1963 |
| | Faeringehavn, Greenland | 60 | Mallet *et al.*, 1963 |
| | Bay of Naples, Italy | 2·2 | Boucart and Mallet, 1965 |
| | North Sea coast, France | traces–40 | Perdriau, 1964 |
| Green algae (marine) | Saint-Briac, France | ND | Perdriau, 1964 |
| Brown algae (marine) | Brest, France | ND | Mallet, 1961 |
| Brown algae (*Fucus seratus*) | Cherbourg, France | 5 | Perdriau, 1964 |
| Freshwater algae | Pskov region, USSR | 5 | Ershova, 1968 |
| Freshwater diatom (*Asterionella formosa*) | Lake Constance, GFR | 2 | Borneff and Fischer, 1962 |
| Sea grass (*Zostera marina*) | Bay of St. Malo, France | 25 | Mallet and Priou, 1967 |
| Marine vascular plants | Clipperton Lagoon, equatorial Pacific | 376 | Niaussat and Auger, 1970 |
| Freshwater vascular plants | Rublevskoye Reservoir, USSR | 0·6–2·7 | Il'nitskii *et al.*, 1972 |
| | Khimkinskoe Reservoir, USSR | 1·7–37·8 | Il'nitskii *et al.*, 1972 |
| 2. *Plankton* | | | |
| | Bay of Disko, Greenland | 5·5 | Mallet *et al.*, 1963 |
| | Bay of Naples, Italy | 6·1–607 | Boucart and Mallet, 1965 |
| | Clipperton Lagoon, equatorial Pacific | 7·3 | Niaussat and Auger, 1970 |
| | Dinard, France | 100–350 | Mallet and Lami, 1964 |
| | French Channel coast | 400 | Mallet and Sardou, 1965 |
| | French Mediteranean coast | ND–5 | Mallet and Sardou, 1965 |
| | Rublevskoye Reservoir, USSR | 0·7–1·8 | Il'nitskii *et al.*, 1972 |
| | Khimkinskoe Reservoir, USSR | 7·3–8·8 | Il'nitskii *et al.*, 1972 |

TABLE 43—*contd.*

| *Organism* | *Location* | *BaP*[a] (μg/kg dry wt) | *Reference* |
|---|---|---|---|
| 3. *Molluscs*[b] | | | |
| Mussels (*Mytilus edulis*) | St. Efflam, France | ND | Mallet, 1961 |
| | | 3·0 (shell) | |
| | Arcachon Basin, France | 5·0 | Mallet, 1961 |
| | | traces (shell) | |
| | North Sea coast, France | ND–230 | Mallet *et al.*, 1963 |
| | Faeringehavn, Greenland | 55 | Mallet *et al.*, 1963 |
| | | 18 (shell) | |
| | Toulon roads, France | 16–22 | Greffard and Meury, 1967 |
| | Bay of Naples, Italy | 128–543 | Boucart and Mallet, 1965 |
| | Les Quatre Vaux, France | 1·0 | Perdriau, 1964 |
| | Seine estuary, France | ND–380 | Perdriau, 1964 |
| | Falmouth, Massachusetts, USA | | |
| | Little Sippewissett | <2 | Pancirov and Brown, 1977 |
| | Wild Harbor | 2 | Pancirov and Brown, 1977 |
| | Tillamook Bay, Oregon, USA | <0·4–67·4 | Mix *et al.*, 1977 |
| | Yaquina Bay, Oregon, USA | 0·48–120·8 | Mix *et al.*, 1977 |
| | Alsea Bay, Oregon, USA | <0·4 | Mix *et al.*, 1977 |
| | Vancouver, British Columbia, Canada | | Dunn and Stich, 1975 |
| | outer harbor | 8 ± 1·2 | |
| | wharf, marina and dock areas | 72 ± 19·6 | |
| | False Creek | 168 ± 24 | |

| | | | |
|---|---|---|---|
| | Vancouver marina area | | |
| | (May '74) | 7·6 ± 0·4 | |
| | (Sept '74) | 7·2 ± 0·4 | |
| | wharf area on rock and cables | | |
| | (May '74) | 52 ± 21·2 | |
| | (Sept '74) | 156 ± 29·2 | |
| | creosoted pilings | | |
| | (May '74) | 272 ± 52 | |
| | (Sept '74) | 532 ± 76 | |
| *M. edulis and M. californeanus* | 25 stations between Bodega Head and San Diego, California, USA | <0·4–32·8 | Dunn and Young, 1976 |
| *M. californeanus* | West coast of Vancouver Island, British Columbia, Canada | 0·4 ± 0·4 | Dunn and Stich, 1975 |
| Oyster (*Ostrea edulis*) | Penze estuary, France | 15 | Mallet, 1961 |
| | Parc d'Marennes, France | 4 | Mallet, 1961 |
| | Arcachon Basin, France | traces | Mallet, 1961 |
| | Bay of St. Malo, France | 30·0 | Mallet and Priou, 1967 |
| *Gryphaea angulata* | Arcachon Basin, France | ND | Mallet, 1961 |
| *Crassostrea virginica* | Norfolk Harbor, Virginia, USA | 20–60 | Cahnmann and Kuratsune, 1957 |
| | Long Island Sound, USA | 8·0 | Pancirov and Brown, 1977 |
| | Chincoteague, Virginia, USA | 1·2 | Pancirov and Brown, 1977 |
| *Crassostrea gigas* | Tillamook Bay, Oregon, USA | <0·4 | Mix *et al.*, 1977 |
| Gaper clam (*Tresus capax*) | Tillamook Bay, Oregon, USA | 0·64–12·8 | Mix *et al.*, 1977 |
| | Netarts Bay, Oregon, USA | <0·4 | Mix *et al.*, 1977 |
| | Yaquina Bay, Oregon, USA | 1·0–2·24 | Mix *et al.*, 1977 |
| | Alsea Bay, Oregon, USA | <0·4 | Mix *et al.*, 1977 |
| | Coos Bay, Oregon, USA | 0·56–2·04 | Mix *et al.*, 1977 |

TABLE 43—*contd.*

| *Organism* | *Location* | *BaP*[a] (*µg/kg dry wt*) | *Reference* |
|---|---|---|---|
| Soft-shell clam (*Mya arenaria*) | Tillamook Bay, Oregon, USA | 1·2 | Mix *et al.*, 1977 |
| | Alsea Bay, Oregon, USA | <0·4 | Mix *et al.*, 1977 |
| | Coos Bay, Oregon, USA | 1·32–26·64 | Mix *et al.*, 1977 |
| Butter clam (*Saxidomas giganteus*) | Coos Bay, Oregon, USA | 1·16–4·20 | Mix *et al.*, 1977 |
| Cockle (*Cardium edule*) | St. Efflam, France | ND | Mallet, 1961 |
| | Seine estuary, France | 220–780 | Perdriau, 1964 |
| Razor clam (*Solen marginatus*) | Penze estuary, France | traces | Mallet, 1961 |
| Scallop (*Chlamys varia*) | Brest, France | 90 | Mallet, 1961 |
| | Brest, France | 4 | Perdriau, 1964 |
| | Bay of St. Malo, France | 3·1 (shell) | Mallet and Priou, 1967 |
| Clam (*Tapes decussatus*) | Arcachon Basin, France | ND | Mallet, 1961 |
| Clam (*unidentified*) | Seine estuary, France | 280–750 | Perdriau, 1964 |
| | Chincoteague, Virginia, USA | 1·2 | Pancirov and Brown, 1977 |
| | Darien, Connecticut, USA | <4·0 | Pancirov and Brown, 1977 |
| Limpet (unidentified) | Cherbourg, France | traces | Perdriau, 1964 |
| Whelk (*Buccinum undatum*) | Bay of St. Malo, France | 4·1 (shell) | Mallet and Priou, 1967 |
| 4. *Crustaceans* | | | |
| Spider crab (*Maia squinado*) | Bay of St. Malo, France | 9·5 (carapace) | Mallet and Priou, 1967 |
| | Brest, France | 3·5 | Perdriau, 1964 |
| Isopods (*Lygia* sp.) | Clipperton Lagoon, equatorial Pacific | 536 | Niaussat and Auger, 1970 |
| Terrestrial crab (*Gecarcinus planatus*) | Clipperton Lagoon | 119 (muscle) | Niaussat and Auger, 1970 |
| Crab (*Callapa hepatica*) | outside Clipperton Lagoon | ND | Niaussat and Auger, 1970 |

| | | | |
|---|---|---|---|
| Crab (unidentified) | North Sea coast of France | ND–40 | Mallet *et al.*, 1963 |
| | Arcachon Basin, France | ND | Mallet, 1961 |
| | Cherbourg, France | 5·0 | Perdriau, 1964 |
| | Le Havre, France | traces | Perdriau, 1964 |
| | Raritan Bay, New Jersey, USA | 12 | Pancirov and Brown, 1977 |
| Crab (*Callinectes sapidus*) | Chesapeake Bay, Virginia, USA | <2·0 | Pancirov and Brown, 1977 |
| Barnacle (unidentified) | St. Cast, France | 0·5 | Perdriau, 1964 |
| Shrimp (*Penaeus aztecus*) | Palacios, Texas, USA | <4 | Pancirov and Brown, 1977 |
| 5. *Echinoderms* | | | |
| Sea urchin (unidentified) | St. Efflam, France | ND | Mallet, 1961 |
| | Morgat, France | ND | Mallet *et al.*, 1963 |
| Star fish (unidentified) | North Sea coast, France | ND–126 | Mallet *et al.*, 1963 |
| Sea cucumber (unidentified) | west coast of Greenland | ND | Mallet *et al.*, 1963 |
| 6. *Annelids* | | | |
| Freshwater oligochaetes (*Tubifex* sp.) | freshwater pool, Italy | 50 | Scaccini-Cicatelli, 1965, 1966 |
| 7. *Fish* | | | |
| Sardine | Bay of Naples, Italy | 65·4 | Boucart and Mallet, 1965 |
| Mullet (*Mugil chelo*) | Bay of St. Malo, France | 17·3 (flesh)<br>155·0 (viscera)<br>22·0 (scales) | Mallet and Priou, 1967 |
| *M. chelo* | Le Havre, France | 12·0 | Perdriau, 1964 |
| Flounder (*Pleuronectes platessa*) | Bay of St. Malo, France | traces | Mallet and Priou, 1967 |
| Flounder (unidentified) | Long Branch, New Jersey, USA | <8 | Pancirov and Brown, 1977 |
| | South of Long Island, New York, USA, 40°27′N 73°06′W | <4 | Pancirov and Brown, 1977 |
| Sole (*Solea solea*) | Bay of St. Malo, France | 10·0 | Mallet and Priou, 1967 |
| | North Sea 51°04·20′N 02°20·10′E | 38 | Perdriau, 1964 |

TABLE 43—*contd.*

| *Organism* | *Location* | *BaP*[a] (*μg/kg dry wt*) | *Reference* |
|---|---|---|---|
| Cod (*Gadus* sp.) | Holsteinborg, Greenland | 15 | Mallet *et al.*, 1963 |
| | Atlantic, 40 km off Toms River, New Jersey, USA | <10 | Pancirov and Brown, 1977 |
| Eel (*Anguilla* sp.) | Dunkerque, France | 30 | Perdriau, 1964 |
| Sand eel (*Amphioxus*) | Roscoff, France | 1·5 | Mallet, 1961 |
| Menhaden | Raritan Bay, New Jersey, USA | 6·0 | Pancirov and Brown, 1977 |
| Anchovy and smelt | San Diego Bay, California, USA | up to 5000 | Lee *et al.*, 1972b |
| Assorted fish | Dunkerque, France | ND | Perdriau, 1964 |
| | Adriatic Coast, Italy | ND–900 | Piccinetti, 1967 |
| *Kulhia taeneura* | Clipperton Lagoon, equatorial Pacific | traces (viscera)<br>4·0 (muscle)<br>26 (integument) | Niaussat and Auger, 1970 |
| *Caranx lugubris* | Clipperton Lagoon | 2·4 (viscera)<br>0·5 (muscle)<br>13·6 (integument)<br>5·5 (intestine) | Niaussat and Auger, 1970 |
| *Kulhia taeneura* | outside Clipperton Lagoon | ND | Niaussat and Auger, 1977 |
| Lake trout | Lake Maskenonge, Ontario, Canada | <4 | Pancirov and Brown, 1977 |

[a] ND, none detected; traces, detected but concentration below limits of quantification.
[b] Unless indicated otherwise, values for molluscs are for the soft tissues only without the shell.

Harbor (Virginia, USA), often but not always contained higher B*a*P concentrations than organisms from more remote areas.

Organisms from the coast of Greenland, presumably a relatively pristine area, contained unusually high concentrations of B*a*P. Mallet *et al.* (1963) attributed this to biosynthesis of B*a*P by sediment-dwelling anaerobic bacteria. A possible alternative source of B*a*P in this environment is the leaching of B*a*P from peat deposits which are extensive in parts of Greenland. Peat is known to contain high concentrations of PAH (see Table 1).

Another case of B*a*P contamination of a seemingly pristine environment was reported in the Clipperton Lagoon in the tropical eastern Pacific Ocean (Niaussat and Auger, 1970). The Clipperton atoll is quite remote from human activities. It lies approximately 1300 km west of Mexico and is the only atoll in the eastern tropical Pacific lying north of the equator (Niaussat, 1970). It is not inhabited and is remote from major shipping and airline routes. The central Lagoon is completely enclosed and the Lagoon water is highly stratified. Surface water is nearly fresh (3–4‰ salinity) while bottom water at a depth of 30–40 m is saline (32‰ salinity). Lagoon waters below the halocline, as well as bottom sediments, are undoubtedly anoxic. Relatively few species of macrofauna live in the Lagoon, but its waters are extremely rich in phytoplankton and aerobic and facultative anaerobic bacteria. Bottom sediments are rich in anaerobic bacteria. These unique conditions plus the high ambient temperatures (28–30 °C) may provide an ideal milieu for the direct or indirect biosynthesis of selected PAH. Niaussat and Auger (1970) determined the concentrations of B*a*P and perylene in the water and biota of the Lagoon, in land plants from the atoll, and in animals from the adjacent ocean. Surface water of the Lagoon contained 1·6 μg B*a*P/liter and 3·05 μg perylene/liter. Of lagoonal biota, the aquatic vascular plants (*Najas* sp., *Zostera* sp., and *Ruppia* sp.) and isopod crustaceans (*Lygia* sp.) contained the highest concentrations of the two PAH (Table 44). Mixed phyto- and zooplankton and fish contained much lower concentrations. Grasses (*Cenchrus* sp. and *Eragrostis* sp.) along the shore of the Lagoon and land crabs (*Gecarcinus planatus*) contained relatively high concentrations of both B*a*P and perylene. In most cases, perylene was more abundant than B*a*P. PAH were undetectable in land plants (*Ipomea pes caproe*) not subject to periodic inundation with lagoon water or in fish and crustaceans collected in the ocean adjacent to the atoll. Thus, PAH in the Lagoon were not derived from exogenous sources such as fallout of airborne particulates or contamination from ocean water.

The origin of PAH in the Clipperton Lagoon is not known.

TABLE 44

CONCENTRATIONS OF BENZO[*a*]PYRENE AND PERYLENE IN WATER AND ORGANISMS FROM THE VICINITY OF THE CLIPPERTON ATOLL IN THE EASTERN TROPICAL PACIFIC OCEAN (FROM NIAUSSAT AND AUGER, 1970, WITH PERMISSION OF THE PUBLISHER)

| *Location and sample* | *BaP* (*μg/kg*) | *Perylene* (*μg/kg*) | *Perylene/ BaP ratio* |
|---|---|---|---|
| *Lagoon* | | | |
| (Seawater | 1·6 μg/liter | 3·05 μg/liter | 1·9) |
| Phyto- and zooplankton | 7·3 | 2·7 | 0·36 |
| Aquatic vascular plants | 376 | 71·6 | 0·19 |
| Isopod crustaceans (*Lygia* sp.) | 536 | 865 | 1·6 |
| Fish (*Kulhia taeneura*) | | | |
| viscera | traces | traces | — |
| muscle | 4·0 | 5·0 | 1·2 |
| integument | 26 | 30 | 1·1 |
| Fish (*Caranx lugubris*) | | | |
| viscera | 2·4 | 10·5 | 4·3 |
| muscle | 0·5 | 1·7 | 3·4 |
| integument | 13·6 | 25 | 1·8 |
| intestine | 5·5 | 88 | 16 |
| *Shore of Lagoon* | | | |
| Marsh grasses (*Cenchrus and Eragrostis*) | 71·5 | 219 | 3·06 |
| Land plants (*Ipomea pes caproe*) | traces | traces | — |
| Land crab (*Gecarcinus planatus*) | | | |
| viscera | ND | ND | — |
| muscle | 119 | 288 | 2·4 |
| carapace | 5·4 | 16·6 | 3·07 |
| *Ocean adjacent to atoll* | | | |
| Crab (*Callapa hepatica*) | | | |
| viscera | traces | traces | — |
| muscle | ND | ND | — |
| carapace | ND | traces | — |
| Fish (*Kulhia taeneura*) | | | |
| viscera | ND | traces | — |
| muscle | ND | ND | — |
| integument | ND | ND | — |

ND, below level of detection.

Predominance of perylene over B*a*P in nearly all samples suggests the possibility of reduction of extended quinones produced by the rich surface microflora as they settle into the anoxic basin of the Lagoon (Aizenshtat, 1973). Direct biosynthesis of PAH by anaerobic microorganisms in deep waters and bottom sediments of the Lagoon is another possibility.

A detailed GC–MS analysis of the complete composition of the PAH assemblage in the Lagoon might provide valuable clues to its origin. It should be stressed that the elevated levels of B*a*P encountered in seemingly pristine areas along the coast of Greenland and in the Clipperton Lagoon represent special and unusual situations.

These elevated B*a*P levels do not represent a normal background of aquatic B*a*P as suggested by Mallet *et al.* (1963) and others. A large number of plant and animal samples collected from coastal and fresh waters of Europe and North America have been found to contain little or no B*a*P. The absence of B*a*P in a great many aquatic biota samples is not adequately reflected in the data summarized in Table 43. For instance, Piccinetti (1967) analyzed 276 individuals representing 53 different species of animals from the Adriatic coast. Only 35 individuals (12·7 per cent) contained detectable levels of B*a*P. Similar results were reported by Mix *et al.* (1977) for molluscs from Oregon bays and by Pancirov and Brown (1977) for invertebrates and fish from the east coast of the USA. Dunn and Stich (1975, 1976a) and Dunn and Young (1976) showed quite convincingly that there was a substantial correlation between the degree of contamination of mussels *Mytilus edulis* and *M. californeanus* and the degree of industrial, urban, and recreational uses of the bodies of water in which these mussels resided. Creosoted wharf pilings and domestic sewage discharges represented major sources of B*a*P contamination.

The authors also observed a sharply falling gradient of B*a*P contamination of mussels with increasing distance from the point-source of B*a*P. Mussels collected from creosoted pilings in Vancouver, British Columbia (Dunn and Stich, 1975, 1976a), from estuaries in Oregon, USA (Mix *et al.* 1977) or at island and coastal stations in southern California (Dunn and Young, 1976) contained the highest concentrations of B*a*P. Mussels and other molluscs from nearby uncreosoted surfaces contained much lower B*a*P concentrations. Dunn and Young (1976) found that levels of B*a*P contamination in mussels taken from locations at least 1 km from piers and wharves in southern California waters were usually at or near the detection limit of 0·4 $\mu$g B*a*P/kg dry weight. The only mussels with elevated levels of B*a*P not growing on creosoted pilings were those collected near large harbors and marinas. Zitko (1975) also identified PAH from creosote oil by fluorescence and mass spectrometry in the tissues of mussels (*Mytilus edulis*), periwinkles (*Littorina littorea*), whelks (*Buccinum undatum*), and clams (*Mya arenaria*) collected from the vicinity of creosoted wharf pilings in Passamaquoddy Bay and at St. Andrews, New Brunswick, Canada.

Dunn and Stich (1976a) reported elevated B*a*P concentrations in mussels collected at five stations along the southern shore of the outer harbor, Vancouver, British Columbia. At all but one station, B*a*P levels in the mussels showed a definite seasonal trend (Fig. 10). B*a*P levels in the mussels were highest in the winter and lowest in the summer. Dunn and Stich attributed these seasonal variations in B*a*P contamination of mussels to overflow discharges from Vancouver's combined sewer and storm drain

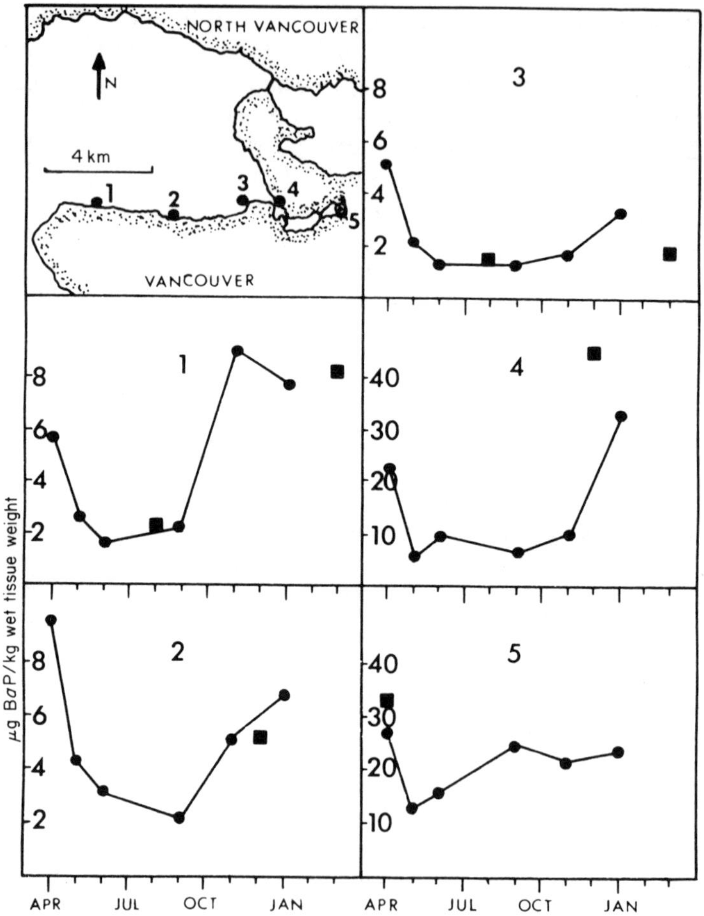


FIG. 10. Seasonal variations in the concentration of benzo[*a*]pyrene (B*a*P) in the tissues of mussels (*Mytilus edulis*) collected at five stations along the southern shore of the outer harbor, Vancouver, British Columbia, Canada. ■, January 1974 to March 1975; ●, April 1975 to January 1976 (from Dunn and Stich, 1976a, reproduced by permission of the Minister of Supply and Services, Canada).

system which occur most frequently during the rainy winter months. Thus, surface runoff from land and municipal sewage can contribute significantly to B*a*P contamination of coastal organisms.

Several other studies point to anthropogenic B*a*P emissions as the main vector for B*a*P contamination of aquatic organisms. B*a*P concentrations were relatively high in molluscs collected in Oregon bays receiving industrial and domestic wastes, but were low or undetectable in molluscs from regions where human activity was less (Mix *et al.*, 1977). Fish from San Diego Bay, California, which is a large shipping port, contained up to 5000 μg B*a*P/kg dry weight (Lee *et al.*, 1972b). On the east coast of the USA animals from areas such as Long Island Sound and Raritan Bay, which receive large amounts of industrial and municipal wastes, contained relatively high concentrations of B*a*P and other PAH (Pancirov and Brown, 1977). Animals from the Atlantic Ocean and fairly pristine areas such as Chincoteague Bay, Virginia, contained little or no B*a*P and other PAH.

There are a few apparent correlations in the data summarized in Table 43 between B*a*P concentrations in aquatic organisms and their phyletic position, trophic level, or habitat preference. Filter-feeding bivalve molluscs often contained lower concentrations of B*a*P than did the algae and plankton upon which they feed. However, there are notable exceptions. Marine algae and mixed plankton from the Bay of Naples, Italy, contained 2·2 and 6·1 to 607 μg B*a*P/kg, respectively, while mussels (*Mytilus edulis*) from the same area contained 128 to 543 μg B*a*P/kg (Boucart and Mallet, 1965). Marine crustaceans, with the exception of

TABLE 45

CONCENTRATIONS OF BENZO[*a*]PYRENE IN THE TISSUES OF BENTHIC MARINE ANIMALS AND IN BOTTOM SEDIMENTS COLLECTED AT THE SAME LOCATION (BASED ON DATA FROM PERDRIAU, 1964)

| *Location* | *BaP concentration (μg/kg dry weight)* | |
|---|---|---|
| | *Benthic animals* | *Sediments* |
| Seine estuary, France | 20 | 80 |
| | 30 | 100 |
| | 40 | 130 |
| Dunkerque, France | 110 | 220 |
| | 40–50 | 180 |
| | 20 | 20 |

those collected from the Clipperton Lagoon, contained low or undetectable levels of B*a*P as did most of the echinoderm samples analyzed. B*a*P concentrations in fish were quite variable. Demersal species such as flounder and sole did not contain notably higher B*a*P concentrations than did pelagic species such as sardines and menhaden, contrary to the conclusion of ZoBell (1971). Perdriau (1964) provided a small amount of data on the concentration of B*a*P in benthic animals and bottom sediments, both collected at the same stations, in two European estuaries (Table 45). With one exception, B*a*P concentrations were higher by a factor of 2–4 in sediments than in the animals living on or in the sediments. Animals living in highly contaminated sediments usually had higher B*a*P concentrations than those residing in relatively clean sediments. This could indicate that the animals accumulated B*a*P from contaminated sediments, or, more likely, that both sediments and animals received B*a*P from a common exogenous source.

### 4.3.2 Other PAH

Comparatively little information is available concerning the presence of PAH other than B*a*P or concerning the PAH homolog distribution in the tissues of aquatic organisms. Such information could provide valuable clues to the sources of PAH in aquatic organisms. Pancirov and Brown (1977) applied low-voltage mass spectral analysis to samples of Long Island Sound oysters. These analyses showed that several PAH homologous series were composed principally of the parent compound with much lower amounts of methyl-substituted homologs. For instance, the ratio of methylpyrene to pyrene was nearly constant at 0·26 for all samples for which data could be obtained. This ratio strongly suggests a pyrolytic rather than a petroleum source for PAH in the animal tissues. Cahnmann and Kuratsune (1957) measured the concentrations of eight PAH in the tissues of oysters, *Crassostrea virginica*, from the harbor of Norfolk, Virginia, an area fairly heavily polluted by domestic, industrial, and shipping wastes (Table 46). Total approximate PAH concentrations ranged from 700 to 1200 μg/kg wet weight. B*a*P represented only 0·2–0·3 per cent of the total. Fazio (1971) measured the concentrations of nine PAH in oysters, *C. virginica*, from Galveston and Aransas Bays, Texas. These oysters were much less contaminated than those from Norfolk, Virginia, despite the extensive oil refinery and port activity around Galveston Bay (Table 47). Interestingly, no B*a*P was detected in any of the oyster samples by a method demonstrated to be sensitive to 2 μg/kg wet weight or less. Samples from the most heavily contaminated station, an area in Galveston

TABLE 46

CONCENTRATIONS OF SELECTED PAH IN THE TISSUES OF OYSTERS, *Crassostrea virginica*, FROM THE HARBOR OF NORFOLK, VIRGINIA, USA (FROM CAHNMANN AND KURATSUNE, 1957, WITH PERMISSION OF THE AMERICAN CHEMICAL SOCIETY)

| *Compound* | *Approximate concentration (μg/kg wet weight)* |
|---|---|
| Fluoranthene | 600–1000 |
| Pyrene | 100–160 |
| Chrysene | 20–40 |
| Benz[*a*]anthracene | <10 |
| Benzo[*k*]fluoranthene | 8–12 |
| Benzo[*a*]pyrene | 2–6 |
| Benzo[*e*]pyrene | <20 |
| Benzo[*ghi*]perylene | 1–5 |

TABLE 47

CONCENTRATIONS OF SELECTED PAH IN THE TISSUES OF OYSTERS, *Crassostrea virginica*, FROM THE TEXAS, USA GULF COAST (FROM FAZIO, 1971, WITH PERMISSION OF THE PUBLISHER)

| *Compound* | *Concentration (μg/kg wet weight)* | | | | | |
|---|---|---|---|---|---|---|
| | *Station A*[a] | | *Station B*[b] | | *Station C*[c] | |
| | *Mean* | *Range* | *Mean* | *Range* | *Mean* | *Range* |
| Phenanthrene | — | | — | | 2·2 | 0·0–6·9 |
| Fluoranthene | 1·7 | 1·7 | 3·0 | 1·6–5·6 | 7·8 | 2·8–14 |
| Pyrene | 0·9 | 0·8–1·0 | 1·8 | 1·3–5·9 | 6·5 | 3·8–13 |
| Chrysene | 0·5 | 0·3–0·6 | 0·2 | 0·0–1·7 | 0·6 | 0·0–2·0 |
| Benzo[*b*]fluoranthene | 0·3 | 0·0–0·5 | 1·2 | 0·5–1·8 | 2·2 | 0·0–4·1 |
| Benzo[*k*]fluoranthene | — | | 0·1 | 0·0–0·6 | 0·4 | 0·0–1·1 |
| Benzo[*a*]pyrene | — | | — | | — | |
| Benzo[*e*]pyrene | 0·2 | 0·0–0·4 | 1·2 | 0·0–4·0 | 2·1 | 0·8–4·2 |
| Perylene | — | | 0·7 | 0·0–3·5 | 0·1 | 0·0–1·5 |

[a] Station A, Aransas Bay oyster reefs relatively uncontaminated from domestic and petroleum operations.
[b] Station B, Galveston Bay oyster reefs approved for commercial harvesting.
[c] Station C, Galveston Bay oyster reefs closed to commercial harvesting owing to domestic sewage contamination.

Bay closed to commercial oyster harvesting owing to elevated sewage contamination, contained a mean total of 21·9 μg/kg wet weight of other PAH. Oysters from uncontaminated regions of Galveston Bay and the relatively unpolluted Aransas Bay contained mean total PAH concentrations of 8·2 and 3·6 μg/kg, respectively. Lower molecular weight PAH were more abundant than higher molecular weight PAH in oysters from both Galveston Bay and Norfolk. In each case, fluoranthene and pyrene were the most abundant PAH in oyster tissues. The predominance of low molecular weight over high molecular weight PAH in oyster tissues probably reflects in part greater bioavailability of the more water-soluble PAH.

Ehrhardt (1972) analyzed the hydrocarbons in oysters, *C. virginica*, from a heavily contaminated reef at the entrance to the Houston ship channel in Galveston Bay. Oyster tissues contained 236 ppm (mg/kg wet weight) total hydrocarbons. Fractions 4 (naphthalenes) and 5 (PAH fraction) of the oyster extract contained 124 ppm of hydrocarbons. The fractions were analyzed by mass spectrometry and revealed a complex mixture of benzene, naphthalene, fluorene, phenanthrene, anthracene, and their alkyl homologs. Higher molecular weight PAH either were not examined or not detected. Because of the high concentration of PAH and the compositional complexity of the PAH mixture in oyster tissues, Ehrhardt concluded that hydrocarbon contamination was of petroleum origin. An analysis of the alkyl homolog distribution, which could have confirmed the source of PAH, was not performed.

Recently, Bravo *et al.* (1978) determined the concentration of total PAH in oysters, *C. virginica*, from ten stations along the Mexican coast of the Gulf of Mexico. PAH from naphthalene to dibenzopyrene were analyzed in the benzene fraction of oyster tissues by capillary-column gas chromatography. Mean total PAH concentration in the oysters ranged from 2·08 to 9·16 ppm (mg/kg wet weight). Highest values were obtained in oysters from the Machona and Carmen Lagoons near petroleum production areas. Nearly all samples were from sparsely populated regions. Although natural oil seepage occurs in several areas along this coastline (Wilson *et al.* 1974) and could contribute to the high levels of PAH reported, the levels appear much too high (by a factor of $10^2$–$10^3$) for a relatively uninhabited area, especially when compared to the values of Fazio (1971) for Texas Gulf coast oysters. This apparent anomaly deserves further investigation. Another report of unusually high concentrations of total PAH in marine animals from a seemingly pristine environment is that of Wong *et al.* (1976). They reported concentrations of total PAH, as measured by GC–MS, of 9–31 μg/kg wet weight in the tissues of the marine fish, arctic cisco, least cisco, and pomfret from the southern Beaufort Sea. A sample of tuna from ocean station PAPA in the north Pacific Ocean contained 184 μg/kg total PAH. Obviously, considerably more information is needed, particularly on the composition of PAH assemblages in the tissues of aquatic organisms from both pristine and heavily impacted areas, if we are to gain an understanding of the sources and fluxes of PAH through the aquatic ecosystem.

A variety of sources has been suggested, and sometimes verified, for B*a*P and other PAH analyzed in aquatic organisms. These include biosynthesis, creosoted structures, runoff from land, domestic and industrial sewage effluents, and petroleum. Dunn and Young (1976) suggested that fallout of airborne particulate PAH was not an important source of B*a*P contamination in southern California mussels, although this is now considered by many investigators to be the major source of the complex PAH assemblages in relatively uncontaminated nearshore marine and freshwater sediments. The authors suggested that particle-bound B*a*P may not be biologically available to filter-feeding bivalves.

CHAPTER 5

# *Degradation of PAH in the Aquatic Environment*

PAH present in the aquatic environment are subject to degradation through a variety of different processes. The most important of these degradative processes are photooxidation, chemical oxidation, and biological transformation by aquatic bacteria, fungi, and animals. Since the majority of natural mechanisms of oxidative transformation of PAH require molecular oxygen, PAH are generally quite stable in hypoxic or anoxic environments. Thus, PAH may persist indefinitely in oxygen-poor water basins or in anoxic sediments.

## 5.1 PHOTOOXIDATION OF PAH

The chemical reactions of photo-induced oxidation of PAH in the aqueous phase by singlet oxygen, ozone, HO · radical, and other oxidizing agents are similar to those involved in photooxidation of airborne PAH and were discussed earlier in the section on airborne PAH. Photooxidation reactions of PAH in solution have been reviewed by the National Academy of Sciences (NAS, 1972); the most common of these reactions is the formation of endoperoxides (Fig. 11A). Photolysis or pyrolysis of the resulting peroxides through ring cleavage and dealkylation produces a variety of products. This process proceeds by a free-radical mechanism (cleavage of the O–O bond) and initiates autooxidation. Diones are often formed when, for steric reasons, no endoperoxide can be formed, so that photooxidation of B*a*P yields a variety of diones (Fig. 11B). Photooxidation involves an energy transfer from the triplet state of PAH, yielding singlet oxygen ($^1O_2$), which reacts with excited PAH to yield the peroxide or dione. Singlet oxygen derived from other sources can also react with PAH.

The photoreactivity of PAH adsorbed to the surface of particulates

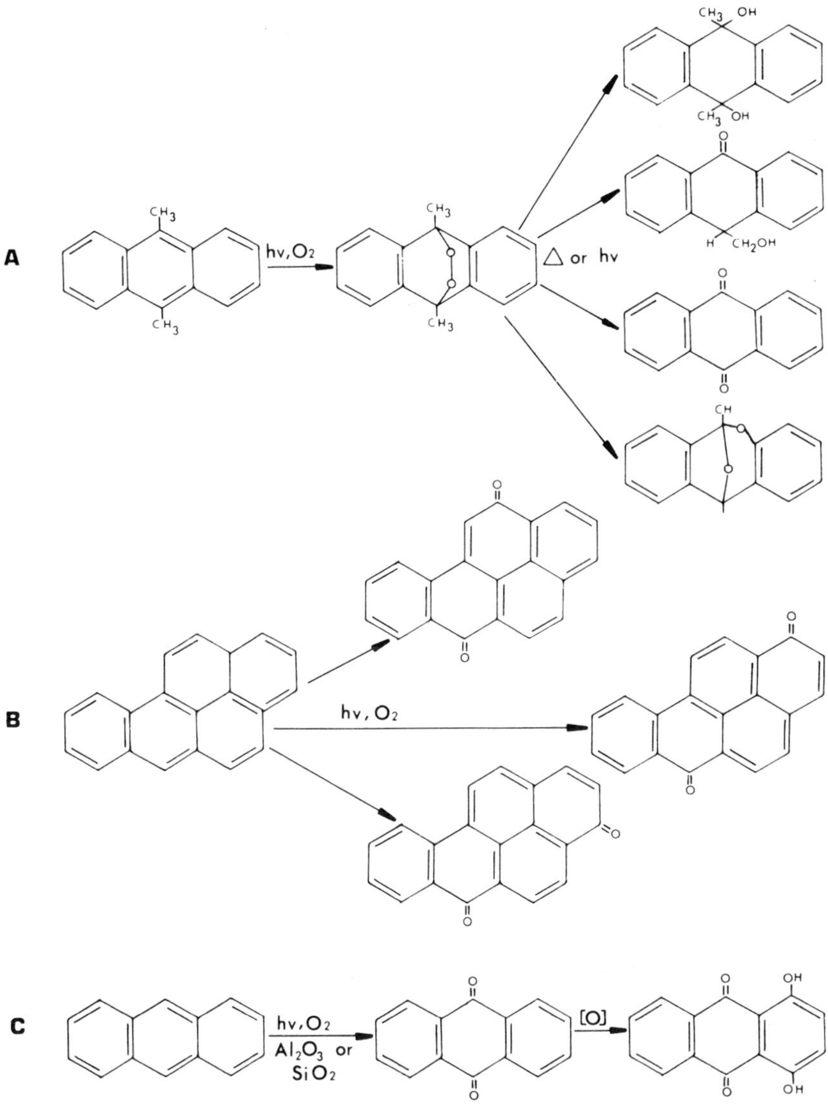


FIG. 11. Types of reactions proposed for the photooxidation of polycyclic aromatic hydrocarbons. A, Photooxidation of 9,10-dimethylanthracene through a 9,10-endoperoxide intermediate to yield several oxygenated products. B, Photooxidation of benzo[*a*]pyrene to the 6,12-, 1,6-, and 3,6-diones. C, Photooxidation of anthracene adsorbed to silica gel or alumina to 1,4-dihydroxy-9,10-anthraquinone (reproduced from *Particulate Polycyclic Organic Matter* (1972), pp. 67–9, with permission of the National Academy of Sciences, Washington, D.C.).

seems to be substantially greater than that of PAH in solution. Under these conditions, the oxidative pathway is different and does not appear to involve an endoperoxide intermediate. Anthracene adsorbed to alumina or silica gel is oxidized in the presence of light to anthraquinone. Further oxidation yields, 1,4-dihydroxy-9,10-anthraquinone (Fig. 11C). Since the majority of PAH in the aquatic environment are adsorbed to particulates, this latter observation has particular significance in the assessment of the environmental fate of PAH.

Suess (1972a) studied the effects of several factors on the photodegradation of B*a*P adsorbed to calcite particles in water. B*a*P–calcite suspensions were exposed to an illumination of 0·31 mW/cm$^2$, which is about one-fourth of the solar radiation at the earth's surface on a winter day. The temperature was 21 °C. Rate of B*a*P degradation was first order with time (an exponential decay). Under these conditions, the half-life of B*a*P was about 15 h when the water sample was exposed to normal atmospheric air. It was reduced to 11 h under an oxygen atmosphere. With a helium atmosphere, the half-life increased to about 35 h. These results show the dependence of B*a*P degradation and its rate on oxygen concentration of the water. No oxidation was observed in the complete absence of illumination. Thus, under the experimental conditions used in this investigation, the dominant mode of PAH degradation in the presence of oxygen was photooxidation; the dominant reaction was probably that of singlet oxygen with B*a*P excited to the triplet state. The influence of temperature on the rate of B*a*P photooxidation was also investigated. Rate of degradation increased exponentially with temperature as described by the Arrhenius equation. The rate constant of B*a*P degradation was not influenced by the concentration of B*a*P adsorbed to calcite between 0·08 and 1·3 $\mu$g/g, by changes in pH between 7 and 10, or by increasing the ionic strength of the medium. Suess (1972a) concluded that B*a*P photooxidation in natural waters will depend on water depth and will vary seasonally owing to changes in solar radiation, temperature, and dissolved oxygen. The degradation rate constant will decrease with depth owing to light absorption by water and suspended matter, and the usual decreases in temperature and dissolved oxygen with depth. Owing to lack of solar radiation and oxygen, photooxidation of B*a*P will be negligible in bottom sediments.

PAH vary substantially in their sensitivity to photooxidation. Il'nitskii *et al.* (1971a) studied the effects of UV radiation on the degradation of seven PAH. B*a*P was the most resistant of the PAH tested and 9,10-dimethylbenz[*a*]anthracene was about nine times more sensitive to photodegradation than B*a*P. More recently, Nagata and Kondo (1977) studied the rate of photodegradation of several PAH in mixed acetone–water or carbon tetrachloride–water solvents (Table 48). All PAH tested were photosensitive but the rates of degradation were quite variable. Anthracene, phenanthrene, and benz[*a*]anthracene were the most sensitive to photodegradation. Chrysene, fluorene, pyrene, and benzo[*a*]pyrene were

TABLE 48

PHOTODEGRADATION OF PAH UNDER NATURAL LIGHT IN MIXED ACETONE–WATER OR CARBON TETRACHLORIDE–WATER SOLUTIONS. INITIAL QUANTITIES OF STARTING MATERIALS ARE NORMALIZED TO 100 mg, EXCEPT IN THE MIXED 1-METHYLNAPHTHALENE–HETEROAROMATIC TESTS WHERE EQUIMOLAR QUANTITIES OF STARTING MATERIALS WERE USED (FROM NAGATA AND KONDO, 1977, WITH PERMISSION OF THE AMERICAN PETROLEUM INSTITUTE)

| *Compound* | *Degree of photodegradation (mg compound remaining) at* | |
|---|---|---|
| | *5 h* | *10 h* |
| Anthracene | 52·9 | 32·6 |
| Phenanthrene | 57·0 | 35·9 |
| Benz[*a*]anthracene | 45·5 | 0·0 |
| Chrysene | 96·0 | 94·0 |
| Fluorene | 94·3 | 91·9 |
| Pyrene | 94·6 | 89·1 |
| Benzo[*a*]pyrene | 93·6 | 90·5 |
| Benzothiophene + | 181·0 | 36·3 |
| 1-methylnaphthalene | 0·0[a] | 0·0[a] |
| Dibenzothiophene + | 20·6 | 16·2 |
| 1-methylnaphthalene | 40·1[a] | 0·0[a] |

[a] Quantity of naphthalene after photoirradiation.

relatively resistant to photodegradation. When a mixture of a heteroaromatic and a PAH was tested, the latter was quickly decomposed followed by a gradual decomposition of the former. This may have been due to formation of an intermolecular $\pi$-type charge-transfer complex between the two compounds. Earlier, Kuratsune and Hirohata (1962) studied the decomposition under UV light of 14 PAH dissolved in purified cyclohexane. Naphthalene was the most sensitive PAH tested. Of the others, only 7,12-dimethylbenz[*a*]anthracene, benzo[*a*]pyrene, 7-methylbenz[*a*]anthracene, and anthracene showed any decomposition, even after 235 h of irradiation. The remaining PAH showed no measurable decomposition. Thus, solvent type may have a marked effect on the photosensitivity of PAH. Predictions of the fate of PAH in natural waters based on studies performed in non-aqueous solvents must be made with extreme caution.

Several studies have been performed on photooxidation of petroleum in water. However, in most cases, the rate of degradation of individual PAH was not reported. Larson *et al.* (1976, 1977) studied the photooxidation of a No. 2 fuel oil containing 23 per cent aromatics. Oil samples were exposed in sample trays to near-UV radiation (280–400 nm) at a temperature of 21–23 °C. UV light energy at the surface of the oil was 0·038 gcal/cm$^2$/min. The total UV irradiation received in 24 h approximated that received on a bright summer day at 40 °N latitude (55 gcal/cm$^2$/day). Major photoreaction products identified were peroxides, phenols, acids, and carbonyls. Peroxides were formed at a roughly linear rate for about 90 h and then slowly declined from the maximum. After an initial decline, phenols increased linearly with time for the duration of the exposure. Acidity increased slowly to a maximum at 91 h. Total carbonyls also increased gradually during the 164 h timecourse of irradiation. There were no significant changes in the concentrations of these materials in oil retained in darkness. Fluorenone and alkylfluorenones were identified by their UV spectra and GC and TLC elution patterns.

The authors suggested a photooxidation scheme that would account for many of the products identified. Fluorene or another reactive PAH could be converted by long-wavelength UV to an excited state and then decay to radical form (Fig. 12). Combination of the radical with oxygen would produce a hydroperoxy radical. Abstraction of hydrogen from various hydrogen donors in the oil would yield a hydroperoxide and a chain-propagating radical also capable of autooxidation. Fluorene hydroperoxide could break down to fluorenone or cleave heterolytically under influence of Lewis acids to an electropositive hydroxyl, an active hydroxylating agent toward electron-rich aromatic hydrocarbons. In this scheme, photooxidation of photosensitive components of oil could lead to the production of oxygenated products which are effective oxidizing agents for other oil components.

Recently, Larson and Hunt (1978) showed that fuel oil contains photosensitizers (compounds which catalyze the production of electronically excited singlet oxygen from ground state $O_2$). Addition of singlet oxygen to reactive hydrocarbons produces hydroperoxides which then decompose thermally or photochemically to radicals which can initiate chain reactions.

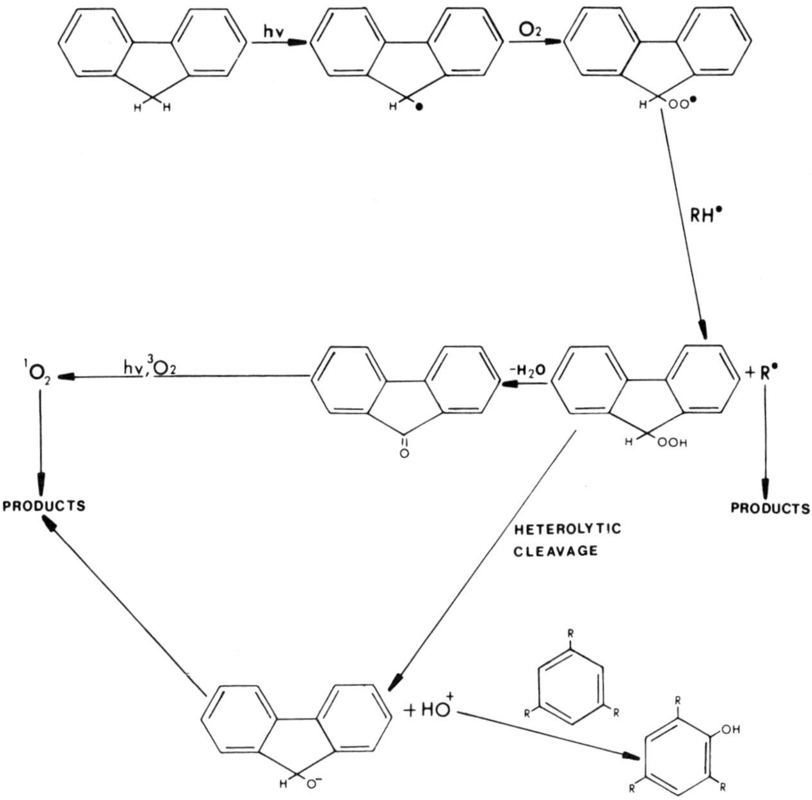


FIG. 12. Proposed scheme for the photooxidation of fluorene contained in No. 2 fuel oil. Fluorene is oxidized through a hydroperoxide intermediate to fluorenone or an electropositive hydroxyl (from Larson *et al.*, 1977, with permission of the American Chemical Society).

Burwood and Speers (1974) studied the photoöxidation of a medium sulfur (about 2 per cent weight) Middle East crude oil during long-term equilibration with sea-water. After four weeks of equilibration, a major portion of the water-soluble fraction was identified by GC–MS techniques as a complex mixture of thiacyclane-1-oxides. Burwood and Speers suggested that the thiacyclane oxides were produced through oxidation of indigenous crude oil thiacyclanes by aromatic hydroperoxides, products of photooxidation of aromatic hydrocarbons in the oil.

It is apparent that the photochemical reactions that take place when petroleum is exposed to solar radiation are extremely complex. The nature of the chemical reactions, as well as the products formed, is likely to vary substantially with variations in the chemical composition of the oil. An environmentally significant consequence of petroleum photooxidation is that the resulting oxygenated products are generally considerably more water soluble than parent compounds. Some of them have been shown to be quite toxic to aquatic organisms (Larson *et al.*, 1976, 1977) whereas others resemble known carcinogens or mutagens.

## 5.2 CHEMICAL OXIDATION OF PAH IN WATER

Chlorination or ozonation is often used for the destruction of pathogens in drinking water and sometimes for the oxidation of organic compounds in industrial wastewater effluents. Several investigations have been performed on the efficiency of these chemicals for use in the destruction of aqueous PAH. Such studies have yielded conflicting results, perhaps owing in part to variations in and lack of control of factors such as chlorine or ozone concentration, pH, and temperature (Harrison *et al.*, 1975).

Harrison *et al.* (1976a,b) have performed careful research dealing with the effects of chlorination on aqueous concentrations of several PAH. Sodium hypochlorite at concentrations of 0–13·2 mg/liter 'free chlorine' was used as the chlorinating agent. Temperature, pH, and the initial concentrations of PAH were carefully controlled. Eight PAH were evaluated. Under standardized conditions (temperature = 20 °C; 'free chlorine' = 2·2 mg/liter; pH = 6·8) the rate of PAH degradation was hyperbolic with time (Fig. 13A). Pyrene was oxidized most rapidly and fluoranthene most slowly. The rate of PAH oxidation during the initial 5 min of treatment increased with increase in 'free chlorine' concentration between 0·44 and 13·2 mg/liter (Fig. 13B). There was only a slight increase in reaction rate when temperature was increased from 5 to 20 °C.

The effect of pH was also studied. Estimated reductions in PAH levels increased considerably with decreasing pH (Fig. 13C), owing to the effect of increased hydrogen ion concentration on the hypochlorous acid dissociation equilibrium. A distinct deflection in the curves at a pH of about 6·5 probably indicated a change in reaction mechanisms. The chemical kinetics of chlorination of the PAH, pyrene, and benzo[*ghi*]perylene in dilute aqueous solutions was also investigated (Harrison *et al.*, 1976b). A first-order dependence of reaction rate upon the concentration of hypochlorous acid (the active chlorinated agent) and upon PAH concentration was found.

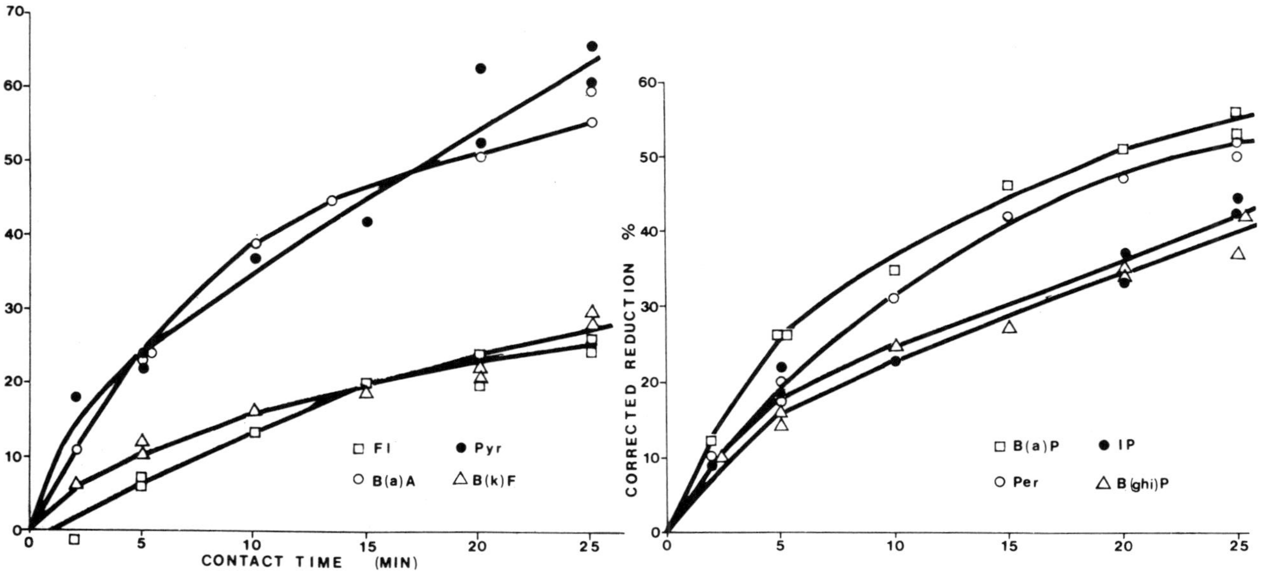


FIG. 13A. Degradation of polycyclic aromatic hydrocarbons by chlorination: rate of degradation of different PAH under standard conditions (temperature = 20 °C; 'free chlorine' = 2·2 mg/liter; pH = 6·8).

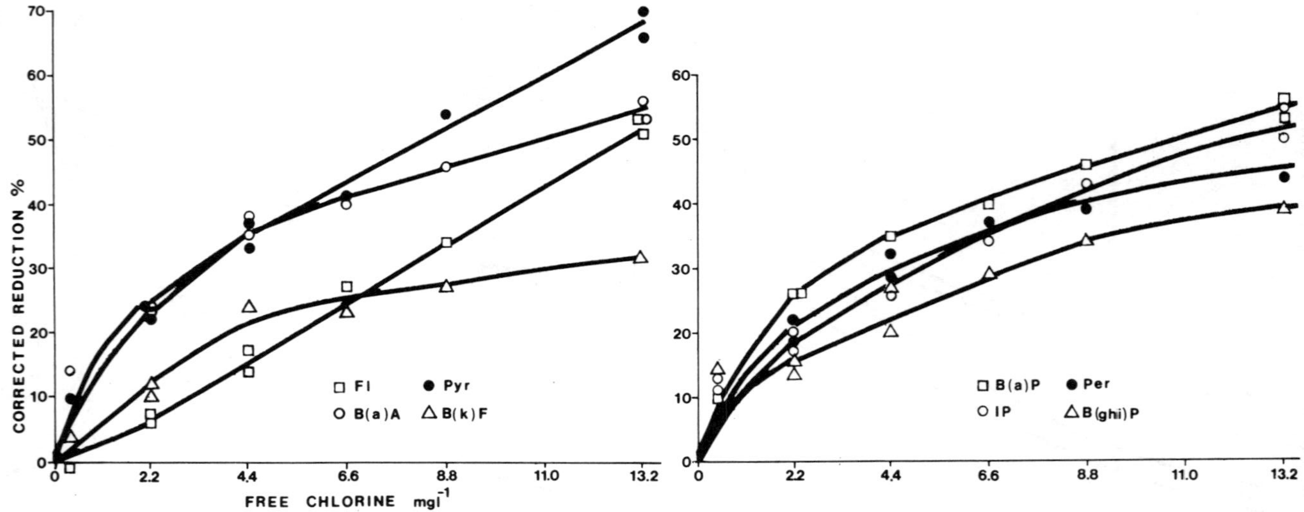


FIG. 13B. Degradation of polycyclic aromatic hydrocarbons by chlorination: effect of 'free chlorine' concentration on the rate of PAH degradation during the initial 5-min contact time of treatment (temperature = 20 °C; pH = 6·8).

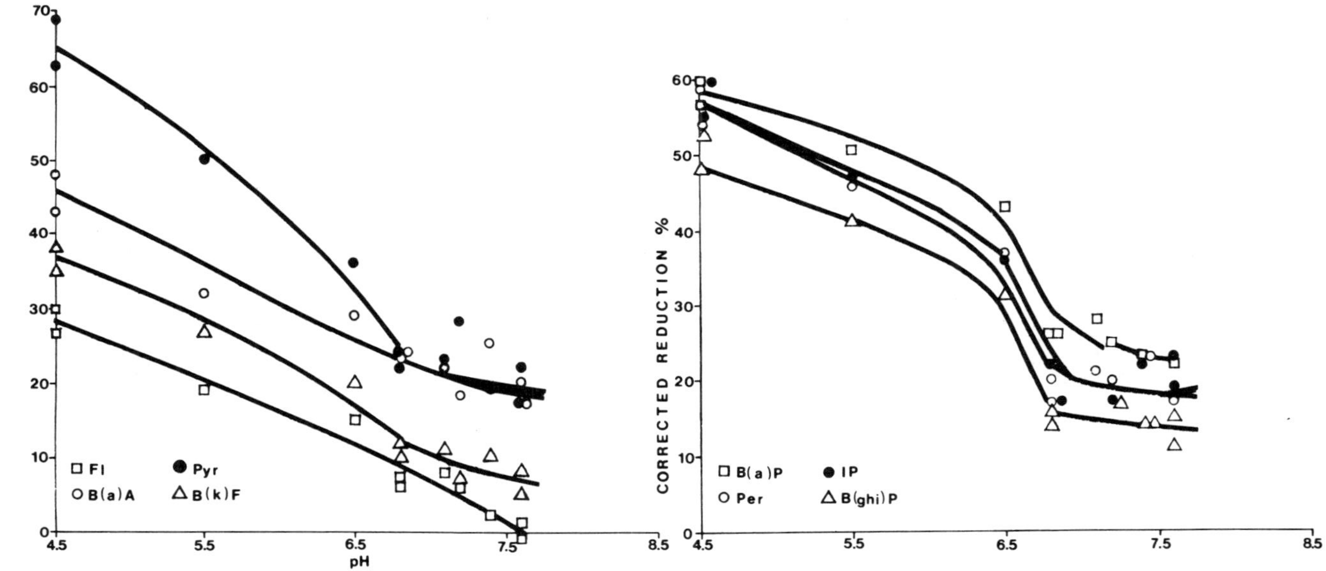


FIG. 13C. Degradation of polycyclic aromatic hydrocarbons by chlorination: effect of initial pH of the reaction mixture on the rate of degradation of PAH (contact time = 5 min; 'free chlorine' = 2·2 mg/liter; temperature = 20°C). (From Harrison *et al.*, 1976a, with permission of the American Chemical Society.)

Harrison *et al.* (1976a) also evaluated the efficiency of water treatment processes, including chlorination, for removal of PAH from drinking water. At the treatment plant studied, water is pumped from a river to a buffer storage reservoir with an average retention time of 2–4 days. Water is treated initially by rapid gravity filtration, followed by slow sand filtration of 12–18 h mean detention time. Finally, the water is chlorinated. Automatic dechlorination by sulfur dioxide is utilized, if necessary. Since PAH present in water will primarily adsorb to suspended particulates, sedimentation of solids in the reservoir caused a reduction in aqueous PAH levels (Table 49). However, since smaller particles and colloidal materials

TABLE 49

CONCENTRATION OF PAH IN WATER AT VARIOUS STAGES IN A WATER PURIFICATION PLANT (FROM HARRISON *et al.*, 1976a, WITH PERMISSION OF THE AMERICAN CHEMICAL SOCIETY)

| *Compound* | *PAH concentration (μg/liter)* | | | |
|---|---|---|---|---|
| | *River intake* | *After reservoir* | *After filtration* | *After chlorination* |
| Fluoranthene | 0·150 | 0·140 | 0·081 | 0·045 |
| Pyrene | 0·100 | 0·075 | 0·045 | 0·018 |
| Benz[*a*]anthracene + chrysene | 0·090 | 0·072 | 0·033 | 0·012 |
| Benzo[*b* + *j* + *k*]fluoranthene | 0·147 | 0·132 | 0·039 | 0·021 |
| Benzo[*a* + *e*]pyrene | [a] | 0·051 | 0·030 | 0·009 |
| Perylene | [a] | 0·039 | 0·024 | 0·006 |
| Indeno[1,2,3-*cd*]pyrene | 0·069 | 0·066 | 0·027 | 0·009 |
| Benzo[*ghi*]perylene | 0·072 | 0·063 | 0·033 | 0·009 |

[a] Measurement impossible owing to background.

contain most of the PAH and have very low sedimentation rates, the PAH reduction was not very large. Filtration removed most of the finer particulates and resulted in nearly a 50 per cent reduction in aqueous PAH levels. Chlorination was performed at 17°C, a pH of 7·5, and an initial chlorine concentration of 5 mg/liter and removed approximately 60 per cent of PAH remaining in the filtered water. Approximately 0·13 μg/liter total PAH remained in the drinking water after treatment.

The products of aqueous chlorination of PAH solutions are not fully known (Jolley *et al.*, 1975). Smith *et al.* (1977) identified several polychlorinated aromatics produced during chlorination of solutions of biphenyl and naphthalene. Chlorination of biphenyl solutions produced several dichlorobiphenyls. Chlorination of naphthalene solutions yielded

1-chloronaphthalene, 1,2-dichloronaphthalene, 1,4-dichloronaphthalene, 1,2,3,4-tetrachloro-1,2,3,4-tetrahydronaphthalene, and several other unidentified chlorinated products. Recently, Oyler *et al.* (1978) identified several reaction products resulting from the chlorination of aqueous solutions of PAH. Chlorination products included anthraquinone, a chlorohydrin of fluoranthene, and monochloro derivatives of fluorene,

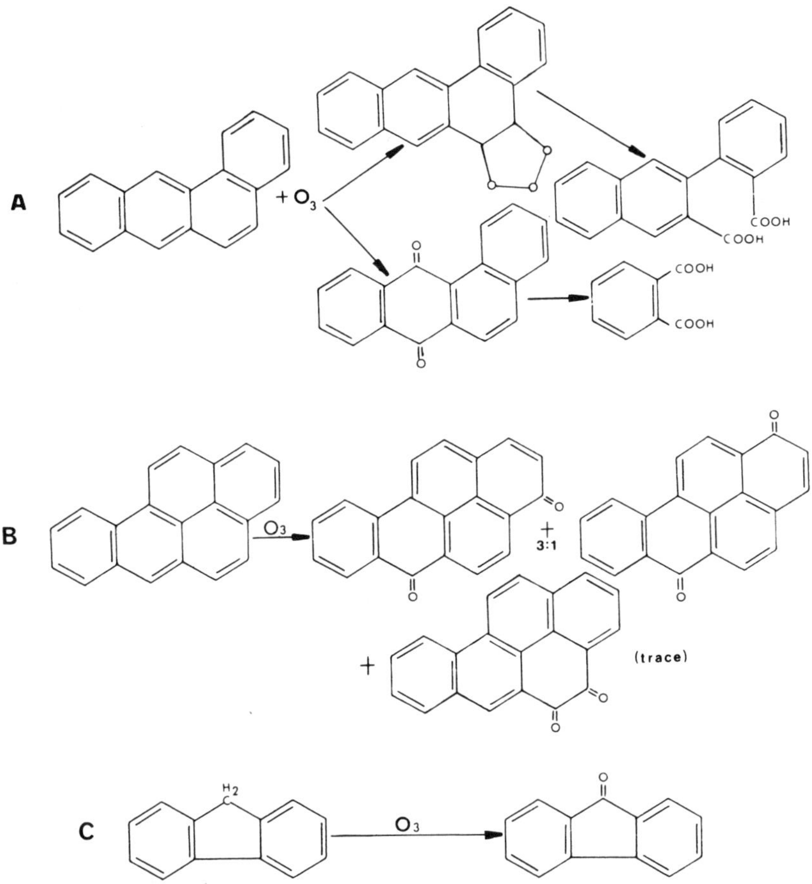


FIG. 14. Proposed reactions of polycyclic aromatic hydrocarbons with ozone. A, Ozonation of benz[*a*]anthracene to the 7,12-quinone and other products. B, Ozonation of benzo[*a*]pyrene to the 3,6-, 1,6-, and 4,5-diones. C, Ozonation of fluorene to fluorenone (reproduced from *Particulate Polycyclic Organic Matter* (1972), pp. 71–3, with permission of the National Academy of Sciences, Washington, D.C.).

phenanthrene, 1-methylphenanthrene, and 1-methylnaphthalene. Since polychlorinated biphenyls (Peakall, 1975), polychlorinated naphthalenes (Green and Neff, 1977), and probably other polychlorinated PAH are highly toxic to aquatic organisms and persist in the environment, special consideration should be given to alternative methods for purification of wastewaters or drinking waters containing high concentrations of PAH.

Ozone is frequently used for water treatment in Europe and, because chlorination may produce toxic byproducts, is beginning to replace chlorination as the purification method for some types of wastewater in the USA. Ozone is usually applied in batches at levels of about 5 mg/liter (Radding *et al.*, 1976). It reacts readily with PAH. The products of ozonolysis have been examined in some detail and the results are summarized by NAS (1972). Several reactions are possible depending on the species of PAH and reaction conditions. These reactions include cleavage of phenanthrene-like bonds producing, under oxidative conditions, various diacids, oxidation of anthracene 9,10-like positions to yield quinones, and oxidation of alkyl side-chains (Fig. 14). Ozonation is generally less efficient than chlorination for the removal of PAH from water. The toxicity of the products of PAH ozonation to aquatic organisms has not been determined.

Radding *et al.* (1976) used data of Il'nitskii *et al.* (1968) to calculate rate constants and PAH half-lives for reaction of 5 mg/liter ozone with PAH (Table 50). The computations show that various PAH generally display similar reactivities toward ozone; alkylation enhances this reactivity considerably. However, reaction rates are sufficiently low that ozonation would be effective in a water-treatment plant only if ozone did not evaporate.

TABLE 50

RATE CONSTANTS AND HALF-LIVES FOR REACTION OF PAH WITH OZONE IN WATER AT 25 °C (FROM RADDING *et al.*, 1976, WITH PERMISSION OF THE ENVIRONMENTAL PROTECTION AGENCY)

| *Compound* | *Rate constant,* $k_2$ *(liter mole$^{-1}$ s$^{-1}$)* | $t_{1/2}$ *(min)*[a] |
|---|---|---|
| Pyrene | 170 | 41 |
| Benzo[*a*]pyrene | 110 | 63 |
| Benz[*a*]anthracene | 260 | 27 |
| Dimethylbenzanthracene | >680 | <10 |
| Dibenzanthracene | 280 | 25 |

[a] Calculated from $t_{1/2} = 0{\cdot}69/k_2[O_3]$, where $[O_3] = 10^{-4}$ M.

PAH can also react with additional organic and inorganic oxidants including various electrophiles, peroxides, nitrogen oxides, and sulfur oxides. These reactions have been extensively studied in relation to oxidation of atmospheric PAH and are reviewed by NAS (1972).

## 5.3 METABOLIC DEGRADATION OF PAH BY AQUATIC BACTERIA AND FUNGI

Bacteria and fungi show a tremendous diversity in their ability to use different organic molecules as a sole carbon source. Since PAH and related compounds have been trace components of the biosphere for billions of years, it is not surprising that many microbes have evolved the capability to metabolize aromatic hydrocarbons. Some organisms are able to oxidize simple aromatic hydrocarbons completely to carbon dioxide and water and use them as a source of energy and as a source of carbon for biomass accretion. Others are not able to carry out the complete oxidation of aromatics, but if an alternative growth substrate is available they are able to partially metabolize aromatic hydrocarbons into various oxygenated metabolites. The latter process is called 'cooxidation' (Gibson, 1977).

Many bacterial species are able to use monocyclic aromatic hydrocarbons (benzene) as the sole carbon source. Initial reactions involve the introduction of two hydroxyl groups, through a dihydrodiol intermediate, into the aromatic nucleus (Gibson, 1976, 1977). The hydroxyl groups may be *ortho* to each other as in catechol or *para* to each other as in gentisic acid. Both oxygens in the hydroxylated product are derived from a single molecule of atmospheric oxygen. Dihydroxylation is apparently a necessary prerequisite for subsequent enzymatic cleavage of the benzene ring. Catechols can undergo enzymatic cleavage between the hydroxyl groups (*ortho*-cleavage) or adjacent to one of the hydroxyl groups (*meta*-cleavage).

Several species of bacteria are able to partially or completely metabolize some PAH. Initial reactions in these cases are similar to the benzene conversions described. Naphthalene is oxidized by *Pseudomonas putida* and certain other species to *cis*-1,2-dihydroxy-1,2-dihydronaphthalene (Jerina *et al.*, 1971) (Fig. 15). The dihydrodiol is next oxidized to 1,2-dihydroxynaphthalene. *Ortho*-cleavage of the dihydroxylated benzene ring yields, by way of intermediates, salicyclic acid, which is degraded through catechol to carbon dioxide and water. An important feature of the bacterial

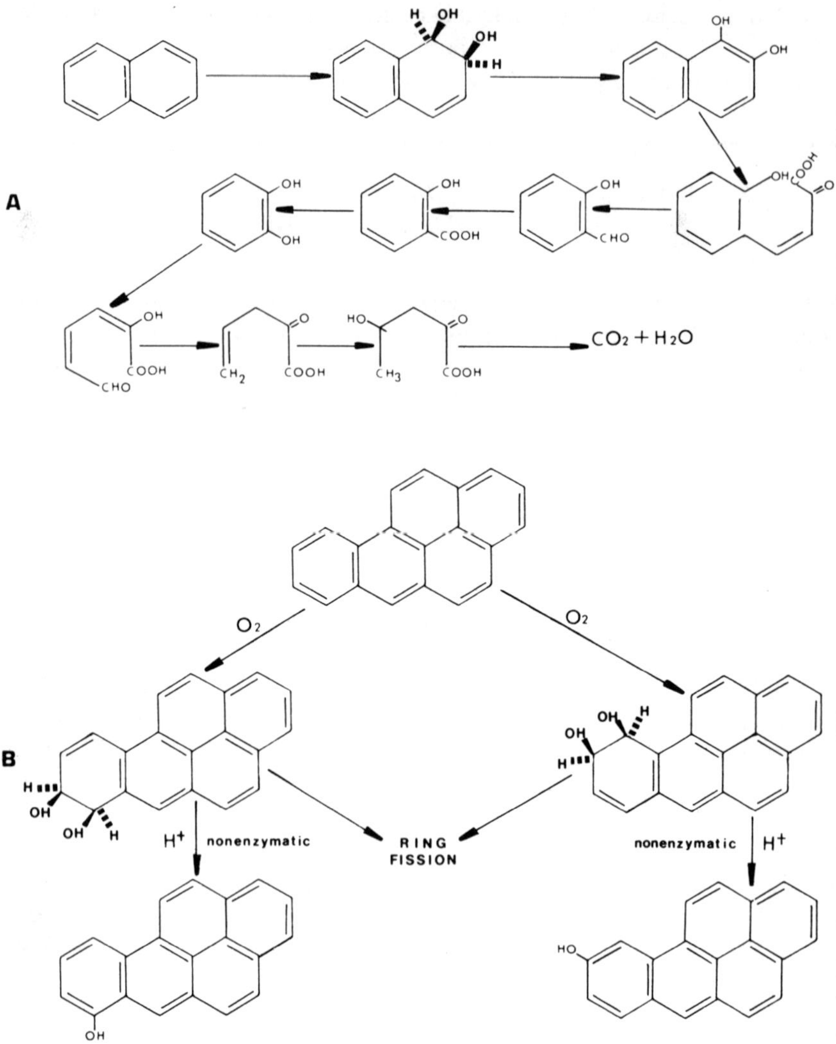


FIG. 15. Metabolism of polycyclic aromatic hydrocarbons by bacteria. A, Metabolism of naphthalene by *Pseudomonas putida* to carbon dioxide and water (from Gibson, 1976, with permission of the American Institute of Biological Sciences). B, Initial reactions in the metabolism of benzo[*a*]pyrene by *Beijerinckia* sp. (from Gibson, 1977, with permission of Pergamon Press).

pathway is that the *cis*-dihydrodiol is apparently produced through a dioxetane intermediate, whereas in the mammalian microsomal system a *trans*-dihydrodiol is produced through an arene oxide intermediate. The arene oxides, or their immediate oxidation products, appear to be responsible for the carcinogenicity or mutagenicity of PAH showing these properties (Jerina and Daly, 1974; Huberman *et al.*, 1976; Heidelberger, 1976).

Several species of bacteria are able to catalyze the oxidation of other PAH including the carcinogens benzo[*a*]pyrene and benz[*a*]anthracene. In most cases, the higher molecular weight PAH are inadequate sole carbon sources, apparently because of the inability of microbes to oxidize them completely to carbon dioxide and water (Dean-Raymond and Bartha, 1975). Instead, various phenolic or acidic metabolites are produced. It appears that *cis*-hydroxylation is the principal mechanism for the commencement of PAH degradation (Gibson, 1977). Gibson *et al.* (1975) isolated a bacterium, tentatively identified as *Beijerinckia* sp., from a polluted freshwater stream which was able to use biphenyl as the carbon source. A mutant strain was produced which oxidized B*a*P to *cis*-9,10-dihydroxy-9,10-dihydrobenzo[*a*]pyrene (major component) and *cis*-7,8-dihydroxy-7,8-dihydrobenzo[*a*]pyrene (10%) (Fig. 15). The dihydrodiols could subsequently be reduced nonenzymatically to 9-hydroxy-B*a*P and 7-hydroxy-B*a*P, respectively. Wild-type *Beijerinckia* sp. were able to further degrade the dihydriols by ring fission. This same mutant strain was able to metabolize phenanthrene to *cis*-3,4-dihydroxy-3,4-dihydrophenanthrene, anthracene to *cis*-1,2-dihydroxy-1,2-dihydroanthracene, and benz-[*a*]anthracene to *cis*-1,2-dihydroxy-1,2-dihydrobenz[*a*]anthracene.

Barnsley (1975) studied the metabolism of fluoranthene and B*a*P by several species of pseudomonads. Degradation occurred most rapidly during the stationary phase of the growth cycle, was heat sensitive, and required oxygen. The presence of cyanide in the culture medium enhanced the degradation of fluoranthrene but not B*a*P. It was hypothesized that cyanide stimulated fluoranthene oxidation by inhibiting the enzyme superoxide dismutase. The superoxide radical is thought to be involved in fluoranthene hydroxylation.

Sisler and ZoBell (1947) studied the utilization of PAH by mixed cultures of marine bacteria. PAH were introduced into seawater cultures adsorbed to ignited sand. The amount of PAH metabolized was determined by measuring the amount of carbon dioxide evolved in hydrocarbon oxidation and subtracting carbon dioxide produced by control cultures (Table 51). The mixed culture of marine bacteria metabolized phenanthrene and

TABLE 51

AMOUNT OF CARBON DIOXIDE PRODUCED IN 4 DAYS BY THE ACTION OF MARINE BACTERIA ON 25 mg OF PAH AT 32 °C (FROM SISLER AND ZOBELL, 1947, WITH PERMISSION OF THE AMERICAN ASSOCIATION FOR THE ADVANCEMENT OF SCIENCE)

| *Substrate* | $CO_2$ *produced (mg)* | *Amount of PAH oxidized (%)* |
|---|---|---|
| Naphthalene | 44·2 | 51 |
| Anthracene | 53·5 | 64 |
| Phenanthrene | 58·5 | 68 |
| Benz[*a*]anthracene | 44·2 | 47 |
| Dibenz[*a*,*h*]anthracene | 11·6 | 13 |

anthracene more rapidly than naphthalene, benz[*a*]anthracene, and dibenz[*a*,*h*]anthracene. No intermediates or metabolites were identified.

Ability of freshwater and marine bacteria to degrade petroleum has been studied extensively. The majority of reports give little information on rates and mechanisms of microbial oxidation of aromatic hydrocarbons from oil. Dean-Raymond and Bartha (1975) isolated six strains of bacteria from the oil-polluted estuarine waters of the Arthur Kill River, New Jersey, USA, which were tentatively identified as types of *Pseudomonas putida* and *P. multivorans*. Isolation was based on ability to culture the strain using naphthalene, 1-methylnaphthalene, or 2-methylnaphthalene as the sole carbon source. None of the organisms could utilize any of several dimethylnaphthalenes tested as sole carbon sources, but all could partially metabolize them. Three of the strains were unable to metabolize any of the higher molecular weight PAH tested; the other three strains could metabolize some of the PAH although they could not use them for growth (Table 52). The latter three strains showed considerable variability in their ability to utilize different PAH. Fluorene, phenanthrene, and anthracene were more readily oxidized than other PAH. The metabolites produced were substrate-dependent and included salicylic acids, naphthoic acids, naphthyl alcohols, and other oxygenated products (Table 53). Lee and Takahashi (1977) studied the degradation of PAH by marine bacteria isolated from assorted depths in a 2 m diameter × 15 m deep controlled ecosystem enclosure in Saanich Inlet, British Columbia. Water in the enclosure was contaminated with No. 2 fuel oil to a concentration of 10–20 $\mu$g/liter total non-volatile petroleum hydrocarbons. Rate of PAH degradation was estimated by measuring rate of $^{14}CO_2$ production following dosing of the cultures with $^{14}$C–PAH. Naphthalene and

TABLE 52

UTILIZATION OF PAH BY *Pseudomonas putida* (N1 AND N2) AND *P. multivorans* (N4) ISOLATED FROM AN OIL-POLLUTED ESTUARY (FROM DEAN-RAYMOND AND BARTHA, 1975, WITH PERMISSION OF THE SOCIETY FOR INDUSTRIAL MICROBIOLOGY)

| *Substrate* | *Amount of PAH utilized* (%) | | |
|---|---|---|---|
| | *N*1 | *N*2 | *N*4 |
| Acenaphthylene | 18·3 | 6·6 | 0·0 |
| Acenaphthene | 1·5 | 0·0 | 0·0 |
| Fluorene | 19·0 | 7·7 | 5·9 |
| Perylene | 5·6 | 3·5 | 0·0 |
| Phenanthrene | 11·8 | 0·0 | 74·2 |
| Anthracene | 8·4 | 7·3 | 12·8 |
| Pyrene | 0·0 | 0·0 | 6·9 |
| Chrysene | 0·0 | 0·0 | 0·0 |
| Benz[*a*]anthracene | 0·0 | 0·0 | 2·3 |
| Benz[*b*]anthracene | 0·0 | 0·0 | 0·0 |

TABLE 53

METABOLITES OF SELECTED PAH PRODUCED BY *Pseudomonas putida* AND *P. multivorans* ISOLATED FROM AN OIL-POLLUTED ESTUARY (FROM DATA OF DEAN-RAYMOND AND BARTHA, 1975)

| *Substrate* | *Metabolites identified* |
|---|---|
| Naphthalene | salicylic acid |
| 1-Methylnaphthalene | 3-methylsalicylic acid |
| 1-Ethylnaphthalene and 2-ethylnaphthalene | ethylsalicylic acids |
| 1,3-Dimethylnaphthalene and 2,4-dimethylnaphthalene | dimethylsalicylic acids |
| 1,5-Dimethylnaphthalene | 1-methyl-5-naphthoic acid<br>1-methyl-5-naphthyl alcohol |
| 2,6-Dimethylnaphthalene | 2-methyl-6-naphthoic acid<br>2-methyl-6-naphthyl alcohol |
| 1,4-Dimethylnaphthalene | methylnaphthyl alcohol<br>methylnaphthoic acid<br>methylnaphthaldehyde<br>dimethylsalicylic acid |
| Anthracene and phenanthrene | derivatives of *o*-hydroxynaphthoic acid |
| Acenaphthylene | a quinone |
| Fluorene | fluorenone |

methylnaphthalene were degraded at a low rate prior to the addition of oil (Table 54). Three days after the addition of oil, metabolism of these PAH was greatly accelerated. Fluorene and B*a*P were not metabolized at all by bacteria isolated from the enclosure prior to oil dosing. Three days later, B*a*P was degraded at a barely detectable rate but fluorene was still unmetabolized.

TABLE 54

MICROBIAL DEGRADATION OF $^{14}$C–PAH IN WATER FROM A CONTROLLED ECOSYSTEM ENCLOSURE TO WHICH No. 2 FUEL OIL HAD BEEN ADDED (FROM LEE AND TAKAHASHI, 1977, WITH PERMISSION OF THE PUBLISHER)

| *PAH (initial concentration, μg/liter)* | *Collection depth (m)* | *Incubation time (h)* | *Time after oil addition (days)* | *Degradation rate (μg/liter/ day ± SD)* | *Turnover time (days)* |
|---|---|---|---|---|---|
| Benzo[*a*]pyrene (16) | 5–10 | 48 | 0 | 0 | — |
| Benzo[*a*]pyrene (16) | 5–10 | 24 | 3 | 1 ± 0·7 | 1 400 |
| Fluorene (30) | 5–10 | 48 | 0 | 0 | — |
| Fluorene (30) | 5–10 | 48 | 3 | 0 | — |
| Methylnaphthalene (50) | 0–5 | 24 | 0 | 10 ± 6 | 500 |
| Methylnaphthalene (50) | 0–5 | 24 | 3 | 26 ± 4 | 200 |
| Naphthalene (50) | 0–5 | 24 | 0 | 10 ± 3 | 500 |
| Naphthalene (50) | 0–5 | 16 | 3 | 250 ± 7 | 22 |
| Naphthalene (50) | 0–5 | 10 | 4 | 100 ± 5 | 57 |
| Naphthalene (50) | 5–10 | 10 | 4 | 500 ± 11 | 10 |

In similar studies, Lee (1977) estimated the rate of degradation of several PAH by natural bacterial populations in marine and estuarine waters of the southeast coast of the USA. Bacterial populations from the oil-polluted Cooper River estuary, South Carolina, rapidly degraded naphthalene and 2-methylnaphthalene but degraded fluorene and anthracene only to a very limited extent, if at all. Similar results were obtained with aquatic bacteria from the relatively unpolluted Skidaway River, Georgia. Rates of degradation of naphthalene and 2-methylnaphthalene were many times lower during January–February than April–October. Only very limited degradation of fluorene and anthracene was observed. Degradation of B*a*P was observed in only one of three incubations and the rate was very low ($2 \pm 1 \times 10^{-3}$ μg/liter/day at an initial B*a*P concentration of 5 μg/liter). Microbial degradation of benz[*a*]anthracene could not be demonstrated. Naphthalene and 2-methylnaphthalene were degraded more rapidly by bacteria from estuarine waters than by those from coastal waters (Table 55).

TABLE 55

MICROBIAL DEGRADATION RATES OF NAPHTHALENE AND 2-METHYLNAPHTHALENE AT AN INITIAL CONCENTRATION OF 30 $\mu$G/LITER IN ESTUARINE, COASTAL AND GULF STREAM WATERS IN THE VICINITY OF THE ATLANTIC COAST OF GEORGIA, USA (FROM LEE, 1977, WITH PERMISSION OF THE AMERICAN PETROLEUM INSTITUTE)

| *Hydrocarbon* | *Sampling site* | *Degradation rate ($\mu$g/liter/day) $\times 10^3 \pm SD$* | *Turnover time (days)* |
|---|---|---|---|
| Naphthalene | estuarine | $870 \pm 20$ | 34 |
| | coastal | $330 \pm 40$ | 230 |
| | Gulf Stream | $12 \pm 7$ | 2 500 |
| 2-Methylnaphthalene | estuarine | $250 \pm 60$ | 120 |
| | coastal | 0 | — |
| | Gulf Stream | 0 | — |

Gulf Stream water contained very little microbial hydrocarboniclastic activity.

Walker *et al.* (1976a,b) showed that bacteria isolated from Atlantic Ocean marine sediments were able to utilize a mixed hydrocarbon substrate. The artificial mixed hydrocarbon substrate contained n-alkanes, cycloalkanes, and aromatic hydrocarbons. All hydrocarbons, including the PAH naphthalene, phenanthrene, benz[*a*]anthracene, perylene and pyrene showed some degradation. Bacteria isolated from the sediment were mainly various species of *Pseudomonas* and *Acinetobacter*.

Activated sludge treatment is frequently used in the purification of domestic and industrial wastewaters. Sludge microbes effectively metabolize and remove many potentially toxic organic compounds. Malaney *et al.* (1967) evaluated the sensitivity of organic carcinogens, including PAH, to oxidation by activated sludge. All PAH tested were quite resistant to treatment with several activated sludges from the Nashville, Tennessee, USA, area. The authors concluded that the sludges were unable to effect any significant removal of organic carcinogens by the oxidative mechanism within normal retention times in activated sludge treatment ponds.

Much less information is available on oxidation of PAH by fungi (yeasts and molds). Fungi unlike bacteria but like mammals, possess a cytochrome P-450 (a heme protein) enzyme system (Duppel *et al.*, 1973; Ferris *et al.*, 1976; Cerniglia and Gibson, 1978). The fungus *Cunninghamella elegans* degrades naphthalene by the arene oxide–*trans*-naphthalene dihydrodiol pathway characteristic of mammals (Ferris *et al.*,

1973). A strain of *C. elegans* was shown to oxidize naphthalene to *trans*-1,2-dihydroxy-1,2-dihydronaphthalene, 1-naphthol, 2-naphthol, 4-hydroxy-1-tetralone, 1,2-naphthoquinone, and 1,4-naphthoquinone (Cerniglia and Gibson, 1978). Activity of the naphthalene-degrading enzymes in the fungus was increased (induced) 2–4-fold by preincubation of the cells in a culture medium containing naphthalene, 3-methylcholanthrene, or phenobarbitol. Thus, initial reactions of metabolic degradation of PAH by fungi resemble those in animals more than in bacteria.

Ahearn *et al.*, (1971) examined the capacity of yeasts from various marine, freshwater, and terrestrial environments to assimilate petroleum hydrocarbons. Many hydrocarboniclastic yeasts were found predominantly in oil-polluted habitats. One of the yeasts isolated from an oil slick in the Gulf of Mexico, *Trichosporon* sp., was able to utilize methylnaphthalene, the only PAH tested, as a growth substrate.

Although several species and strains of marine and freshwater bacteria and some species of marine and freshwater fungi have been shown to be capable of degrading PAH, their quantitative role in the removal of PAH from the aquatic environment is poorly understood. The concentrations of hydrocarboniclastic bacteria and fungi in all but heavily polluted fresh and marine waters are extremely low. Since most species cannot use PAH as a sole carbon source, a deficiency of alternative organic carbon growth substrates may limit population increases in areas of elevated PAH inputs. Marine and freshwater sediments usually contain much larger numbers of hydrocarboniclastic microorganisms (Ahearn *et al.*, 1971; Dean-Raymond and Bartha, 1975; Walker *et al.*, 1976a) and higher concentrations of organic carbon. Thus, microbial degradation of PAH in sediments may be significant as long as the sediments remain oxygenated. Since microbial oxidation of PAH requires molecular oxygen, it will not proceed to any appreciable extent in anoxic sediments or water. As an example, Lee *et al.* (1978a) showed that $^{14}$C-benz[*a*]anthracene near the surface of a marine sediment was degraded rapidly while that in deeper, presumably anoxic layers, was degraded very slowly.

## 5.4 BIOTRANSFORMATION OF PAH BY AQUATIC ANIMALS

### 5.4.1 Hepatic Microsomal MFO Activity

In mammals, an enzyme system variously known as the cytochrome P-450-dependent mixed-function oxidase, mixed-function oxygenase, aryl hydrocarbon hydroxylase, or drug metabolizing system is responsible for initiating the metabolism of various lipophilic organic compounds, including xenobiotics (foreign organic compounds such as alkanes, PAH, pesticides, and drugs) and endogenous compounds such as steroid hormones and bile salts. The primary function of this system appears to be to render poorly water-soluble lipophilic materials more water soluble and therefore more available for excretion. Although this system effectively detoxifies certain xenobiotics, others, such as certain PAH and alkenes, are transformed to intermediates which are highly toxic, mutagenic, or carcinogenic to the host. Oxidative metabolism of PAH in this system proceeds via highly electrophilic intermediate arene oxides, some of which bind covalently to cellular macromolecules such as DNA, RNA, and protein. It is now generally agreed that metabolic activation by the mixed-function oxygenase (MFO) system is a necessary prerequisite for PAH-induced carcinogenesis and mutagenesis (Jerina and Daly, 1974; Huberman *et al.*, 1976; Freudenthal and Jones, 1976). For instance, the carcinogenicity of B*a*P is thought to be due primarily to the isomeric 9,10-epoxy-7,8-dihydro-7,8-dihydroxybenzo[*a*]pyrenes (Sims *et al.*, 1974; Lehr and Jerina, 1977; Yang *et al.*, 1977).

Active epoxides may be converted to less toxic products by various enzymatic or nonenzymatic reactions. Major enzymatic biotransformation pathways include microsomal epoxide hydrase which converts epoxides to diols, glutathione S-transferases which catalyze the formation of glutathione (GSH) conjugates, the initial step in mercapturic acid biosynthesis, and sulfokinases which convert phenols to sulfates. Nonenzymatic pathways include reactions with GSH, solvolysis to yield diols, and isomerization of arene oxides to form phenols (Corner, 1975; Bend *et al.*, 1977). Epoxide hydrase may either increase or decrease the mutagenicity of B*a*P (Bentley *et al.*, 1977). This seeming paradox is probably due to the fact that both the 7,8-epoxide and the diol-epoxide (active carcinogen) of B*a*P can serve as the substrate for epoxide hydrase. Which reaction will take precedence apparently depends on the initial concentration of B*a*P or the activity of the MFO system (Bentley *et al.*, 1977).

The MFO system is located in the liver and sometimes in several other organs of many vertebrates and invertebrates. Within the cell it is localized primarily in association with the endoplasmic reticulum (microsomal fraction) and consists of several components including a nicotinamide adenine dinucleotide phosphate (NADPH) cytochrome reductase, a phospholipid, and the heme protein called cytochrome P-450 because of its characteristic carbon monoxide absorption spectrum (Fig. 16). These

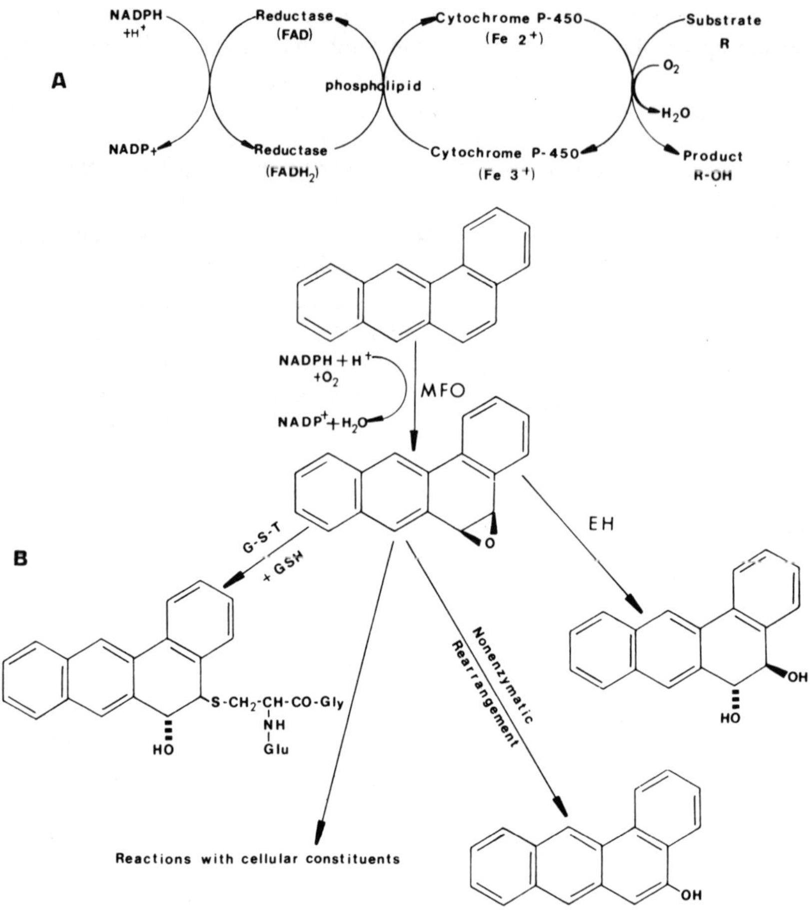


FIG. 16. Hepatic cytochrome P-450–mixed-function oxidase (MFO) system of animals. A, Components of the MFO system (FAD, flavine-adenine dinucleotide) (from Stegeman, 1977, with permission of *Oceanus* magazine). B, Several possible degradative pathways of benz[*a*]anthracene (MFO, mixed-function oxidase; EH, epoxide hydrase; G-S-T, glutathione S-transferase) (from Swaisland *et al.*, 1973, with permission of Pergamon Press).

components function as an electron-transport system when combined with NADPH as a source of reducing power, molecular oxygen, and an appropriate substrate such as PAH. The system catalyzes the incorporation of one atom of molecular oxygen into water and one atom into the substrate.

Several possible reactions in the biotransformation of benz[*a*]anthracene are summarized in Fig. 16. Benz[*a*]anthracene is converted to the 4,5-epoxide by the MFO system. The epoxide can then be converted to several alternative products. Microsomal epoxide hydrase converts it stereospecifically to the *trans*-4,5-dihydro-4,5-dihydroxybenz[*a*]anthracene. The second oxygen is derived from water. Cytoplasmic glutathione S-transferase may catalyze the conjugation of the epoxide to glutathione. Alternatively, the epoxide may undergo nonenzymatic isomerization to produce 4-hydroxybenz[*a*]anthracene or it may bind to cellular macromolecules. The biotransformation of B*a*P by mammalian hepatic microsomes has been more extensively studied. In the scheme summarized in Fig. 17 (Yang *et al.*, 1977), B*a*P is converted to 7,8-epoxide by the MFO system. Epoxide hydrase catalyzes conversion of the epoxide almost exclusively to the (−)*trans*-7,8-diol enantiomer. Oxygen of the $C_7OH$ group is derived exclusively from atmospheric oxygen while oxygen at the $C_8$ position is derived from water. The (−)*trans*-7,8-diol is further metabolized stereoselectively by MFO to a major diol epoxide I and a minor diol epoxide II in a ratio of approximately 10 to 1. These metabolically formed diol epoxides are unstable in aqueous media and each is hydrolyzed nonenzymatically to a pair of tetrahydrotetrahydroxybenzo[*a*]pyrenes (tetrols) or reduced nonenzymatically by NADH or NADPH to a tetrahydrotrihydroxybenzo[*a*]pyrene (triol). The reactions proceed through carbonium ion intermediates.

Diol epoxide I has been demonstrated to be the major species formed metabolically by mammalian microsomes and to be the most mutagenic B*a*P metabolite in cultured mammalian cells. Many other reactions and products are possible (DePierre and Ernster, 1978; Yang *et al.*, 1978). For instance, initial attack by MFO on the B*a*P molecule can also occur at the 1,2, 2,3, 4,5 (K region), or 9,10 positions. MFO can also attack the diol at several positions. The epoxide may isomerize nonenzymatically to the phenol and both the epoxide and the diol epoxide may be conjugated enzymatically to GSH, sulfate, or carbohydrate. Dihydroxy-PAH could be oxidized to quinones. There may be large interspecies differences in the preferred pathways and in the dominant metabolites produced. In all cases where the stereochemistry has been established, the PAH-diols produced by eukaryotic cells are in the *trans* configuration (Gibson, 1977). However, relatively little comparative research has been done in these areas.

A good deal of effort has been expended in recent years to identify the cytochrome P-450–MFO system in nonmammalian species, including many aquatic animals. Two strategies have been used. In the first, animals

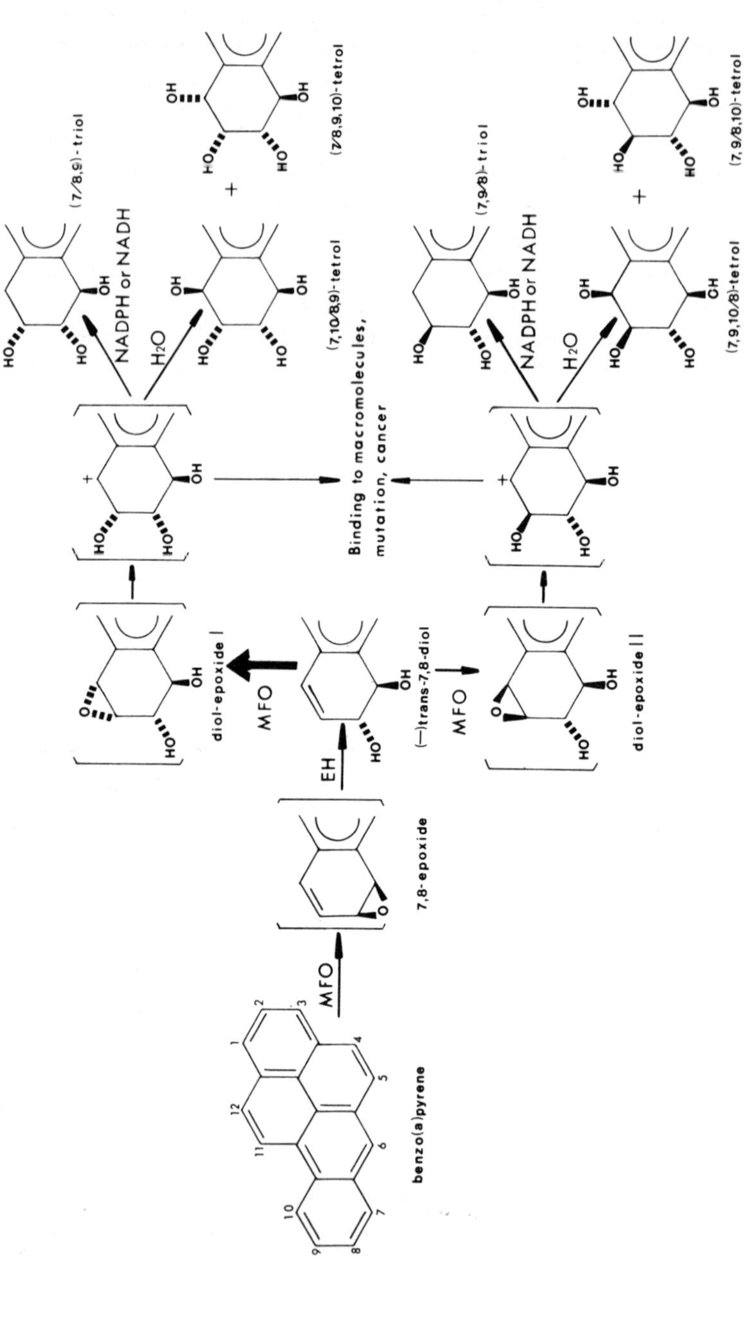


FIG. 17. Metabolic activation and detoxification of benzo[*a*]pyrene by mammalian hepatic microsomes. MFO, mixed-function oxidase; EH, epoxide hydrase (from Yang *et al.*, 1977, with permission of AAAS, copyright 1977 by the American Association for the Advancement of Science).

are exposed to a PAH or other MFO substrate and an attempt is made to isolate and identify metabolites in the animal's tissues or in the ambient medium. In the second approach, MFO enzymatic activity of homogenates or microsomal preparations of animal tissues is assessed. Using these methods, a large number of aquatic species has been identified which possess at least a limited ability to metabolize PAH to more polar metabolites. Table 56 summarizes much of the available data on PAH metabolite identification in aquatic species. Among invertebrates, the ability to produce polar metabolites from PAH seems to be restricted primarily to some members of the phyla Arthropoda and Annelida. The Cnidaria and Ctenophora as well as most members of the phylum Mollusca studied so far seem to lack this ability. Only four species of annelid worms have been investigated, but all possess at least a limited ability to transform PAH. According to Payne (1976, as cited by Malins, 1977b and Varanasi and Malins, 1977), the squid *Ilex illecebrosus* (Mollusca) and the sea star *Asterias* sp. (Enchinodermata) are able to transform PAH. Most of the boney fish (Teleostei) so far studied are able to transform PAH. Among the Amphibia, only American toad embryos and an aquatic salamander have been studied, and both are able to transform PAH. Recently, Payne (1977) reported that aryl hydrocarbon hydroxylase (AHH) activity was not present in the marine algae, *Fucus* sp., *Ascophyllum* sp., and an unidentified colonial green alga, in the sea anemone *Metridium* sp., and the sea urchin *Strongylocentratus drobrachiensis*.

Distribution of AHH activity and cytochrome P-450 in the liver or hepato-pancreas of aquatic species (Table 57) tends to coroborate the metabolite production data. Unfortunately, AHH activity is expressed by various researchers in different units which are not strictly comparable or interconvertible. Therefore it is difficult to compare AHH activities among diverse species. Cytochrome P-450 concentrations have been converted to the same units to facilitate comparisons. The marine worm *Nereis virens* has a highly active AHH activity localized in its lower intestinal epithelium. Cytochrome P-450 concentration is substantially less in this species than in most of the teleosts studied. The American oyster *Crassostrea virginica* is the only mollusc in which AHH activity has been quantified; it is low in this species and is measurable only after long periods of induction by exposure to PAH (Anderson, 1978). Since one enzyme unit as defined by Lee *et al.* (1978b) is equal to approximately 1 pmole 3-OH-B*a*P/mg protein/10 min, AHH activity in the worm is approximately 3·5 nmole 3-OH-B*a*P/mg protein/10 min or 21·5 times the AHH activity in the oyster's digestive gland.

TABLE 56

IDENTIFICATION OF PAH METABOLITES PRODUCED BY AQUATIC ORGANISMS FOLLOWING EXPOSURE TO DIFFERENT PAH

| *Species* | *Substrate(s)*[a] | *Metabolites*[b] *identified* | *Nature of metabolites*[c] | *Reference* |
|---|---|---|---|---|
| Cnidaria | | | | |
| jellyfish (unidentified) | N, B*a*P, MC | 0 | | Lee, 1975 |
| Ctenophora | | | | |
| *Pleurobrachia pileus* | N, B*a*P, MC | 0 | | Lee, 1975 |
| Annelida | | | | |
| *Nereis succinea*, *N. virens*, and *Capitella capitata* | B*a*A | + | 5,6-dihydrodiol | Lee *et al.*, 1978b |
| *Neanthes arenaceodentata* | N | + | PM | Rossi, 1977 |
| Mollusca | | | | |
| *Crassostrea virginica* (oyster) | B*a*P | + | 9,10-dihydrodiol; 4,5(or 7,8) dihydrodiol- 1,6(or 3,6)-quinone; 3-OH-B*a*P and unknown OH-B*a*P | Anderson, 1978 |
| *Mytilus edulis* (mussel) | N, B*a*P | 0 | | Lee *et al.*, 1972a |
| *Physa* sp. (snail) | B*a*P | + | PM | Lu *et al.*, 1977 |
| Arthropoda | | | | |
| *Maia squinado* (spider crab) | N | + | 1,2-dihydrodiol; 1,2-dihydrodiol-glucoside; 1-naphthyl-$SO_4$; 1-naphthylglucoside | Corner *et al.*, 1973 |
| *Callinectes sapidus* (blue crab) | N | + | naphthol; PM | Lee *et al.*, 1976 |
| | 2-MN | + | hydroxymethyl-N; PM | |
| | B*a*P | + | OH-B*a*P; PM | |
| | F | + | OH-F; PM | |
| *Pandalus platyceros* larvae (spot shrimp) | N | + | PM | Sanborn and Malins, 1977 |
| *Calanus plumchrus* (copepod) | N, B*a*P, MC | + | hydroxides; PM | Lee, 1975 |
| *C. helgolandicus* (copepod) | N, B*a*P, MC | + | hydroxides; PM | Lee, 1975 |
| *C. hyperboreus* (copepod) | N, B*a*P, MC | + | hydroxides; PM | Lee, 1975 |

| | | | | |
|---|---|---|---|---|
| *Euchaeta japonica* (copepod) | N, B*a*P, MC | + | hydroxides; PM | Lee, 1975 |
| *Calanus helgolandicus* | N | + | PM | Corner *et al.*, 1976a |
| *Daphnia magna* (water flea) | B*a*P | + | PM | Lu *et al.*, 1977 |
| *D. magna* | A | + | PM | Herbes and Risi, 1978 |
| *Parthemista pacifica* (amphipod) | N, B*a*P, MC | + | hydroxides; PM | Lee, 1975 |
| *Cyphocaris challengeri* (amphipod) | N, B*a*P, MC | + | hydroxides; PM | Lee, 1975 |
| *Uca pugilator* (fiddler crab) | N | + | 1-naphthol; PM | Burns, 1976b |
| *Culex pipiens quinque-fasciatus* (mosquito larva) | B*a*P | + | PM | Lu *et al.*, 1977 |
| Chordata | | | | |
| Teleostei (bony fishes) | | | | |
| *Salmo gairdneri* (rainbow trout) | 2-MN | + | PM | Statham *et al.*, 1978 |
| *Salmo trutta lacustris* (lake trout) | B*a*P | + | dihydriols; 3-,6-quinone; 3-OH-B*a*P | Ahokas *et al.*, 1975 |
| *Oncorhynchus kisutch* (coho salmon) | N | + | 1-naphthol; 1-naphthylglucuronate; 1-naphthyl-α-glycoside; mercapturic acid; 1-naphthyl-$SO_4$; 1,2-dihydrodiol | Roubal *et al.*, 1977b |
| *Gambusia affinis* (mosquito fish) | B*a*P | + | PM | Lu *et al.*, 1977 |
| *Gillichthys mirabilis* (mudsucker) | N | + | 1,2-dihydrodiol | Lee *et al.*, 1972b |
| | B*a*P | + | 7,8-dihydrodiol | |
| Amphibia | | | | |
| *Bufo americanus* embryos (American toad) | B*a*P | + | PM | Clark and Diamond, 1971 |

[a] A, Anthracene; B*a*A, benz[*a*]anthracene; B*a*P, benzo[*a*]pyrene; F, fluorene; MC, methylcholanthrene; 2-MN, 2-methylnaphthalene; N, naphthalene.

[b] +, Metabolites were identified qualitatively or quantitatively; 0, the production of metabolites could not be established.

[c] PM, metabolites identified only as polar or water-soluble materials.

TABLE 57

*In vitro* ARYL HYDROCARBON HYDROXYLASE (AHH) ACTIVITY AND CYTOCHROME P-450 CONCENTRATIONS IN THE LIVER OF FISH OR ANALOGOUS ORGAN OF AQUATIC INVERTEBRATES

| *Species* | *AHH*[a] | *Cytochrome P-450*[b,c] (*nmole/mg microsomal P*) | *Reference* |
|---|---|---|---|
| Annelida | | | |
| *Nereis virens* (clam worm) (intestine) | 21 000 EU/mg microsomal P | 0·0066 | Lee *et al.*, 1978b |
| Mollusca | | | |
| *Crassostrea virginica* (oyster) (digestive gland) | 0·163 nmole/mg P/10 min | — | Anderson, 1978 |
| *Mya arenaria* (soft-shell clam) (whole) | ND | — | Vandermeulen *et al.*, 1977 |
| *Mytilus edulis* (mussel) (whole) | ND | — | Vandermeulen *et al.*, 1977 |
| *Ostrea edulis* (oyster) (whole) | ND | — | Vandermeulen *et al.*, 1977 |
| *Placopecten* sp. (scallop) | ND | — | Varanasi and Malins, 1977 |
| *Littorina littorea* (snail) | ND | — | Varanasi and Malins, 1977 |
| Arthropoda | | | |
| *Panulirus argus* (spiny lobster) | ND–0·04 FU/mg P/min | 0·88 ± 0·52 | Bend *et al.*, 1977 |
| *Homarus americanus* (lobster) | ND | 0·04–0·13 | Bend *et al.*, 1977 |
| *Callinectes sapidus* (blue crab) | ND–0·01 FU/mg P/min | ND–0·29 | Bend *et al.*, 1977 |
| *C. sapidus* | | | |
| male | 10 ± 12 EU/mg microsomal P | — | Singer and Lee, 1977 |
| female | 9·5 ± 6·5 EU/mg microsomal P | — | Singer and Lee, 1977 |
| *Cancer irroratus* (rock crab) | <0·01 FU/mg P/min | — | Pohl *et al.*, 1974 |
| Chordata | | | |
| *Microcosmos sulcata* (protochordate) | ND | — | Kurelec *et al.*, 1977 |
| Cyclostomata | | | |
| *Myxine glutinosa* (hagfish) | 0·18 ± 0·02 FU/mg P/min | — | Pohl *et al.*, 1974 |
| Elasmobranchii | | | |
| *Dasyatis sabina* (Atlantic stingray) | 0·86 ± 0·36 FU/mg P/min | 0·43 ± 0·07 | Bend *et al.*, 1977 |

| | | | |
|---|---|---|---|
| *D. sayi* (bluntnose stingray) | 0·17 ± 0·05 FU/mg P/min | 0·32 ± 0·12 | Bend *et al.*, 1977 |
| *Raja ocellata* (large skate) | 0·30 ± 0·09 FU/mg P/min | 0·36 ± 0·41 | Bend *et al.*, 1977 |
| *R. erinacea* (little skate) | 0·17 ± 0·10 FU/mg P/min | 0·32 ± 0·12 | Bend *et al.*, 1977 |
| *R. radiata* (thorny skate) | 0·12 ± 0·02 FU/mg P/min | — | Pohl *et al.*, 1974 |
| *Squalus acanthus* (dogfish shark) | 0·07 + 0·02 FU/mg P/min | 0·23 ± 0·29 | Bend *et al.*, 1977 |
| Teleostei | | | |
| *Archosargus probatocephalus* (sheep's head) | 1·38 ± 0·47 FU/mg P/min | 0·29 ± 0·12 | Bend *et al.*, 1977 |
| *Lutjanus griseus* (mangrove snapper) | 6·64 ± 0·48 FU/mg P/min | 0·25 ± 0·05 | Bend *et al.*, 1977 |
| *Pseudopleuronectes americanus* (winter flounder) | 2·54 ± 1·65 FU/mg P/min | 0·17 ± 0·01 | Bend *et al.*, 1977 |
| *Pogonias cromis* (black drum) | 1·62 ± 1·91 FU/mg P/min | 0·14 ± 0·03 | Bend *et al.*, 1977 |
| *Anguilla rostrata* (eel) | 0·21 FU/mg P/min | — | Pohl *et al.*, 1974 |
| *Hemitripterus americanus* (king of Norway) | 0·02–0·04 FU/mg P/min | — | Pohl *et al.*, 1974 |
| *Scomber scombrus* (mackerel) | <0·07 FU/mg P/min | — | Pohl *et al.*, 1974 |
| *Fundulus heteroclitus* (mummichog) | 4·10 ± 2·10 FU/mg P/min | — | Pohl *et al.*, 1974 |
| *F. heteroclitus* | — | <0·009 | Burns, 1976b |
| *F. heteroclitus* | 34·1–75·9 pmole 3-OH-B*a*p/mg microsomal P/min | — | Stegeman and Sabo, 1976 |
| *Esox lucius* (pike) | 0·76 ± 0·5 pmole prod./mg microsomal P/min | 0·12 ± 0·02 | Ahokas *et al.*, 1976a |
| *Clupea pilchardus* (sardine) | 20–110 AU | — | Kurelec *et al.*, 1977 |
| *Mugil cephalus* (striped mullet) | — | 0·282 ± 0·061 | Yarbrough and Chambers, 1977 |
| *Tautogolabrus adspersus* (cunner) | 16–19·9 FU/mg P | — | Payne, 1976 |
| *Salmo gairdneri* (rainbow trout) | 2·404 μmole 3-OH-B*a*P/g P/min | — | Pedersen *et al.*, 1974 |
| *S. gairdneri* | 0·04–2·28 μmole 3-OH-B*a*P/g P/min | — | Pedersen *et al.*, 1976 |
| *S. gairdneri* | 0·03 nmole/mg microsomal P/min | 0·21–0·22 | Statham *et al.*, 1978 |

TABLE 57—*contd.*

| *Species* | *AHH*[a] | *Cytochrome P-450*[b,c] *(nmole/mg microsomal P)* | *Reference* |
|---|---|---|---|
| *S. trutta* (brown trout) | 68 ± 14 FU/mg P | — | Payne and Penrose, 1975 |
| *S. trutta lacustris* (lake trout) | — | 0·20–0·40 | Ahokas *et al.*, 1976b 1977 |
| *S. trutta lacustris* | 0·14 nmole/mg microsomal P/min | 0·22 | Ahokas *et al.*, 1975 |
| *Salvelinus fontinalis* (brook trout) | 0·23 FU/mg P | — | Vandermeulen *et al.*, 1977 |
| *Oncorhynchus tshawytscha* (chinook salmon) | 0·7–1·75 nmole prod./mg P/20 min | — | Gruger *et al.*, 1977b |
| *O. kisutch* (coho salmon) | 0·58–0·72 nmole prod./mg P/20 min | — | Gruger *et al.*, 1977b |
| *Stenotomus versicolor* (scup) | 126 pmole 3-OH-B*a*P/mg microsomal P/min | 0·352 | Stegeman and Sabo, 1976 |
| Amphibia | | | |
| *Ambystoma tigrinum* (tiger salamander) | 0·2 nmole/mg P/min | — | Busbee *et al.*, 1975 |

[a] Units of AHH activity vary and are not strictly comparable. EU/mg microsomal P (P = protein), the fluorescence equivalent to $1 \times 10^{-12}$ moles of 3-hydroxybenzo[*a*]pyrene produced in 60 min at 30 °C; nmole/mg P/10 min, nmoles $^{14}$C-B*a*P metabolized/mg protein/10 min; FU/mg P/min, fluorescent units/mg protein/min. One fluorescent unit equals the fluorescence intensity of 3 μg quinine sulfate/ml in 0·1 N sulfuric acid (excitation 400 nm; emission 525 nm); pmole 3-OH-B*a*P/mg microsomal P/min, picomoles 3-hydroxy-B*a*P equivalents produced/mg microsomal protein/min; AU, fluorescence units/100 μliter supernatant, containing 210 ± 37 μg protein (no time given); FU/mg P, alkali-extractable fluorescence (excitation 360 nm; emission >460 nm)/mg protein; μmole 3-OH-B*a*P/g P/min, μmoles 3-hydroxy-B*a*P equivalents produced/g protein/min; nmole prod./mg P/20 min, nanomoles polar metabolites formed/mg protein/20 min; pmole 3-OH-B*a*P/mg microsomal P/min, picomoles 3-hydroxy-B*a*P produced/mg microsomal protein/min. ND, none detected.

[b] Concentrations of cytochrome P-450 expressed as nanomoles/milligram microsomal protein (nmole/mg microsomal P).

[c] —, Not determined.

Crustaceans investigated have low and quite variable levels of hepato-pancreatic AHH activity. *In vitro* activities listed may greatly under-estimate the true *in vivo* activity of this enzyme system. The crustacean hepato-pancreas is a digestive gland containing high concentrations of powerful hydrolytic enzymes. Homogenization releases these enzymes and inhibits *in vitro* AHH activity (Pohl *et al.*, 1974). Metabolite identification experiments with several marine crustaceans have shown that the intact animal produces PAH metabolites very rapidly (Corner *et al.*, 1973; Lee, 1975; Lee *et al.*, 1976; Sanborn and Malins, 1977). In addition, cytochrome P-450 concentrations in crustacean hepato-pancreas are in the same range as those measured in fish liver. The information on AHH activity and cytochrome P-450 concentrations in aquatic invertebrates is so scanty that few generalizations can be made at this time. Often, the *in vitro* assay conditions used for invertebrate tissues were designed for use with mammalian microsomal preparations and therefore may not represent optimal conditions for assaying the invertebrate enzymes.

Khan *et al.* (1974) measured accumulation of the insecticide aldrin by 12 species of freshwater invertebrates and its subsequent epoxidation to dieldrin (Table 58). Epoxidation of aldrin is catalyzed by the MFO system, therefore dieldrin production is a good index of MFO activity. All species studied, representing five invertebrate phyla, were able to catalyze expoxidation of aldrin, although in some activity was slight. Particularly noteworthy is the relatively high rate of aldrin epoxidation exhibited by the two molluscan species. Aldrin epoxidation has also been detected in fiddler crab (*Uca pugilator*) hepato-pancreas (Burns, 1976a) and the gut and fat body of caddis-fly larvae, *Limnephilus* sp. (Kreiger and Lee, 1973). Thus, the MFO system may be more widespread among invertebrates than is implied in Tables 56 and 57. As will be discussed below, in many invertebrates the MFO system may not be localized primarily in hepatic or analogous tissues as it is in vertebrates.

The only nonvertebrate chordate studied, *Microcosmos sulcata*, did not exhibit measurable AHH activity even after exposure for one month to seawater saturated with diesel oil (Kurelec *et al.*, 1977). Among fish investigated (cyclostomes, elasmobranchs, and teleosts) all but the mackerel *Scomber scombrus* displayed detectable levels of AHH activity. Others, such as the dogfish shark *Squalus acanthus* and king of Norway *Hemitripterus americanus*, showed barely detectable levels of AHH activity. The most interesting aspect of these data is the wide inter- and intraspecific variations in levels of AHH activity among fish. These variations do not seem to be related to habitat preference or feeding habits

TABLE 58

ACCUMULATION OF ALDRIN AND ITS SUBSEQUENT EPOXIDATION TO DIELDRIN BY FRESHWATER INVERTEBRATES (FROM KHAN *et al.*, 1974, WITH PERMISSION OF ACADEMIC PRESS)

| *Species* | *Insecticide absorbed (ng/animal)* | | | *Aldrin epoxidation (% dieldrin/animal)* |
|---|---|---|---|---|
| | *Aldrin* | *Dieldrin* | *Total* | |
| Coelenterata | | | | |
| *Hydra littoralis* (hydroid) | 45·02 | 1·20 | 46·25 | 2·59 |
| Platyhelminthes | | | | |
| *Dugesia* sp. (flatworm) | 23·01 | 0·24 | 23·25 | 1·03 |
| Annelida | | | | |
| Leech | 34·14 | 0·54 | 35·65 | 1·51 |
| Mollusca | | | | |
| *Anodonta* sp. (clam) | 533·89 | 130·81 | 684·70 | 19·10 |
| *Lymnaea palustis* (snail) | 69·15 | 13·95 | 83·10 | 16·79 |
| Arthropoda | | | | |
| *Asellus* sp. (isopod) | 25·02 | 0·59 | 25·61 | 2·30 |
| *Gammarus* sp. (amphipod) | 19·98 | 1·52 | 21·50 | 7·07 |
| *Daphnia pulex* (water flea) | 27·16 | 1·64 | 28·80 | 5·69 |
| *Cyclops* sp. (copepod) | 26·23 | 1·13 | 27·36 | 4·13 |
| *Cambarus* sp. (crayfish) | 86·43 | 8·01 | 94·44 | 8·48 |
| *Aeschna* sp. (dragon-fly larva) | 48·59 | 16·11 | 64·70 | 24·90 |
| *Aedes aegypti* (mosquito larva) | 95·06 | 69·82 | 164·88 | 42·35 |

of the fish. There is little difference in the range of AHH activity between freshwater and marine species, or between species collected from Maine and Florida, USA (Philpot *et al.*, 1976). Pedersen *et al.* (1976) noted a range of mean uninduced AHH activity of 0·04–2·28 μmole 3-OH-B*a*P produced/g protein/min in six strains of rainbow trout *Salmo gairdneri*. Similar wide variations were seen within other species.

Hepatic microsomal cytochrome P-450 concentrations showed much less deviation among fish. Values ranged from less than 0·009 in the mummichog to 0·43 nmoles/mg microsomal protein in the Atlantic sting-ray. Elasmobranchs generally displayed higher hepatic cytochrome P-450 concentrations and lower hepatic AHH activities than teleost fish (Philpot et al., 1976; Bend *et al.*, 1977). The significance of the relationship between AHH activity and cytochrome P-450 concentration in a tissue is not clearly understood. It is apparent from the data in Tables 56 and 57 that many marine and freshwater animals possess a cytochrome P-450–MFO scheme

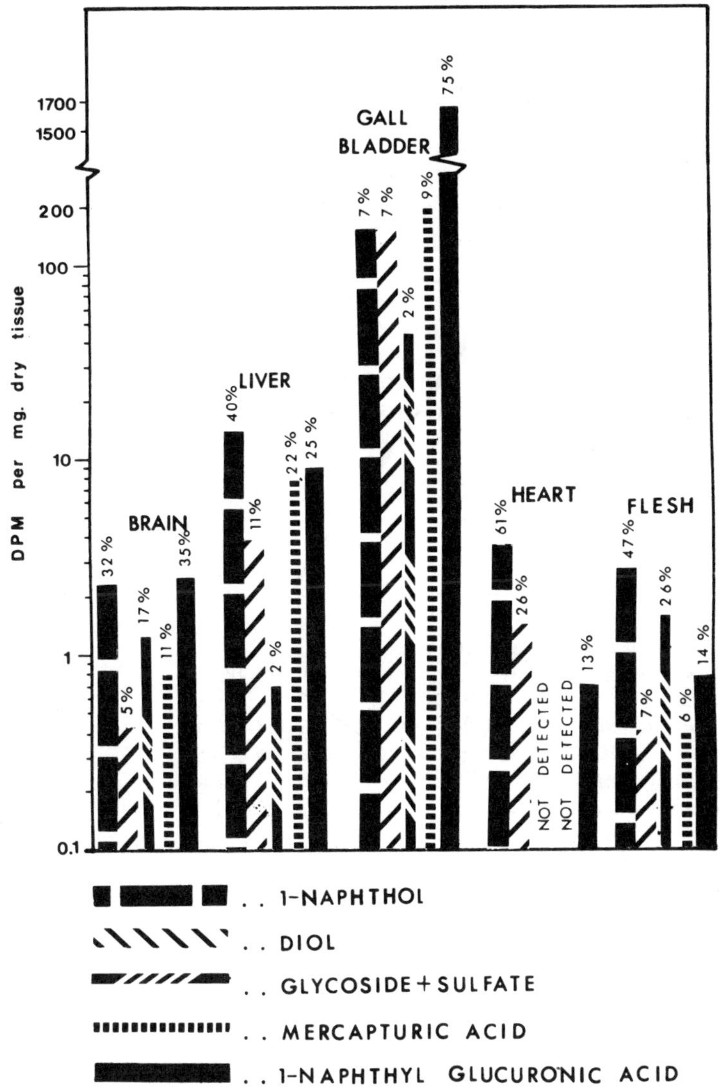


FIG. 18. Composition and distribution in the tissues of solvent-extractable aromatic metabolites in coho salmon (*Oncorhynchus kisutch*) 24 h after intraperitoneal injection of 2·5 $\mu$Ci naphthalene-1-$^{14}$C. DPM, disintegrations per minute (from Roubal *et al.*, 1977b, with permission of Springer-Verlag).

similar in many respects to the well-characterized system in mammalian liver.

The types of PAH metabolites produced by marine and freshwater animals are also analogous to those which have been identified in mammals. Metabolites include PAH diols, phenols, and quinones as well as PAH conjugates with sulfate, monosaccharide, glucuronate, and glutathione (Table 56). The major metabolite of B*a*P in the oyster *Crassostrea virginica* was the 1,6- or 3,6-quinone (Anderson, 1978). Dihydrodiols represented 17·6 per cent and monophenols 32·7 per cent of the total PAH metabolites

TABLE 59

MICROSOMAL EPOXIDE HYDRASE AND 176 000 *g* SUPERNATANT FRACTION GLUTATHIONE S-TRANSFERASE ACTIVITY (±SD) FROM THE LIVER OR HEPATO-PANCREAS OF SEVERAL MARINE SPECIES. BENZO[*a*]PYRENE-4,5-OXIDE WAS USED AS A SUBSTRATE FOR BOTH ENZYMES (FROM BEND *et al.*, 1977, WITH PERMISSION OF THE NEW YORK ACADEMY OF SCIENCES)

| *Species* | *Epoxide hydrase (nmole/mg protein/min)* | *Glutathione S-transferase (nmole/mg protein/min)* |
|---|---|---|
| Molluscs | | |
| *Mya arenaria* (soft-shell clam) | 2·1 | 0·3 |
| *Mytilus edulis* (mussel) | 0·05 | 0·3 |
| Crustaceans | | |
| *Panulirus argus* (spiny lobster) | 16·7 ± 7·4 | 0·4 ± 0·2 |
| *Callinectes sapidus* (blue crab) | 2·6 ± 0·5 | — |
| *Cancer irroratus* (rock crab) | 2·2 ± 1·4 | 0·2 ± 0·1 |
| *Homarus americanus* (lobster) | 6·3 ± 5·3 | 3·3 ± 1·8 |
| Elasmobranchs | | |
| *Dasyatis sabina* (Atlantic stingray) | — | 1·7 ± 0·6 |
| *Raja erinacea* (little skate) | 0·09 ± 0·05 | 23·7 ± 6·5 |
| *Squalus acanthias* (dogfish shark) | 3·2 ± 0·9 | 7·8 ± 1·0 |
| Teleosts | | |
| *Archosargus probatocephalus* (sheep's head) | 3·02 ± 0·85 | 52·7 ± 23·2 |
| *Pogonias cromis* (black drum) | — | 9·2 ± 3·0 |
| *Pseudopleuronectes americanus* (winter flounder) | — | 5·3 ± 2·5 |
| *Anguilla rostrata* (eel) | 2·2 | 2·9, 3·8 |
| *Lutjanus griseus* (mangrove snapper) | — | 2·2 ± 0·3 |

— Not determined.

measured in this species. Roubal *et al.* (1977b) measured distribution of naphthalene metabolites in the tissues of young coho salmon *Oncorhynchus kisutch* 24 h after intraperitoneal injection of naphthalene-1-$^{14}$C (Fig. 18). Ratios and relative concentrations of different metabolites varied from one tissue to another. The gall bladder, one of the main organs through which poorly soluble xenobiotics are excreted, (Statham *et al.*, 1976) contained the highest concentration of all metabolites with 1-naphthylglucuronic acid predominant. In the liver, heart, and flesh 1-naphthol was most abundant. Principal metabolites in the brain were 1-naphthol and 1-naphthylglucuronic acid. Similar naphthalene metabolites were detected in urine of the spider crab *Maia squinado* when the animal was dosed with naphthalene (Corner *et al.*, 1973).

Enzymes involved in the metabolism of PAH epoxides have not been extensively studied in aquatic animals. Activity of epoxide hydrase and glutathione S-transferase in hepatic tissues of several marine animals (Bend *et al.*, 1977) is summarised in Table 59. There are no apparent phylogenetic trends. The two molluscs tested had low glutathione S-transferase activity. However, the soft-shell clam had high epoxide hydrase activity while *Mytilus edulis* had very low epoxide hydrase activity. Neither of these molluscs demonstrated detectable levels of AHH activity. Crustaceans had high epoxide hydrase activity and low glutathione S-transferase activity, while results for elasmobranchs and teleosts were mixed. Relative activities of these two enzymes will in large part determine the ultimate metabolic fate of PAH. Thus, we can predict that dominant PAH metabolites in the little skate and the sheep's head would be glutathione conjugates or their metabolites, mercapturic acids. In crustaceans on the other hand, one would expect primarily dihydrodiols and phenols.

### 5.4.2 Tissue Distribution of MFO Activity

In fish, as in mammals, most MFO activity is localized in the liver (Table 60). The only other tissues of rainbow trout *Salmo gairdneri* which displayed AHH activity, as described by Pedersen *et al.* (1974), were the heart and posterior kidney. However, AHH activity has been detected in the gills of brown trout *Salmo trutta* and capelin *Mallotus villosus* (Payne and Penrose, 1975), and cunner *Tautogolabrus adspersus* (Payne, 1976). AHH activity (Philpot *et al.*, 1976) and glutathione S-transferase activity (Bend *et al.*, 1977) were assayed in several tissues of the little skate *Raja erinacea*. AHH activity was highest in the liver and very low in all other tissues examined. It was not detected in the spiral valve mucosa. Glutathione S-transferase activity was also highest in the liver and kidney.

The heart and testes had moderate activity, while other tissues had relatively low activity. The gill yielded lowest glutathione S-transferase activity of the organs assayed.

Among the invertebrates the pattern is quite different. Singer and Lee (1977) measured AHH activity in several tissues of adult male and female

TABLE 60

DISTRIBUTION OF B*a*P HYDROXYLASE (=AHH) IN THE TISSUES OF THE RAINBOW TROUT *Salmo gairdneri* (FROM PEDERSEN *et al.*, 1974, WITH PERMISSION OF SPRINGER-VERLAG)

| *Tissue* | *μmoles* 8-*OH-BaP formed*/g *protein*/*min*[a] |
|---|---|
| Blood | ND |
| Gill | ND |
| Heart | 0·006 |
| Posterior kidney | 0·026 |
| Liver | 2·404 |
| Muscle | ND |

[a] ND, not detectable.

blue crabs (*Callinectes sapidus*) (Table 61). The hepato-pancreas contained lower AHH specific activity than all other tissues except the blood of males and females and heart of males. Highest activity was located in the green gland (excretory organ) and pyloric stomach, especially of the females. AHH specific activity of the green gland was more than 3 600 times higher than that of female hepato-pancreas. The gill, eyestalk, and testes also contained relatively high AHH specific activity. As stated previously, most of the AHH activity in *Nereis virens* was localized in the lower intestinal epithelium (Lee, 1976; Lee *et al.*, 1977); the only other tissue containing AHH activity was the stomach. Kreiger and Lee (1973) detected aldrin epoxidase activity in the gut and fat body but not in the head, thorax or silk gland of the caddis-fly larva *Limnephilus* sp. In lobsters (*Homarus americanus*), epoxide hydrase activity was highest in the hepato-pancreas (Bend *et al.*, 1977). Activity in the green gland, egg masses, and gill was much lower. Thus, hepato-pancreatic MFO and related enzyme activity does not, in most cases, reflect the true metabolic capabilities of aquatic invertebrates.

TABLE 61

DISTRIBUTION OF ARYL HYDROCARBON HYDROXYLASE (AHH) ACTIVITY IN THE TISSUES OF ADULT MALE AND FEMALE BLUE CRABS (*Callinectes sapidus*). ONE UNIT OF ENZYME ACTIVITY EQUALS THE FLUORESCENCE EQUIVALENT TO $1 \times 10^{-12}$ MOLE OF 3-HYDROXYBENZO[*a*]PYRENE PRODUCED IN A 60 MIN INCUBATION AT 28 °C. SPECIFIC ACTIVITY GIVEN AS MEAN ± SD WITH THE NUMBER OF ASSAYS GIVEN IN PARENTHESES (FROM SINGER AND LEE, 1977, WITH PERMISSION OF THE PUBLISHER)

| *Tissue* | *Total activity (units)* | *Specific activity (units/mg protein)* | |
|---|---|---|---|
| *A. Mature males* | | | |
| green gland | 950 | 450 ± 270 | (8) |
| hepato-pancreas | 560 | 10 ± 12 | (13) |
| testes | 1 300 | 150 ± 110 | (13) |
| median vas deferens | 300 | 18 ± 13 | (3) |
| proximal vas deferens | 1 500 | 15 ± 15 | (3) |
| gill | 6 900 | 140 ± 80 | (12) |
| blood | 5 000 | 3·1 ± 1·1 | (2) |
| pyloric stomach | 20 000 | 1 400 ± 170 | (4) |
| eyestalk | 43 | 140 ± 26 | (26) |
| heart | 0 | 0 | |
| *B. Mature females* | | | |
| green gland | 109 000 | 34 000 ± 2 700 | (7) |
| hepato-pancreas | 1 400 | 9·4 ± 6·5 | (6) |
| ovary | 170 | 7·7 ± 7·6 | (4) |
| seminal receptacle | 800 | 97 ± 91 | (5) |
| gill | 1 900 | 140 ± 124 | (3) |
| blood | 3 000 | 2·6 | (1) |
| pyloric stomach | 49 000 | 2 500 ± 910 | (6) |
| eyestalk | 90 | 87 | (1) |
| heart | 500 | 21 | (1) |

### 5.4.3 Effect of Endogenous and Exogenous Factors on MFO Activity

A wide variety of endogenous and exogenous factors may profoundly influence the levels of MFO system and related enzyme activities in tissues of aquatic animals. Dewaide and Henderson (1970) measured seasonal variations in hepatic aminopyrene-N-demethylase and aniline-*p*-hydroxylase activities (both reactions are mediated by the MFO system) of wild populations of the roach *Leuciscus rutilus* (a freshwater teleost) from the River Rhine near Nijmegen, The Netherlands. Highest enzyme activities were measured during August and September (Fig. 19). Activity of both enzymes was significantly lower statistically during the winter,

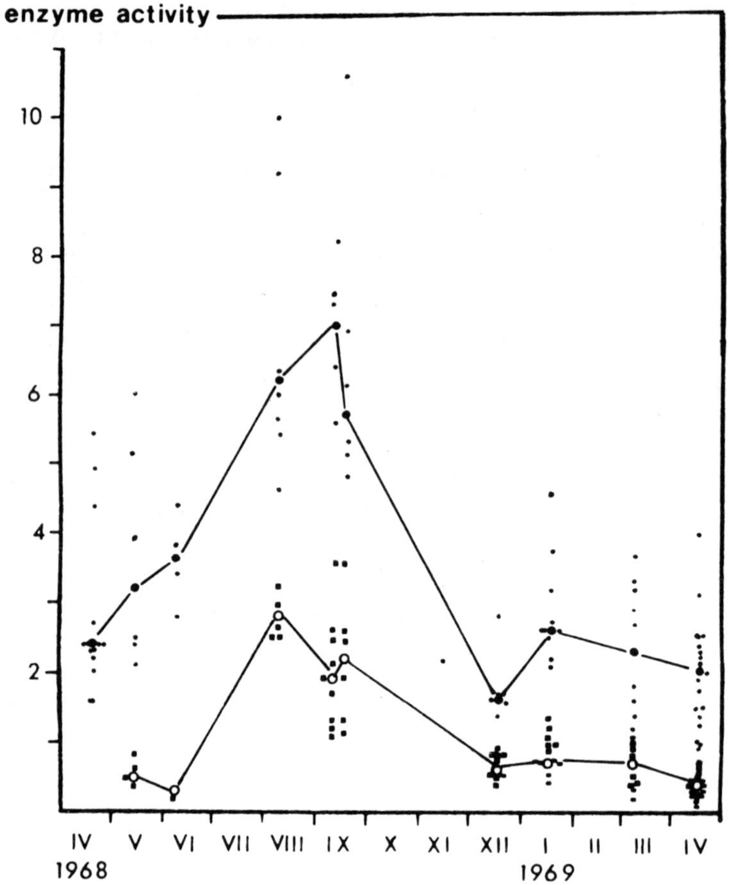


FIG. 19. Seasonal variations in hepatic microsomal aminopyrene-N-demethylase (●—●) and aniline-*p*-hydroxylase (○—○) activities in wild populations of the roach *Leucisus rutilus* (a freshwater teleost) (from Dewaide and Henderson, 1970, with permission of Pergamon Press).

December–April, and spring, May–June. Subsequently, Dewaide (1970) acclimated *L. rutilus* and the trout *Salmo irideus* to laboratory temperatures of 5 and 18 °C over a two-week period. Both species showed significantly higher hepatic levels of aminopyrene-N-demethylase and aniline-*p*-hydroxylase activities at the lower temperature. Thus, the response to seasonal temperature acclimation in the natural environment contrasted short-term laboratory thermal acclimation. Since the fish for the laboratory experiment were collected during the wintertime, the results suggest that acclimation of the MFO system to elevated ambient temperatures is a relatively slow process requiring more than two weeks in fish. Starved trout had lower activity of both enzymes, particularly at the higher acclimation temperature, than fed trout (Dewaide, 1970) suggesting that nutritional status has an influence on MFO activity and particularly on the rate of acclimation to new environmental regimes. Since feeding activity is depressed during cold winter months in most species of fish from temperate climates, the apparent effect of nutritional status on MFO activity may partially explain depressed levels of enzyme activity during the winter in natural populations of roach.

Dewaide and Henderson (1970) also observed that maintenance of roach in running tap-water depressed MFO activity by up to 50 per cent in 40–55 days, while roach maintained in river water showed no significant change in enzyme activity during the same period. They suggested that this decrease could be due to termination of exposure of the fish to inducing agents by removal of roach from river water to cleaner water, or to stress-induced effects of laboratory confinement.

Singer and Lee (1977) showed that there were drastic changes in AHH

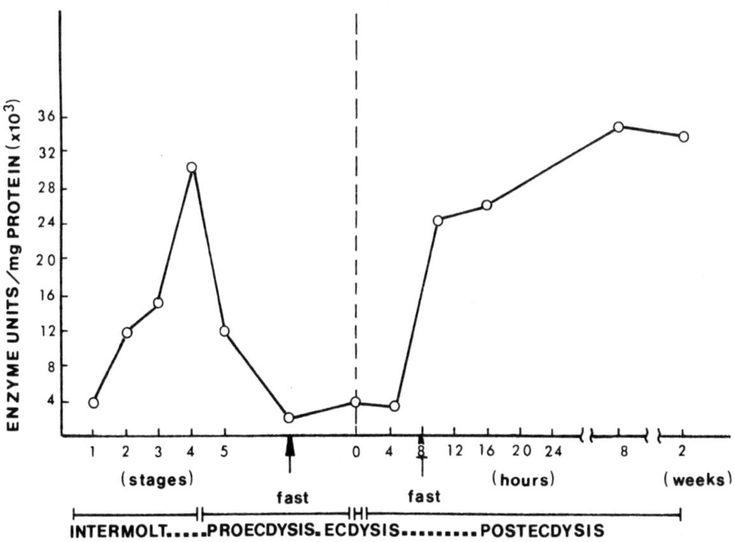


FIG. 20. Variations in aryl hydroxylase activity in the green gland of female blue crabs (*Callinectes sapidus*) during the molt cycle (from Singer and Lee, 1977, with permission of the publisher).

activity in the green gland of the female blue crab *Callinectes sapidus* during the molting cycle (Fig. 20). AHH activity was high during late intermolt and dropped to very low levels just before and during the molt (ecdysis). It rose sharply to its highest level about 48 h after the molt. These dramatic changes in AHH activity during the normal molting cycle of a crustacean strongly suggest that green-gland AHH activity functions in controlling the concentration of steroid molting hormones (crustecdysones) in the hemolymph of the crab. High levels of AHH activity during the intermolt and postmolt periods would insure that ecdysteroid titers would remain low, preventing untimely initiation of the molting process. Reduction in AHH activity during proecdysis would allow ecdysteroid titers to rise to levels high enough to initiate molting. It is interesting to note that induction of the MFO system by chlorinated hydrocarbons has been implicated in the increased rate of catabolism of cortisol, a steroid hormone, in mammals (Kupfer and Bugler, 1976).

### 5.4.4 Induction of MFO Activity

Exposure to various environmental xenobiotics can dramatically affect activity of MFO and related enzymes in aquatic animals. Several investigators have shown that exposure to PAH or complex mixtures of hydrocarbons such as petroleum results in rapid induction of elevated levels of MFO activity in some aquatic species. Pedersen *et al.* (1976) were able to induce AHH activity in five strains of rainbow trout *Salmo gairdneri* by intraperitoneal injection of 3-methylcholanthrene (Table 62). The degree of induction (induced/control ratio) varied substantially and was highest in the Chester Morse strain (47·75). Similarly, AHH activity was increased nearly 15-fold in the liver of the sheep's head *Archosargus probatocephalus* within six days of receiving injections of 3-methylcholanthrene (Table 63) (Bend *et al.*, 1977). There was also a rising trend in concentration of hepatic microsomal cytochrome P-450 in the dosed fish, but because of large standard deviations, differences were not statistically significant. Of major significance was the observation that activities of epoxide hydrase and glutathione S-transferase were not increased by treatment with 3-methylcholanthrene. Induction of MFO without concomitant induction of epoxide detoxifying enzymes could lead to an accumulation of the highly carcinogenic and mutagenic epoxides. Busbee *et al.* (1975, 1978) and Colvin (1974; as cited by Rose, 1977) found that hepatic AHH in the tiger salamander *Ambystoma tigrinum* was inducible by 3-methylcholanthrene administered through intraperitoneal injection or by exposure in the ambient water. Salamanders from a PAH-contaminated lagoon at Reese Air

TABLE 62

INDUCTION OF HEPATIC ARYL HYDROCARBON HYDROLASE ACTIVITY IN FIVE STRAINS OF RAINBOW TROUT *Salmo gairdneri*. FOR INDUCTION, FISH WERE INJECTED INTRAPERITONEALLY 48 h PRIOR TO SACRIFICE WITH APPROXIMATELY 75 mg 3-METHYLCHOLANTHRENE/kg BODY WEIGHT IN CORN OIL. CONTROLS WERE INJECTED WITH CORN OIL. VALUES INDICATE MEAN SPECIFIC ACTIVITIES ± 1 SD. NUMBERS OF DETERMINATIONS ARE INDICATED IN PARENTHESES (FROM PEDERSEN *et al.*, 1976, REPRODUCED BY PERMISSION OF THE MINISTER OF SUPPLY AND SERVICES, CANADA)

| *Strain* | *μmole 3-hydroxybenzo[a]pyrene formed/g protein/min* | | |
|---|---|---|---|
| | *Control* | *Induced* | *Induced/control ratio* |
| Chambers Creek | 2·28 ± 2·82 (25) | 6·73 ± 4·05 (10) | 2·95 |
| Spokane | 1·51 ± 0·38 (16) | 5·39 ± 1·24 (6) | 3·57 |
| Mt. Whitney | 1·34 ± 0·43 (18) | 3·36 ± 0·78 (8) | 2·51 |
| Chester Morse | 0·04 ± 0·04 (12) | 1·91 ± 0·74 (5) | 47·75 |
| Donaldson | 0·55 ± 0·18 (92) | 1·27 ± 0·61 (21) | 2·31 |

TABLE 63

INDUCTION OF VARIOUS COMPONENTS OF THE PAH-METABOLIZING ENZYME SYSTEM IN THE LIVER OF THE SHEEP'S HEAD *Archosargus probatocephalus* BY INTRAPERITONEAL INJECTION OF 20 mg 3-METHYLCHOLANTHRENE/kg BODY WEIGHT IN CORN OIL ON DAYS 1 AND 3 OF THE EXPERIMENT (FROM BEND *et al.*, 1977, WITH PERMISSION OF THE NEW YORK ACADEMY OF SCIENCES)

| *Treatment* | *Number of samples* | *Cytochrome P-450 (nmole/mg protein)* | *BaP hydroxylase (AHH) activity (FU/mg protein/min)*[a] | *Epoxide hydrase activity with styrene 7,8-oxide (nmole/mg protein/min)* | *Glutathione S-transferase (GSH) activity with BaP-4,5-oxide (nmole GSH conjugated/mg protein/min)* |
|---|---|---|---|---|---|
| Control | | | | | |
| Corn oil only | 7 | 0·28 ± 0·09 | 1·41 ± 0·49 | 6·5 ± 4·0 | 49·5 ± 7·9 |
| *Test* | | | | | |
| Day 6 | 6 | 0·44 ± 0·17 | 15·21 ± 8·19[b] | 5·8 ± 2·2 | 44·6 ± 10·2 |
| Day 9 | 4 | 0·54 ± 0·28 | 10·98 ± 3·76[b] | 4·4 ± 2·1 | 57·3 ± 6·8 |
| Day 13 | 5 | 0·37 ± 0·13 | 14·48 ± 5·63[b] | 4·6 ± 1·5 | 37·2 ± 6·6 |

[a] FU, Fluorescent units.
[b] Significantly different from controls; $P < 0{\cdot}05$.

Force Base, Texas, displayed elevated AHH activity but the 4,5-epoxide of B*a*P was not converted to dihydrodiol. The lack, or low activity, of epoxide detoxifying enzymes may have accounted for the high incidence of cancer in this population of salamanders (Rose, 1977).

Bend *et al.* (1977) also reported that type-II inducers such as 3-methylcholanthrene, 2,3,7,8-tetrachlorodibenzo-*p*-dioxin, and dibenz[*ac*]anthracene induced AHH activity in the liver of the little skate *Raja erinacea* and winter flounder *Pseudopleuronectes americanus*. However, 3-methylcholanthrene failed to induce elevated levels of AHH activity or cytochrome P-450 concentration in the liver of the Atlantic sting-ray *Dasyatis sabina*. Stegeman and Sabo (1976) induced a nearly two-fold increase in hepatic AHH activity in the scup *Stenostomus versicolor* by intraperitoneal injection of 3-methylcholanthrene and 5,6-benzoflavone; hepatic cytochrome P-450 concentrations increased only slightly. Injection of 3-methylcholanthrene, benz[*a*]anthracene, and *β*-naphthoflavone increased hepatic AHH activity without altering glucuronyl transferase activity in the rainbow trout *Salmo gairdneri* (Statham *et al.*, 1978). Cytochrome P-450 concentrations were increased in treated fish by approximately 50 per cent. Pretreatment of trout with benzanthracene resulted in an increase in rate of metabolism and biliary excretion of 2-methylnaphthalene *in vivo*.

Much less work has been done on the induction of MFO in aquatic invertebrates. Attempts by Payne (1977) to induce AHH in several species of algae, molluscs, crustaceans and echinoderms failed. Vandermeulen *et al.* (1977) failed to induce AHH activity in three species of bivalve molluscs—*Mya arenaria*, *Mytilus edulis*, and *Ostrea edulis*—by exposure for four days to Kuwait crude oil or Bunker C residual oil. Molluscs chronically exposed to Bunker C oil in Chedabucto Bay, Canada, also showed no detectable AHH activity. However, Anderson (1978) was able to induce AHH in the oyster *Crassostrea virginica* following exposure for 5–9 months to B*a*P or methylcholanthrene in the ambient medium at a concentration of 1 μg/liter (ppb) (Table 64). Lee *et al.* (1978*b*) induced AHH activity in the marine polychaete worm *Capitella capitata* by exposing the worm to benz[*a*]anthracene and Kuwait crude oil in its sand substrate (Table 65). Unexposed worm tissues did not yield detectable levels of AHH activity. Maximal activity was obtained following 10 weeks' exposure to the inducers. Burns (1976a) detected low levels of aldrin epoxidase activity in several tissues of the fiddler crab *Uca pugnax*. Crabs from contaminated environments had similar MFO activities suggesting a lack of induction in this species.

TABLE 64

INDUCTION OF ARYL HYDROCARBON HYDROXYLASE (AHH) ACTIVITY IN THE OYSTER *Crassostrea virginica* BY BENZO[*a*]PYRENE AND METHYLCHOLANTHRENE AT AN EXPOSURE CONCENTRATION OF 1 μG/LITER (PPB) (BASED ON DATA FROM ANDERSON, 1978)

| *Inducing substance* | *Exposure time (months)* | *AHH activity (nmole BaP products formed/mg/10 min)* |
|---|---|---|
| Control | 5 | 0 |
| Benzo[*a*]pyrene | 5 | 0·183 |
| Methylcholanthrene | 5 | 0·206 |
| Control | 9 | 0·035 |
| Benzo[*a*]pyrene | 9 | 0·107 |
| Methylcholanthrene | 9 | 0·081 |
| Benzo[*a*]pyrene and methylcholanthrene | 9 | 0·121 |

Several investigators have demonstrated AHH induction in fish by petroleum under both laboratory and field conditions. Gruger *et al.* (1977a) exposed young coho salmon (*Oncorhynchus kisutch*) to dilute water-soluble fractions (WSF) of Prudhoe Bay crude oil in seawater for six days. AHH activity in females was induced by exposure concentrations as

TABLE 65

INDUCTION OF MIXED-FUNCTION OXYGENASE (MFO) (= ARYL HYDROCARBON HYDROXYLASE, AHH, ACTIVITY) IN THE ESTUARINE POLYCHAETE WORM *Capitella capitata* (FROM LEE *et al.*, 1978b, WITH PERMISSION OF ACADEMIC PRESS)

| *Inducing substance* | *Exposure time (weeks)* | *AHH activity (enzyme units/mg protein)*[a] |
|---|---|---|
| Benz[*a*]anthracene | 0 | 0 |
| (150 mg/kg sand substrate) | 3 | 200 |
| | 6 | 200 |
| | 10 | 500 |
| Kuwait crude oil | 0 | 0 |
| (2·5 g oil/kg sand substrate | 3 | 0 |
| | 6 | 10 |
| | 10 | 100 |

[a] One enzyme unit equals the fluorescence equivalent to $1 \times 10^{-12}$ mole of 3-hydroxybenzo[*a*]pyrene produced in a 60-min incubation at 30 °C.

low as 15 ppb WSF, while in males it was only induced at a concentration of 150 ppb WSF (Table 66). Subsequently, Gruger *et al.* (1977b) fed saltwater-adapted coho salmon (*Oncorhynchus kisutch*) and chinook salmon (*O. tshawytscha*) test diets containing polychlorinated biphenyls (PCB), petroleum hydrocarbons, or both. Induction of AHH activity occurred in hepatic microsomes of the coho salmon within the first two

TABLE 66

HEPATIC ARYL HYDROCARBON HYDROXYLASE (AHH) ACTIVITY IN YOUNG COHO SALMON (*Oncorhynchus kisutch*) EXPOSED IN SEAWATER FOR 6 DAYS TO A DILUTE WATER-SOLUBLE FRACTION (WSF) OF PRUDHOE BAY CRUDE OIL. FISH IN EACH TREATMENT GROUP DID NOT VARY SIGNIFICANTLY IN WEIGHT AND LENGTH (FROM GRUGER *et al.*, 1977a, WITH PERMISSION OF SPRINGER-VERLAG)

| *Treatment* | *Number of fish* | *AHH activity*[a] *females* | *males* | *combined* |
|---|---|---|---|---|
| 150 ppb WSF | 10 | 0·85 ± 0·27 | 0·71 ± 0·22 | 0·78 ± 0·24[b] |
| 15 ppb WSF | 9 | 0·78 ± 0·37 | 0·47 ± 0·15 | 0·54 ± 0·23 |
| Control | 9 | 0·41 ± 0·06 | 0·45 ± 0·23 | 0·46 ± 0·17 |

[a] Per centage relative fluorescence of phenolic or hydroxylated products formed/mg 15 000 × *g* supernate protein/30 min incubation.
[b] Significant ($0{\cdot}01 < P < 0{\cdot}05$) difference relative to controls.

weeks of exposure to hydrocarbons. This induction was potentiated by presence of PCB in the diet. PCB alone had no effect on hepatic AHH activity. Results were quite different with chinook salmon. Hydrocarbons, PCB, or both in the diet depressed hepatic microsomal AHH activity. Thus, the response to dietary xenobiotics was quite dissimilar in two closely related species. Payne and Penrose (1975) reported a highly significant induction of AHH activity in liver and gill microsomes from brown trout *Salmo trutta* and capelin *Mallotus villosus* following exposure to approximately 1 ppm dispersed Tia Juana medium Venezuela crude oil for 7–16 days. After 7–8 days, only gill microsomal AHH activity in capelin was not significantly affected.

Payne (1976) compared liver and gill microsomal AHH activity in the cunner *Tautogolabrus adspersus* from uncontaminated oil-polluted stations on the coast of the Avalon Peninsula, Newfoundland, Canada. AHH activity was significantly higher in liver and gill microsomes in fish collected at oil-polluted sites than those from clean sites (Table 67). Payne (1976) recommended the use of fish hepatic AHH activity as a monitor of oil pollution. Similar results were reported by Kurelec *et al.* (1977). Hepatic microsomal AHH activity in a natural population of blennies (*Blennius pavo*) was dramatically increased within seven days after an oil spill near Rovinj, Yugoslavia (Fig. 21). AHH activity in the fish reached a maximum less than three weeks after the spill and subsequently dropped about three times the prespill levels a further three weeks later. It remained fairly

TABLE 67

ARYL HYDROCARBON HYDROXYLASE (AHH) ACTIVITY IN LIVER AND GILLS OF THE CUNNER *Tautogolabrus adspersus* FROM OIL-CONTAMINATED (1 AND 2) AND UNCONTAMINATED (3 AND 4) SITES ON THE COAST OF THE AVALON PENINSULA, NEWFOUNDLAND, CANADA (FROM PAYNE, 1976, WITH PERMISSION OF AAAS. COPYRIGHT 1976 BY THE AMERICAN ASSOCIATION FOR THE ADVANCEMENT OF SCIENCE)

| *Site* | *Number of samples* | *Mean AHH activity (FU/mg protein)[a] ± SD* | |
|---|---|---|---|
| | | *Liver* | *Gill* |
| 1. (Refinery) | 10 | 53·2 ± 25·2[b] | 0·27 ± 0·23[b] |
| 2. (Harbor) | 8 | 46·6 ± 6·7[b] | — |
| 3. (Control) | 10 | 16·0 ± 7·6 | 0·0 |
| 4. (Control) | 9 | 19·9 ± 6·6 | — |

[a] Specific activity refers to arbitrary units of fluorescence of alkali-soluble reaction product (3-hydroxybenzo[*a*]pyrene) per milligram protein.

[b] Significantly different from control (Student's *t*-test; $P < 0{\cdot}005$ for liver activity between sites 1 and 4; $P < 0{\cdot}025$ for gill activity between sites 1 and 3).

constant at this elevated level for at least five months. The authors were able to induce AHH activity in the laboratory by subjecting blennies to as little as 170 ppb diesel oil. Maximal activity was reached following 12 days of exposure. Burns (1976b) reported higher levels of hepatic aldrin epoxidase activity and cytochrome P-450 concentrations in mummichogs (*Fundulus heteroclitus*) from oil- or pesticide-contaminated areas than in those from uncontaminated areas along the Massachusetts coast. Similarly, Stegeman and Sabo (1976) reported higher levels of hepatic AHH activity in mummichogs from marsh areas contaminated with petroleum or chlorinated hydrocarbons than in those from uncontaminated areas. The extent of induction seemed to be greater in males than in females.

Yarbrough and Chambers (1977) reported a highly significant increase in cytochrome P-450 and cytochrome $b_5$ concentrations in mullet (*Mugil cephalus*) liver following exposure for four days to 75 ppm emulsified Empire Mix or Saudi Arabian crude oil in seawater. In contrast, Ahokas *et al.* (1976a) found lower levels of AHH activity and cytochrome P-450 concentrations in liver microsomes from pike (*Esox lucius*) collected from

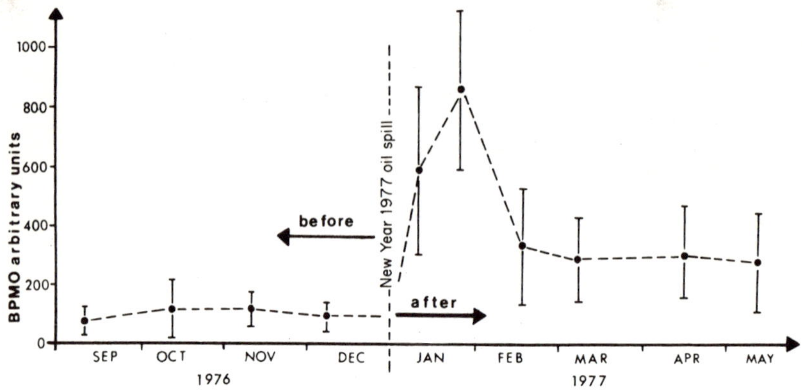


FIG. 21. Hepatic aryl hydrocarbon hydroxylase activity in a natural population of blennies (*Blennius pavo*) before and after an oil spill. BPMO, benzo[*a*]pyrene monooxygenase activity in arbitrary fluorescence units (based on a quinine sulfate standard) per 100 $\mu$liter of the supernatant, containing 210 $\pm$ 37 $\mu$g protein. (From Kurelec *et al.*, 1977, with permission of Springer-Verlag.)

Lake Vatianjarvi, Finland, heavily polluted with mixed effluents from pulp and chemical factories, than in those collected from uncontaminated Lake Paijanne, Finland.

It is apparent that a great many endogenous and exogenous factors can influence levels of AHH and related enzyme activity and concentrations of MFO components in the tissues of aquatic animals. Each species seems to respond somewhat differently to these factors and there are even significant intraspecific (probably genetic) differences. Endogenous factors may include age, sex, nutritional status, period of the molt cycle (in arthropods), etc. Primary exogenous factors influencing rate of PAH metabolism include temperature, season, possibly salinity (though this has not been studied), and current and previous history of exposure to potential inducers or inhibitors of different components of the microsomal PAH-metabolizing system.

## 5.5 INTERACTION OF DEGRADATIVE PROCESSES IN THE REMOVAL OF PAH FROM THE AQUATIC ENVIRONMENT

In the aquatic environment, the assorted oxidative processes that are responsible for degrading PAH do not proceed independently of one another. The nature and extent of interactions between different processes are often hard to ascertain.

Lee *et al.* (1978a) introduced a dispersion of Prudhoe Bay crude oil enriched with a number of PAH into a controlled ecosystem enclosure (ca. 2-m diameter and 15-m deep) and estimated rate of loss of different PAH due to evaporation, photochemical oxidation, microbial degradation, and sedimentation. Initial concentrations of PAH in surface waters varied from 19 $\mu$g/liter methylnaphthalenes to 1.1 $\mu$g/liter benzo[*a*]pyrene. The time required for PAH concentrations to decrease by 50 per cent was 3–4 days in water at 3 m and 4–6 days in water at 7 m. By day 17 all of the spiked hydrocarbons were below the level of detection in water at 3 m. Photooxidation appeared to be important in B*a*P degradation in the upper 5 m. On day 4, 30 per cent B*a*P in the upper 5 m was in the form of the parent compound. The remainder was in the form of quinones and more polar B*a*P degradation products. Approximately 40 per cent B*a*P introduced into the enclosed water was recovered in bottom sediments, suggesting that no more than 50 per cent B*a*P could have been photooxidized.

Rate of microbial degradation was evaluated by addition of $^{14}$C-labeled PAH to water samples from the enclosures followed by measurement of $^{14}CO_2$ production after different incubation periods. Benz[*a*]anthracene and B*a*P were not degraded by bacteria from either the control or oil-treated enclosures. Anthracene was not degraded in water from the uncontaminated enclosure but was degraded at a rate of 0·02 $\mu$g/liter/day in water from the treated enclosure. Approximately 5 per cent of the naphthalene and methylnaphthalenes in the treated enclosure was degraded each day by bacterial activity.

Naphthalenes, with a predominance of dimethylnaphthalenes, were rapidly deposited in bottom sediments during the first seven days of the study, while higher molecular weight PAH accumulated in sediments later in the study. The authors suggested that naphthalenes were rapidly accumulated by phytoplankton which subsequently sank to the bottom. Higher molecular weight PAH became associated primarily with colloidal and detrital material which sank more slowly.

Only naphthalene and methylnaphthalenes were lost from the water column by evaporation. Limited volatilization of anthracene and fluoranthene was proposed. Zooplankton in the enclosure, predominantly the copepod *Pseudocalanus minutus*, accumulated anthracene, fluoranthene, B*a*P, and benz[*a*]anthracene during the first four days but after nine days, no PAH were detected in copepod tissues. Presumably some of the accumulated PAH were degraded by zooplankton.

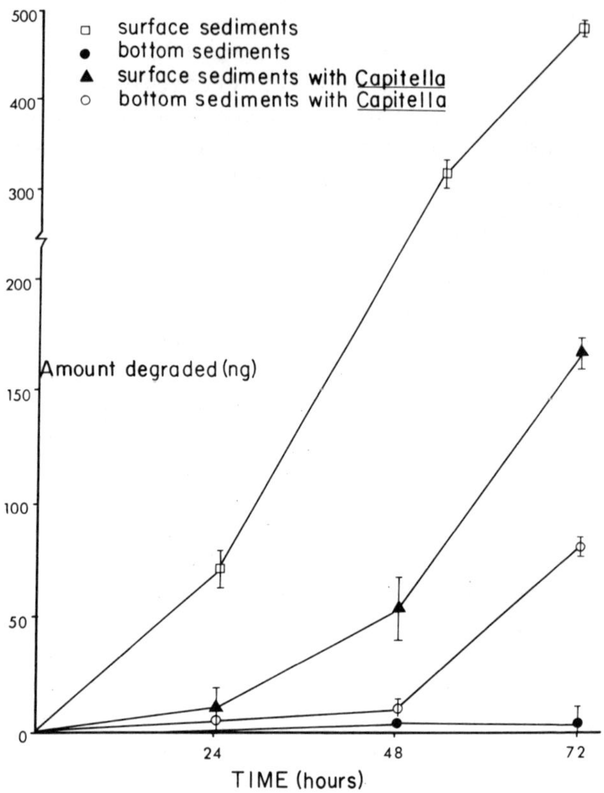


FIG. 22. Effect of the presence of the marine polychaete worm *Capitella capitata* on the rate of degradation of benz[*a*]anthracene in sediments (from Lee *et al.*, 1978b, with permission of Academic Press).

These studies illustrate the brief residence time of PAH in marine waters. Lower molecular weight aromatics (benzene–phenanthrene) are removed primarily by evaporation and microbial activity. Higher molecular weight PAH are removed mainly by sedimentation and photooxidation. Anthracene is also rapidly oxidized. Degradation by animals in the water column is of minor importance. The relative importance of various removal processes is characteristic of each PAH.

Lee *et al.* (1978b) reported that vital activity of the marine worm *Capitella capitata* increased the rate of degradation of benz[*a*]anthracene below the sediment surface (Fig. 22). In sediment trays without worms, microbes were able to degrade PAH at the surface but not in the subsurface sediment layers. With worms present, microbial degradation of PAH in subsurface sediments dramatically increased, suggesting that worms may provide aeration or nutrients to hydrocarbon-degrading microbes in subsurface sediments, promoting greater microbial activity. Lee (1976) discussed the interactions of the microfauna (mainly bacteria and fungi), meiofauna (small animals), and macrofauna in degradation of petroleum hydrocarbons in aquatic sediments. Population density of hydrocarbon-degrading microbes is high in sediments from oil-polluted areas, resulting in rapid degradation of alkanes first with slower attack on isoalkanes, cycloalkanes, and aromatic hydrocarbons. Feeding activity of meio- and macrofauna causes mixing of sediments and exposure of subsurface sediments to the water–sediment interface, where microbial activity is highest. Also, some macrofauna (e.g., polychaete worms) and possibly also some meiofauna ingest sediment and are able to degrade PAH adsorbed to it.

*CHAPTER 6*

# *Accumulation and Release of PAH by Aquatic Organisms*

The presence of PAH in tissues of a wide variety of freshwater and marine organisms, as discussed in Section 4.3, strongly indicates that these organisms are able to accumulate PAH present at low concentrations in the ambient medium, food, or sediments. This is not surprising, since PAH are highly hydrophobic and lipophilic. Intrinsic lipid/water partition coefficients favor their rapid transfer from the aqueous phase into lipophilic compartments such as biological membranes, macromolecules, and depot lipid stores in organisms (Leo *et al.*, 1971; Neely *et al.*, 1974). Release of PAH from tissues of contaminated organisms may be passive, reflecting an equilibrium distribution between the aqueous phase and lipophilic compartments in contact with it, or it may be active and involve metabolic transformation of PAH to polar water-soluble metabolites which are more readily excreted, as described in Section 5.5. A large amount of research has been conducted in recent years on accumulation and release of petroleum hydrocarbons by aquatic (mainly marine) organisms (see reviews of Anderson *et al.*, 1974a; Varanasi and Malins, 1977). Only literature dealing specifically with accumulation and release of PAH by aquatic organisms will be reviewed below.

## 6.1 ACCUMULATION AND RELEASE OF PAH FROM WATER

Kauss *et al.* (1973) and Soto *et al.* (1975) studied the accumulation and release of $^{14}$C-naphthalene by the freshwater green alga *Chlamydomonas angulosa*. Algae were exposed to saturated solutions of naphthalene ($\sim$30 ppm) for up to 7 days in closed glass flasks to prevent naphthalene evaporation. The cells accumulated $^{14}$C-naphthalene to high levels during the 7-day incubation period. When the flasks were opened and naphthalene allowed to evaporate, algal cells retained naphthalene and remained non-motile until cell division commenced (after about 4 days). When cell division began, a decrease in $^{14}$C-naphthalene/cell was observed which corresponded to the increase in cell numbers. If naphthalene-contaminated cells were re-suspended in a clean medium, they began to release $^{14}$C-naphthalene immediately and regained motility within 24 h. In neither case was there evidence of production of polar metabolites. Thus, the algae possessed very limited ability to release or degrade naphthalene.

When polychaete worms (*Neanthes arenaceodentata*) were exposed for 24 h to seawater containing 0·15 ppm $^{14}$C-naphthalene, they accumulated naphthalene to a maximum of 6 μg/g tissue (ppm) in 3–24 h (Rossi, 1977). Upon return to isotope-free seawater, $^{14}$C-naphthalene was rapidly released and reduced to nondetectable levels within 300 h. Approximately one-third of the radioactivity released by *Neanthes* during the first 24 h of depuration was in the form of unmetabolized naphthalene. The remainder was in the form of polar metabolites. Significant levels of $^{14}$C-naphthalene metabolites remained in tissue material after 504 h (21 days) of depuration suggesting that some metabolic products were covalently bound to tissue macromolecules. The freshwater oligochaete *Tubifex* sp. was able to accumulate as much as 88·2 μg B*a*P/kg tissue during exposure for 11 days to a concentration of 100 μg B*a*P/liter freshwater (Scaccini-Cicatelli, 1966). Unexposed control worms contained approximately 72 μg B*a*P/kg tissue. Worms containing 350 μg B*a*P/kg released 75 per cent B*a*P during a 40-day depuration period in clean water. There was no evidence of B*a*P metabolism.

Several studies have been performed on accumulation and release of PAH in solution by marine bivalve molluscs. Lee *et al.* (1972a) demonstrated accumulation by the mussel *Mytilus edulis* of $^{14}$C-naphthalene and $^{3}$H-benzo[*a*]pyrene dissolved in seawater. The highest levels of activity were recorded in the gill and the authors hypothesized that gill tissue of mussels has a micellar layer which absorbs hydrocarbons and then passes them to other tissues. Although no evidence of MFO activity could be detected in the mussels they rapidly released accumulated PAH when returned to isotope-free seawater. Neff *et al.* (1976a) measured relative rates of accumulation and release of four PAH from seawater by the estuarine clam *Rangia cuneata* (Table 68). Phenanthrene was accumulated most rapidly and released most slowly. The rapid release of naphthalene from clam tissues probably masked a similarly rapid uptake during exposure, since both influx and efflux of this compound undoubtedly occurred simultaneously. These results can best be explained

TABLE 68

ACCUMULATION FROM SEAWATER AND RELEASE OF PAH BY THE ESTUARINE CLAM *Rangia cuneata* (FROM NEFF *et al.*, 1976a, WITH PERMISSION OF THE AMERICAN INSTITUTE OF BIOLOGICAL SCIENCES)

| *Parameter* | *PAH* Naphthalene | *Phenanthrene* | *Chrysene* | *Benzo[a]pyrene* |
|---|---|---|---|---|
| Exposure concentration (ppm) | 0·071 | 0·089 | 0·066 | 0·052 |
| Tissue concentration after 24-h exposure (ppm) ± SD | 0·43 ± 0·1 | 2·85 ± 1·1 | 0·54 ± 0·3 | 0·45 ± 0·1 |
| Bioaccumulation factor [tissue]/[water] | 6·1 | 32·0 | 8·2 | 8·7 |
| Tissue concentration after 24-h depuration (ppm) ± SD | 0·15 ± 0·02 | 2·47 ± 1·2 | 0·40 ± 0·15 | 0·38[a] |
| Per cent released in 24 h | 66 | 13 | 26 | 16 |

[a] Only one sample analyzed.

in terms of relative aqueous solubilities and lipid/water partition coefficients of the four PAH. Naphthalene is the most water soluble of the PAH tested rendering it more readily bioavailable. Although it has a high affinity for lipids, its lipid/water partition coefficient favors rapid release to water when naphthalene concentrations in the medium are reduced. B*a*P, at the other extreme, has a very low aqueous solubility ($\sim$1 ppb) so that most B*a*P in the exposure water was in colloidal or particulate form decreasing its bioavailability. Once absorbed, it would not readily partition back into the aqueous phase even when B*a*P concentrations in the medium were extremely low. In other experiments *R. cuneata* accumulated up to 7·2 ppm B*a*P during 24-h exposure to 0·03 ppm $^{14}$C-B*a*P in seawater (Neff and Anderson, 1975). Nearly 75 per cent of the activity was located in the viscera (digestive gland, gonad, etc.). When returned to isotope-free seawater, clams released B*a*P to undetectable levels in 58 days. Tissue distribution of B*a*P remained relatively constant during the depuration period. Dunn and Stich (1976b) measured rate of release of B*a*P from *Mytilus edulis* naturally contaminated with PAH. Concentration of B*a*P in mussel tissues was approximately 45 μg/kg wet weight at the beginning of the experiment. When placed in clean seawater, mussels released B*a*P at an approximately exponential rate over the six-week depuration period. The overall half-life of B*a*P in mussel tissues was approximately 16 days.

Lee *et al.* (1978a) suspended oysters (*Crassostrea virginica*) at a depth of 7 m in a controlled ecosystem enclosure, which was dosed with Prudhoe Bay crude oil enriched with several PAH. Some oysters were subsequently removed to hydrocarbon-free seawater for depuration studies. Naphthalene and alkylnaphthalenes were accumulated most rapidly by the oysters (Table 69). Rate of accumulation of other PAH decreased with increasing molecular weight. When oysters were returned to clean seawater, naphthalenes were released rapidly and reached undetectable levels in 23 days. Anthracene, fluoranthene, benz[*a*]anthracene, and B*a*P were released much more slowly. Based on the depuration experiments, calculated half-lives of the naphthalenes, anthracene, fluoranthene, benz[*a*]anthracene, and B*a*P were 2, 3, 5, 9, and 18 days, respectively. Thus, despite the very limited ability of molluscs to metabolize PAH, all species so far studied are able to release the majority of accumulated PAH from their tissues in periods varying from a few days to several weeks.

Patterns of accumulation and release of PAH in solution by aquatic crustaceans are somewhat different, probably reflecting their more active mode of life and greater PAH-metabolizing abilities. The copepod *Calanus helgolandicus* was able to accumulate naphthalene during exposure for 24 h to seawater solutions containing as little as 0·1 $\mu$g naphthalene/liter (ppb) (Corner *et al.*, 1976a). When returned to naphthalene-free seawater, copepods released naphthalene rapidly (half-line $\sim$1·5 days). Depuration was slightly more rapid in copepods feeding on algal cells than in starved individuals. Subsequently, Harris *et al.* (1977b) studied accumulation of $^{14}$C-naphthalene by *C. helgolandicus* and *Eurytemora affinis* during exposure to concentrations of 0·2 to 992 $\mu$g $^{14}$C-naphthalene/liter seawater for up to 15 days. Initial uptake was rapid, but after exposure for 7–8 days to seawater containing 50 $\mu$g/liter, an equilibrium condition was approached. At an exposure concentration of 1 $\mu$g $^{14}$C-naphthalene/liter, the quantity of radioactivity accumulated in 10 days was nearly 50 times greater in the smaller estuarine species *E. affinis* than in the larger oceanic species *C. helgolandicus*, when expressed in terms of body weight. The difference in uptake rate was not due to differences in lipid content between the two species. When returned to hydrocarbon-free seawater, *C. helgolandicus* released up to 90 per cent of the accumulated radioactivity in 24 h. However, 5 per cent of the accumulated radioactivity remained in the copepods after 11 days. Much of the radioactivity in the copepods and depuration water was identified as naphthalene metabolites.

Lee (1975) obtained similar results when he exposed several species of marine zooplankton to seawater solutions of radiolabeled B*a*P,

TABLE 69

ACCUMULATION AND RELEASE OF PAH BY OYSTERS (*Crassostrea virginica*) EXPOSED TO CRUDE OIL SPIKED WITH SEVERAL PAH IN A CONTROLLED ECOSYSTEM ENCLOSURE (FROM LEE *et al.*, 1978a, WITH PERMISSION OF THE AMERICAN CHEMICAL SOCIETY)

| Duration of exposure (days) | Duration of depuration (days) | Naphthalene | | Methylnaphthalenes | | Dimethylnaphthalenes | | Anthracene | | Fluoranthene | | Benz[a]anthracene | | Benzo[a]pyrene | |
|---|---|---|---|---|---|---|---|---|---|---|---|---|---|---|---|
| | | Oyster (μg/g) | Water (μg/liter) | Oyster (μg/g) | Water (μg/liter) | Oyster (μg/g) | Water (μg/liter) | Oyster (μg/g) | Water (μg/liter) | Oyster (μg/g) | Water (μg/liter) | Oyster (μg/g) | Water (μg/liter) | Oyster (μg/g) | Water (μg/liter) |
| 2 | | 30 | 5 | 58 | 8 | 84 | 10 | 5·6 | 13 | 5·0 | 7·2 | 2·8 | 5·3 | 0·36 | 1·9 |
| 8 | | 12 | 3 | 36 | 3 | 72 | 2 | 2·5 | 1 | 4·0 | 0·4 | 1·8 | 0·1 | 0·30 | 0·1 |
| 2 | 7 | 1 | — | 1 | — | 8 | — | 1·2 | — | 1·7 | — | 1·9 | — | 0·40 | — |
| 8 | 7 | 2 | — | 2 | — | 4 | — | 0·4 | — | 1·4 | — | 1·0 | — | 0·20 | — |
| 8 | 23 | ND[a] | — | ND | — | ND | — | 0·1 | — | 0·4 | — | 0·3 | — | 0·12 | — |

[a] ND, not detected, less than 0·5 μg/g.

methylcholanthrene, and naphthalene. The copepod *Calanus plumchrus* accumulated up to $22 \times 10^{-4}$ $\mu$g B*a*P/individual during a 3-day exposure to 1 $\mu$g B*a*P/liter in seawater (Fig. 23). When returned to isotope-free seawater, the copepods released most of the accumulated radioactivity in 17 days. When depuration was continued beyond 17 days, no further hydrocarbon loss was observed. PAH metabolism contributed significantly to release of PAH by all the crustaceans studied.

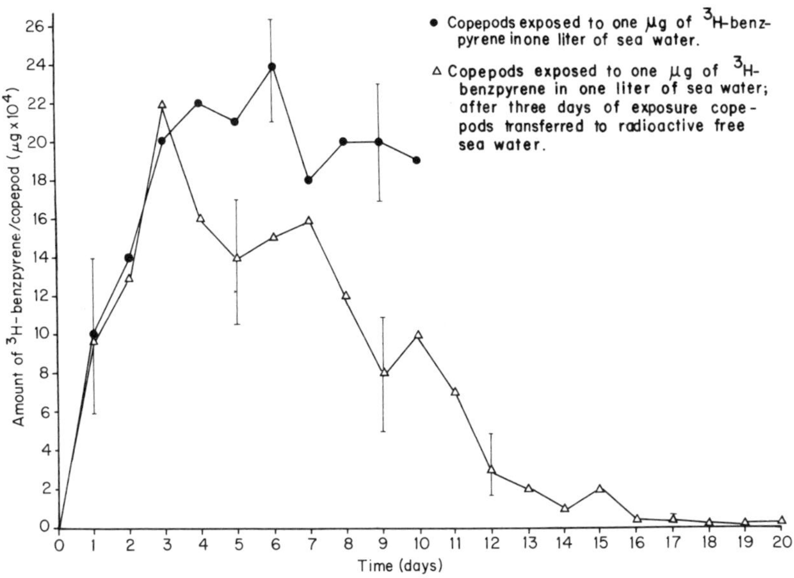


FIG. 23. Accumulation of benzo[*a*]pyrene by the copepod *Calanus plumchrus* during exposure to 1 $\mu$g B*a*P/liter seawater and B*a*P release following return to clean seawater (from Lee, 1975, with permission of the American Petroleum Institute).

Pink shrimp (*Penaeus-duorarum*) exposed to 1 or 5 $\mu$g chrysene/liter seawater accumulated the PAH in both the cephalothorax and abdomen (Miller *et al.*, 1978). At an exposure concentration of 5 ppb, shrimp accumulated approximately 1·8$\mu$g chrysene/g tissue in the cephalothorax and 0·4 $\mu$g chrysene/g tissue in the abdomen in 28 days. When returned to clean seawater, they released most of the chrysene in 10 days. However, a small but measurable amount of chrysene remained in shrimp tissues after 28 days of depuration. No attempt was made to measure metabolite formation.

Juvenile blue crabs (*Callinectes sapidus*) accumulated isotopically labeled B*a*P, methylcholanthrene, and fluorene during exposure to seawater containing 2·5, 1·0, and 30 μg/liter of these compounds, respectively (Lee *et al.*, 1976). Maximum radioactivity in the crabs was reached after 2 days, although uptake from the water continued past this time. After 2 days' exposure, rapid discharge of PAH and metabolites from the crabs balanced uptake from the medium. Initial uptake took place via gill tissue from which radioactivity was transferred primarily to the blood and hepato-pancreas. The pattern of accumulation, distribution in the body, and metabolism and release of $^3$H-B*a*P by crabs is summarized in Table 70. Main sites of B*a*P accumulation were the hepato-pancreas and gill. During both the exposure and depuration periods the fraction of total radioactivity present as unmetabolized B*a*P decreased with time in all tissues analyzed. At every sampling period, more than 50 per cent of accumulated radioactivity was in the hepato-pancreas and, in all but the day-1 samples, more than 50 per cent hepato-pancreatic activity was present as hydrocarbon metabolites. During depuration, concentrations of B*a*P and its metabolites in the crab tissues dropped rapidly so that after 20 days only small amounts of radioactivity remained, primarily in the hepato-pancreas, as polar metabolites.

B*a*P and metabolites were recovered from depuration water. Throughout the first 4 days of depuration, the major excretory product was unmetabolized B*a*P. After longer depuration times, major excretory products were various polar metabolites. More than 50 per cent accumulated B*a*P and its metabolites were excreted in 6 days. More than 70 per cent radioactivity excreted was in particulate form, suggesting that fecal material was the main route of PAH excretion. A fraction of the total radioactivity in crab tissues (varying from 0·3 to 50 per cent of the total) occurred in a form not extractable with the solvents used. This material may have consisted of highly polar compounds or B*a*P metabolites that were covalently bound to tissue macromolecules. The relative proportion of this non-extractable radioactivity in the tissues increased during the depuration period, implying that it was not as readily excreted as B*a*P or its major metabolites. Similar conclusions were obtained with fluorene and methylcholanthrene. All results dramatically demonstrate the importance of PAH metabolism in elimination of these materials from tissues of PAH contaminated crustaceans.

The observations reported in several of the studies discussed above suggest that some of the products of PAH metabolism may be retained in animal tissues longer than unmetabolized PAH. A recent study by Sanborn

TABLE 70

FATE OF $^{3}$H-BENZO[*a*]PYRENE ACCUMULATED FROM WATER (2·5 $\mu$g B*a*P/liter) BY JUVENILE BLUE CRABS (*Callinectes sapidus*). AFTER 2 DAYS' EXPOSURE, CRABS WERE TRANSFERRED TO HYDROCARBON-FREE SEAWATER FOR DEPURATION. VALUES ARE MEANS FOR THREE CRABS SEPARATELY ANALYZED $\pm$1 SD (LEE *et al.*, 1976, WITH PERMISSION OF SPRINGER-VERLAG)

| *Product* | *Time (days)* | *Quantity in tissue ($\mu$g) × $10^4$* | | | | |
|---|---|---|---|---|---|---|
| | | *Gill* | *Blood* | *Hepato-pancreas* | *Stomach* | *Muscle* |
| Total | 1 | 150 ± 70 | 90 ± 17 | 270 ± 90 | 14 ± 2 | 10 ± 3 |
| | 2 | 410 ± 85 | 85 ± 20 | 590 ± 210 | 11 ± 6 | 18 ± 7 |
| | 4 | 100 ± 32 | 80 ± 52 | 300 ± 18 | 10 ± 4 | 8 ± 5 |
| | 8 | 70 ± 55 | 65 ± 21 | 210 ± 16 | 4 ± 5 | 4 ± 3 |
| | 12 | 35 ± 12 | 20 ± 13 | 320 ± 74 | 6 ± 2 | 2 ± 2 |
| | 20 | 1 ± 2 | 6 ± 5 | 40 ± 26 | 2 ± 1 | 2 ± 3 |
| Benzo[*a*]pyrene | 1 | 110 | 50 | 160 | 3 | 4 |
| | 2 | 290 | 30 | 280 | 2 | 2 |
| | 4 | 70 | 12 | 40 | 1 | 2 |
| | 8 | 22 | 8 | 21 | [a] | 1 |
| | 12 | 10 | 1 | 32 | [a] | [a] |
| | 20 | [a] | 1 | 2 | [a] | [a] |
| Hydroxybenzo[*a*]pyrene | 1 | 10 | 27 | 20 | 2 | 2 |
| | 2 | 15 | 12 | 84 | 2 | 2 |
| | 4 | 5 | 21 | 90 | 2 | 1 |
| | 8 | 11 | 9 | 40 | [a] | [a] |
| | 12 | 7 | 6 | 24 | 1 | [a] |
| | 20 | [a] | 2 | 5 | [a] | [a] |
| Polar metabolites | 1 | 15 | 10 | 89 | 8 | 2 |
| | 2 | 3 | 26 | 210 | 6 | 8 |
| | 4 | 12 | 32 | 140 | 7 | 4 |
| | 8 | 17 | 36 | 130 | 2 | 2 |
| | 12 | 11 | 10 | 224 | 2 | 1 |
| | 20 | [a] | 2 | 29 | 1 | 1 |

[a] Detectable but less than $0{\cdot}5 \times 10^{-4}$ $\mu$g.

and Malins (1977) supports this view. Stage-V spot shrimp larvae (*Pandalus platyceros*) accumulated large amounts of radioactivity during exposure to 8–12 μg $^{14}C$-naphthalene/liter in water or complexed with bovine serum albumen. Metabolic products of naphthalene accounted for up to 21 per cent of the radioactivity in larval tissues. $^{14}C$-Naphthalene was almost completely depurated from the tissues during 24–36 h in isotope-free seawater. However, metabolic products were strongly resistant to depuration.

The only detailed study of PAH uptake and release by a freshwater crustacean is that of Herbes and Risi (1978) using the water flea *Daphnia pulex*. When exposed to an aqueous solution of 0·02 μg $^{14}C$-anthracene/liter, *Daphnia* rapidly accumulated anthracene to an equilibrium tissue concentration 760 times the $^{14}C$ concentration in the water. Upon return to isotope-free water, 30–35 per cent of the accumulated radioactivity was released after 5–10 min by starved animals and after about 50 min by animals fed *Enterobacter* cells. Release rates then slowed considerably and asymptotically approached 8 per cent of the initial level in the tissues with a half-life of 3·3 h. Eight per cent of the radioactivity originally accumulated remained in the tissues after 48 h. Anthracene metabolites accounted for only 6 per cent of radioactivity excreted by the water fleas. Thus, most of the accumulated anthracene was excreted unmetabolized and a small fraction of the total accumulated radioactivity remained tightly bound to tissues as in marine crustaceans.

As might be expected from data on their PAH-metabolizing abilities, most fish are able to metabolize and excrete PAH accumulated from the medium even more rapidly than crustaceans. Anderson *et al.* (1974c) reported that accumulation and release of naphthalene and 1-methylnaphthalene were very rapid in the estuarine sheep's-head minnow *Cyprinodon variegatis*. When exposed for 4 h to 1 ppm of each compound in seawater, the fish accumulated 60 ppm naphthalane and 210 ppm 1-methylnaphthalene. Nearly 90 per cent of the accumulated hydrocarbons were released after 29 h in hydrocarbon-free seawater.

Lee *et al.* (1972b) studied accumulation and release of $^{14}C$-naphthalene and $^{3}H$-B*a*P by three species of marine fish. All three species rapidly accumulated the PAH from dilute solution in seawater. Equilibrium levels in the tissues were reached in about 1 h. The main route of uptake appeared to be through the gills. Both naphthalene and B*a*P tended to accumulate primarily in the liver and gall bladder, with the latter organ containing the majority of the activity after several hours' depuration. Following a 24-h period in clean seawater, more than 90 per cent of radioactivity accumulated as $^{14}C$-naphthalene had been lost from most of the fish tissues. B*a*P was released more slowly with losses of 50, 50, 90, and 20 per cent radioactivity in liver, gut, gill, and flesh, respectively, after 24 h in clean seawater. A significant portion of the radioactivity excreted following exposure of fish to both $^{14}C$-naphthalene and $^{3}H$-B*a*P was in the form of polar metabolites. The main routes of excretion of PAH and their metabolites were the gall bladder and urine.

Statham *et al.* (1976) showed that the gall bladder was a principal site of accumulation of several hydrophobic xenobiotics in the trout *Salmo gairdneri*. However, the highest concentration of radioactivity detected in mummichogs (*Fundulus heteroclitus*) following exposure for 4 hours to $^{14}C$-naphthalene was in the spleen (34–105 times the exposure concentration) (DiMichele and Taylor, 1978). Liver, brain, anterior kidney, and gall bladder also contain high concentrations of radioactivity. No attempt was made to identify metabolites. On the other hand, when mangrove snapper (*Lutjanus griseus*) were exposed to 1 or 5 μg chrysene/liter seawater, they accumulated PAH only in the liver and not in other tissues examined (gall bladder, white muscle, intestine) (Miller *et al.*, 1978).

Recently, Melancon and Lech (1978) studied the rate of accumulation, distribution in the body, and rate of release following return to clean water of $^{14}C$-naphthalene and $^{14}C$-2-methylnaphthalene by the rainbow trout *Salmo gairdneri*. During exposure for 8 h to 5 or 23 μg $^{14}C$-naphthalene/liter, the fish accumulated high concentrations of radioactivity in several tissues. Visceral fat contained the highest level of $^{14}C$ while blood and muscle had the lowest levels of the tissues studied. Biliary fluid contained an average of 3·1 μg/ml (ppm) of naphthalene and naphthalene metabolites. When the fish were returned to isotope-free water, all tissues released radioactivity rapidly. Half-lives for $^{14}C$ during the depuration period varied from 6·5 h for muscle to 31·3 h for fat. When the fish were exposed for 26–27 days to $^{14}C$-naphthalene or $^{14}C$-2-methylnaphthalene at aqueous concentrations of 23 or 17 μg/liter, respectively, they accumulated $^{14}C$ to higher levels in the three tissues studied (muscle, blood, and liver) than during the short-term exposure. When the fish were returned to clean water, they released $^{14}C$ much more slowly than following the short-term exposure. Half-lives for $^{14}C$ as naphthalene and its metabolites ranged from 341 h for liver to 909 h for muscle. Release of 2-methylnaphthalene and its metabolites was more rapid. Half-lives ranged from 22·6 h for blood to 724 h for muscle. Release of $^{14}C$ from muscle was biphasic with an initial rapid release phase ($t_{1/2} = 13$ h) followed by a long slow-release phase

($t_{1/2} = 711$ h). The authors showed that the metabolites of naphthalene and 2-methylnaphthalene were released more slowly than the unmetabolized parent compounds. These disparate results suggest either that there are interspecific differences in distribution in fish tissues of PAH accumulated from water or that different PAH have different distribution patterns in the tissues.

The role of the MFO system in excretion of PAH by the rainbow trout *Salmo gairdneri* was recently investigated by Statham *et al.* (1978). When trout were pretreated (induced) with benzanthracene and exposed to $^{14}$C-2-methylnaphthalene in solution, they exhibited significantly elevated rates of biliary excretion of accumulated $^{14}$C-2-methylnaphthalene and metabolites. The bile of induced fish contained a higher proportion of polar metabolites of 2-methylnaphthalene than that of controls. Initial levels of $^{14}$C were higher in livers of induced trout than in uninduced controls, and radioactive material appeared to be retained longer in livers of induced fish during the depuration period. Pretreatment with benzanthracene had little effect on the rate of disappearance of radioactivity from blood and muscle. The greater retention of $^{14}$C in livers of induced fish may be due to the metabolic production of electrophilic metabolites which became bound to tissue macromolecules.

A slightly different approach to the same problem was taken by Lu *et al.* (1977). Accumulation from water and subsequent metabolism of $^{14}$C-B*a*P by three species of freshwater animals—the mosquito fish *Gambusia affinis*, the mosquito larva *Culex pipiens quinquefasciatus*, and the pond snail *Physa* sp.—in the presence and absence of a five-fold excess of piperonyl butoxide, a powerful mixed-function oxidase inhibitor, was measured. In the absence of piperonyl butoxide, none of the radioactive material in the fish occurred as unmetabolized B*a*P at any sampling time (Table 71). In the mosquito larvae and snails, 22 and 88 per cent, respectively, of the total $^{14}$C accumulated after 3 days still existed as the parent compound. In the presence of piperonyl butoxide, the relative percentage of parent compound was increased to 21 per cent in fish and 86 per cent in mosquito larvae but remained about the same (85 per cent) in snails. Therefore, the fish and to a lesser extent mosquito larvae were able to metabolize and excrete B*a*P almost as rapidly as it was taken up from the water, while pond snails, characterized by extremely limited MFO activity, tended to accumulate B*a*P and excrete it and its metabolites very slowly. In all three species, a greater percentage of total radioactivity was retained, particularly after a 3-day exposure, in individuals subjected simultaneously to the MFO inhibitor and B*a*P than in those exposed to B*a*P alone; this provided further

TABLE 71

ACCUMULATION AND METABOLISM OF $^{14}$C-BENZO[*a*]PYRENE (B*a*P) BY FRESHWATER ANIMALS IN THE PRESENCE AND ABSENCE OF PIPERONYL BUTOXIDE. INITIAL B*a*P CONCENTRATION WAS 2·5 μg/liter (ppb) (FROM LU *et al.*, 1977, WITH PERMISSION OF SPRINGER-VERLAG)

| Treatment | *Concentration of $^{14}$C-BaP in μg/g wet weight (ppm)* | | | | | | | | |
|---|---|---|---|---|---|---|---|---|---|
| | *Mosquito fish (Gambusia affinis)* | | | *Mosquito larva (Culex pipiens quinquefasciatus)* | | | *Snail (Physa sp.)* | | |
| | 1 *day* | 2 *days* | 3 *days* | 1 *day* | 2 *days* | 3 *days* | 1 *day* | 2 *days* | 3 *days* |
| *A. $^{14}$C-BaP alone* | | | | | | | | | |
| B*a*P | — | — | — | 0·27 | 0·23 | 0·09 | 4·20 | 5·15 | 5·44 |
| Metabolites | 0·09 | 0·10 | 0·04 | 0·34 | 0·55 | 0·32 | 0·41 | 0·46 | 0·69 |
| Total extractable $^{14}$C | 0·09 | 0·10 | 0·04 | 0·61 | 0·78 | 0·41 | 4·61 | 5·61 | 6·14 |
| Unextractable $^{14}$C | 1·08 | 0·85 | 0·29 | 1·90 | 1·79 | 0·97 | 0·09 | 0·13 | 0·14 |
| Total $^{14}$C | 1·17 | 0·95 | 0·33 | 2·51 | 2·57 | 1·38 | 4·70 | 5·74 | 6·28 |
| *B. $^{14}$C-BaP with 5-fold excess of piperonyl butoxide* | | | | | | | | | |
| B*a*P | 0·01 | 0·02 | 0·06 | 1·07 | 1·49 | 1·13 | 5·44 | 6·10 | 7·63 |
| Metabolites | 0·13 | 0·16 | 0·21 | 0·56 | 0·50 | 0·42 | 0·38 | 0·73 | 1·20 |
| Total extractable $^{14}$C | 0·14 | 0·18 | 0·27 | 1·63 | 1·99 | 1·55 | 5·82 | 6·83 | 8·83 |
| Unextractable $^{14}$C | 1·28 | 0·84 | 1·13 | 1·53 | 0·80 | 0·60 | 0·07 | 0·19 | 0·17 |
| Total $^{14}$C | 1·42 | 1·02 | 1·40 | 3·16 | 2·79 | 2·15 | 5·89 | 7·02 | 9·00 |

indication of importance of the MFO system in excretion of PAH. The experimental results also suggest that, in the absence of piperonyl butoxide, elevated levels of MFO activity were induced within one day in fish and within two days in mosquito larvae, but not at all in snails.

Sharp *et al.* (1978) showed that rates of uptake and release of $^{14}C$-naphthalene changed during the timecourse of embryonic development of the mummichog *Fundulus heteroclitus*. The uptake rate of naphthalene decreased markedly from 1265 dpm $^{14}C$/embryo/h on day 2 of development to 195 dpm $^{14}C$/embryo/h on day 10 of development. Water influx rates varied only slightly during the same time period. Rate of $^{14}C$-naphthalene efflux from the embryos, when returned to isotope-free seawater, varied from 8·65 to 9·95 per cent/h between day 2 and day 6 of development and then dropped to 5·85 per cent/h on day 12 of development. Embryos were unable to metabolize either naphthalene or chrysene to polar metabolites at any stage of embryonic development, possibly accounting for the relatively slow depuration rate.

Kuhnhold and Busch (1978) studied the accumulation of $^{14}C$-naphthalene and $^{14}C$-B*a*P by eggs of the salmon *Salmo salar*. Both naphthalene and B*a*P were accumulated rapidly from the water. The PAH readily penetrated the protective chorion of the eggs and accumulated primarily in the yolk and egg fluid.

## 6.2 ACCUMULATION OF PAH FROM FOOD

Considerably less research has been done on the ability of aquatic animals to accumulate PAH from food sources. Such information is, of course, necessary if we are to assess the potential for biomagnification of PAH in aquatic food chains. Rossi (1977) fed young adult polychaete worms (*Neanthes arenaceodentata*) powdered alfalfa (their normal diet in laboratory culture) contaminated with 10–15 ppm $^{14}C$-2-methylnaphthalene for 16 days in succession. The labeled compound was detected in the worms after feeding for 192 and 384 h on contaminated food. If worms were given uncontaminated fish food for 24 h to allow purging of labeled unassimilated food, no radioactivity was detected in tissue material at any sampling time. In addition, more than 85 per cent of radioactive material recovered from the water of the exposure chambers occurred as the unmetabolized parent compound. Thus, *N. arenaceodentata* had little, if any, ability to accumulate 2-methylnaphthalene from its food.

The situation is quite different in marine crustaceans and fish. Corner *et al.* (1976b) studied dietary uptake of $^{14}$C-naphthalene by the marine copepod *Calanus helgolandicus*. When *Calanus* were fed $^{14}$C-naphthalene-contaminated living or dead copepod nauplii (*Elminius* sp.) or living algae (*Biddulphia*), $^{14}$C-naphthalene was accumulated. Uptake from food was much more efficient than uptake from solution. Approximately 35 per cent of radioactive material taken from the food source was lost during 24 h of depuration, after which 94 per cent of the radioactive compound remaining in the copepods could still be identified as unmetabolized hydrocarbon. However, less than one-third of the radioactive material in depuration water existed as unmetabolized naphthalene. Similar results were obtained by Harris *et al.* (1977a). They compared $^{14}$C-naphthalene uptake from solution versus food supply (*Biddulphia sinensis* cells). Compared to the quantity of $^{14}$C-naphthalene present as suspended food, the amount in solution alone required to give the same increase in hydrocarbon level in copepods was much greater in experiments with female *Calanus* than when males were used (Table 72). The discrepancy between male and female responses was attributed to the higher feeding rate of females in comparison to the smaller males. Approximately 60 per cent of naphthalene ingested with the food was assimilated by both male and female copepods (Table 73). Of the assimilated naphthalene, almost one-half was retained in

TABLE 72

QUANTITATIVE IMPORTANCE OF THE DIETARY PATHWAY IN THE ACCUMULATION OF $^{14}$C-NAPHTHALENE BY THE MARINE COPEPOD *Calanus helgolandicus* (FROM HARRIS *et al.*, 1977a, WITH PERMISSION OF PERGAMON PRESS)

| *Hydrocarbon concentration* | | *Radioactivity as pg hydrocarbon/animal* | | *Hydrocarbon in solution alone equivalent to solution and food level (µg/liter)* C | *Ratio of (C − A):B* |
|---|---|---|---|---|---|
| *In solution (µg/liter)* A | *As food (ng/liter)* B | *Solution alone* | *Solution and food* | | |
| *Experiments with females* | | | | | |
| 0·96 | 0·66 | 53 | 106 | 2·40 | 2 182 |
| 4·69 | 2·38 | 179 | 386 | 11·55 | 2 882 |
| 25·52 | 122·40 | 759 | 4 538 | 230·80 | 1 677 |
| 26·00 | 29·00 | 749 | 1 830 | 76·54 | 1 743 |
| 134·10 | 61·40 | 2 997 | 6 774 | 375·40 | 3 930 |
| *Experiments with males* | | | | | |
| 0·93 | 1·10 | 75 | 88 | 1·24 | 282 |

TABLE 73

FATE OF $^{14}C$-NAPHTHALENE ACCUMULATED BY *Calanus helgolandicus* FROM INGESTION OF CONTAMINATED *Biddulphia* CELLS (FROM HARRIS *et al.*, 1977a, WITH PERMISSION OF PERGAMON PRESS)

| *Sex* | *% Ration* | | | | *Retained*:*Assimilated* (%) | *Soluble release*:*Assimilated* (%) |
|---|---|---|---|---|---|---|
| | *Feces* | *Assimilated* | *Retained* | *Soluble release* | | |
| Female | 41·9 | 58·1 | 31·2 | 26·9 | 53·7 | 46·3 |
| Male | 39·3 | 60·7 | 26·8 | 33·9 | 44·1 | 55·9 |

copepod tissues while the other half was released as either unmetabolized naphthalene or as its metabolites.

Lee *et al.* (1976) showed that juvenile blue crabs (*Callinectes sapidus*) were able to accumulate naphthalene, methylnaphthalene, fluorene, and B*a*P from food. When crabs were fed shrimp or oysters containing radiolabeled PAH, between seven and ten per cent of the radioactive compound was transferred from the stomach to other body tissues. Most of the remainder was unabsorbed and lost in the feces during the first two days after feeding. Radioactive material appeared in the hepato-pancreas after 6 h and in the blood after 18 h. Gills, muscles, and gonads also contained significant levels of radioactivity. The pattern of PAH release from the tissues following ingestion of contaminated food was qualitatively similar to that described earlier for depuration of PAH accumulated from water. A significant fraction of radioactivity was released as polar metabolites.

Dixit and Anderson (1977) administered $^{14}C$-naphthalene to the Gulf killifish *Fundulus similis* by stomach tube. After 2 h, 34 per cent of the administered radioactivity was recovered in tissue material and 34 per cent of this in stomach tissue. Thus, approximately 12 per cent of the administered $^{14}C$-naphthalene was assimilated by *Fundulus*. The majority of this activity was localized in the liver and gall bladder after 2 h. Significant amounts of activity were also present in the heart and lateral body musculature. After 8 h, 79 per cent of the radioactivity recovered was present in the gall bladder. These results strongly suggest that PAH absorbed from the gut are transported to the liver where they are rapidly metabolized and excreted in bile.

Roughly similar results were obtained when $^{14}C$-B*a*P-contaminated squid were fed to young cod (*Gadus morrhua*) (Corner *et al.*, 1976b) or juvenile herring (*Clupea harengus*) (Whittle *et al.*, 1977). Forty-eight hours after feeding, 83·5 per cent of the radioactivity was still present in the stomach and 12 per cent in intestinal contents of the cod (Table 74). The

TABLE 74

DISTRIBUTION OF $^{14}C$ ACTIVITY (AS A PERCENTAGE OF TOTAL ACTIVITY RECOVERED) AT INTERVALS AFTER FEEDING SQUID CONTAINING $^{14}C$-B*a*P AND $^{14}C$-HEXADECANE TO YOUNG CODFISH (*Gadus morrhua*) (FROM CORNER *et al.*, 1976b, WITH PERMISSION OF CAMBRIDGE UNIVERSITY PRESS)

| *Sample* | *Percentage $^{14}C$ activity recovered after:* | | |
|---|---|---|---|
| | 48 *h* | 72 *h* | 96 *h* |
| Stomach | 83·5 | 32·9 | 37·1 |
| Liver | 2·2 | 3·6 | 3·4 |
| Bile fluid | 0·6 | 12·5 | 1·9 |
| Intestinal contents | 12·9 | 8·6[a] | 36·3 |
| Urine | 0 | trace | 0·03 |
| Aquarium residue (mainly feces) | 0·5 | 41·4 | 20·7 |
| Aquarium water | 0 | 0·2 | 0·12 |
| Plasma[b] | 0·03 | 0·08 | 0·05 |
| Blood[b] | 0 | 0·06 | 0·04 |
| Gills | 0·2 | 0·5 | 0·4 |
| Spleen | 0·06 | 0·1 | 0·1 |

[a] A loss of at least 50 per cent occurred during dissection.
[b] Expressed on a per gram basis.

liver, bile fluid, and gills were the only other tissues containing significant activity. After 72 h, the bile fluid contained 12·5 per cent of the total activity; and after 96 h the intestinal contents and feces contained 57 per cent of the total recovered radioactivity, most of the remainder (37·1 per cent) was associated with the stomach. In herring, nearly 80 per cent of the recovered $^{14}C$ remained in the lipid fraction (unmetabolized B*a*P) of the stomach 43 h after ingestion of $^{14}C$-B*a*P-contaminated squid. The largest fraction of the remaining activity was found in the lipid fraction of the intestine (10·3 per cent) and the residual fraction (unextractable B*a*P metabolites) of the stomach, pyloric caecae and intestine (4·9 per cent). Most of the activity recovered in bile was water soluble, indicating polar metabolites. Although the digestive tract of fish represents the major site of both uptake and excretion of orally administered PAH, the fact that more than 98 per cent of radioactivity recovered from the fish 43 h after feeding was in the digestive tract strongly suggests that there was very little assimilation of ingested

B*a*P, and that the small amount of B*a*P that was assimilated was rapidly metabolized and excreted via the gall bladder into the intestine. The authors concluded that retention of B*a*P in the stomach implies strong adsorption or binding to the stomach wall. This binding prevents subsequent absorption and assimilation of ingested B*a*P.

Twenty-four hours after fingerling coho salmon (*Oncorhynchus kisutch*) were fed food containing $^{14}$C-naphthalene or $^{14}$C-anthracene, 0·03 and 0·17 per cent., respectively, of the administered radioactive material was associated with the liver, brain, and flesh of the fish (Roubal *et al.*, 1977b). The specific activity in the liver and brain was higher than in flesh, but flesh contained more total activity. Anthracene and its metabolites were retained longer in fish tissues than were naphthalene and its metabolites.

From the limited data available, it would appear that there are large interspecific differences in ability to absorb and assimilate PAH from food. Polychaete worms have a very limited ability to do so, while in fish adsorption of PAH from the gut is limited and variable depending on species of fish, the PAH, and possibly the food matrix in which PAH is administered. Crustaceans, on the other hand, apparently readily assimilate PAH from contaminated food. In all cases where assimilation of ingested PAH was demonstrated, metabolism and excretion of PAH was rapid. Thus, the potential for food-chain biomagnification of PAH seems to be limited. For such biomagnification to occur, the material must be readily absorbed from food, and once assimilated, it must be relatively resistant to metabolism or excretion.

Lu *et al.* (1977) conducted two types of laboratory freshwater model ecosystem studies to evaluate environmental fate, degradation pathways, bioaccumulation, and food-chain biomagnification potential of B*a*P. In the first, $^{14}$C-B*a*P was applied directly to water in the model ecosystem at an initial concentration of approximately 2 ppb. The $^{14}$C-B*a*P was allowed to pass through a freshwater food chain of plankton, filamentous green algae (*Oedogonium cardiacum*), water fleas (*Daphnia magna*), mosquito larvae (*Culex pipiens*), snails (*Physa* sp.), and mosquito fish (*Gambusia affinis*) for three days at 26·7 °C in an environmental chamber under 750 foot-candle illumination. B*a*P was accumulated and retained in substantial amounts by all organisms in the model ecosystem. The amount of total extractable $^{14}$C still identifiable as B*a*P varied from 46 per cent in mosquito larvae to 90 per cent in *Daphnia*. Bioaccumulation factors ([B*a*P]tissues/[B*a*P]water) ranged from 930 in fish to 134 248 in *Daphnia*. All organisms contained some radioactive material in the form of polar metabolites. The authors reasoned that, since the fish bioaccumulated substantial amounts of B*a*P in this study but did not when they were exposed to B*a*P in water alone, food-chain biomagnification of B*a*P was demonstrated. However, the observations discussed above on accumulation of B*a*P from food by fish would lead one to speculate that much of the radioactivity demonstrated in *Gambusia* in the food-chain study was unassimilated B*a*P in the digestive tract. Lu *et al.* (1977) did not investigate this possibility.

The second study utilized a terrestrial–aquatic model ecosystem. In this experiment, 0·2 mg of $^{14}$C-B*a*P either with or without 1·0 mg piperonyl butoxide was topically applied in acetone solution to *Sorghum vulgare* seedlings at the terrestrial end, simulating atmospheric chemical fallout. The plants were consumed by fourth instar salt-marsh caterpillar larvae (*Estigmene acrea*). $^{14}$C-Labeled products entered the terrestrial and aquatic phases of the system as fecal products, leaf frass, etc. Food-chain organisms in the aquatic component of the system were the same as described above. Radioactive products were allowed to interact in the system for 33 days at 26·7 °C under a 12-h light: dark cycle. Radioactivity in the water reached a maximum level of 5 ppb after 14 days and then declined to 3·44 ppb at 33 days. Unmetabolized B*a*P accounted for 7·1 to 34 per cent of the total $^{14}$C extracted from aquatic organisms at the end of the experiment. Various polar metabolites and unextractable material accounted for the remainder of radioactivity in the organisms. Incorporation of piperonyl butoxide into the model ecosystem increased the total uptake of $^{14}$C by three- to four-fold and increased the percentage of unmetabolized B*a*P in the total extractable radioactivity in all the aquatic organisms. B*a*P was sufficiently persistent in the model ecosystem, particularly in the presence of piperonyl butoxide (which is often added to insecticide formulations), to indicate at least a limited potential for food-chain magnification of this and other high molecular weight PAH. More research is definitely required in this area.

## 6.3 ACCUMULATION OF PAH FROM PETROLEUM

When petroleum is spilled in water, PAH in the oil may enter the water column as PAH in solution, in dispersed form (micro- and macro-oil droplets), or adsorbed to organic or inorganic particles. The physical behavior of oil in natural waters is extremely complex and poorly understood. Dispersed oil droplets have a composition similar to that of the spilled oil. The degree to which different hydrocarbons leave the oil droplets or surface oil slick and enter solution in the water column depends on their concentration in the oil, aqueous solubility, and water/oil partition coefficient. Solubility and water/oil partition coefficients may be influenced by endogenous surfactants in the oil and by the presence in the water of co-solutes and solubilizing agents.

The concentration of PAH in aqueous solution under an oil slick or in the presence of dispersed oil droplets represents an equilibrium between the two phases, oil and water. Because the water/oil partition coefficients of PAH greatly favor their retention in the oil phase, concentrations of PAH in a solution in equilibrium with an oil phase are always substantially lower than saturation concentrations. If a water-soluble fraction of oil is prepared by gently mixing one part oil with nine parts seawater for 20 h (Anderson *et al.*, 1974b), the concentration of individual PAH in the resulting aqueous phase decreases exponentially with increasing ring number (Fig. 24). If the curves are extrapolated to the five-ring PAH such as B*a*P, it can be inferred that they would be present in the water phase in equilibrium with oil at a concentration of approximately 0·01 $\mu$g/liter (ppb).

The physical form in which oil derived PAH occur in the water column may significantly affect the rate at which they are accumulated by aquatic organisms. Several investigators have studied accumulation of petroleum hydrocarbons, including aromatics, by aquatic animals during exposure to petroleum in different aqueous forms.

Stegeman and Teal (1973) and Stegeman (1974) exposed oysters (*Crassostrea virginica*) to 106 $\mu$g No. 2 fuel oil (probably present as both dispersed and water-soluble hydrocarbons) per liter seawater in a flow-through system for 50 days. Oysters with a high lipid content reached an equilibrium tissue concentration of 334 $\mu$g total hydrocarbons/g wet weight in 49 days, while those with a lower lipid content reached an equilibrium tissue concentration of 161 $\mu$g/g wet weight after about 35 days. The accumulated hydrocarbons were rich in aromatics relative to the contaminating oil. In oysters sampled during the exposure period, approximately 40 per cent of the total hydrocarbons present were in the two chromatographic fractions containing mixed aromatic hydrocarbons. By comparison, only 14 per cent of the No. 2 fuel oil was contained in these fractions. When returned to cleaner seawater (11 $\mu$g total hydrocarbons/liter), the oysters rapidly released most of the accumulated hydrocarbons to 34 $\mu$g/g wet weight after 4 weeks. After 28 days of depuration, 16 per cent of the hydrocarbons remaining in oyster tissues were aromatic hydrocarbons. Thus, oysters accumulated aromatic hydrocarbons from oil more rapidly

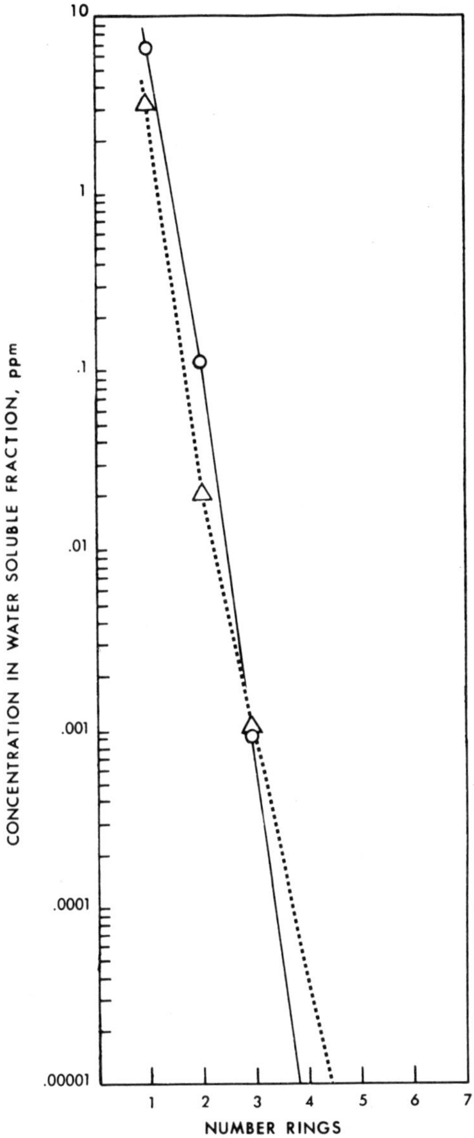


FIG. 24. Relationship between ring number and concentration of aromatic hydrocarbons in the water-soluble fraction prepared by equilibrating one part of either South Louisiana crude oil (○) or Kuwait crude oil (△) with nine parts artificial seawater at 20 °C for 20 h (based on data from Anderson *et al.*, 1974b).

than aliphatics and also released them somewhat more rapidly during depuration.

When mussels (*Mytilus edulis*) were transferred from an unpolluted area to an oil-polluted station in San Francisco Bay, California, for 11 weeks, both aromatic and aliphatic hydrocarbons were taken up (DiSalvo *et al.*, 1975). Aromatic hydrocarbons were accumulated to a much lesser extent than alkanes. When returned to an unpolluted station, approximately 90 per cent of the accumulated hydrocarbons were released in 5 weeks. Final tissue burdens of alkanes and aromatics were similar. When mussels from a chronically oil-contaminated area were transferred to a clean area, 50–65 per cent of the aromatic hydrocarbon burden was lost but little or none of the aliphatic burden over a 10-week span. Bories *et al.* (1976) measured accumulation and release of n-paraffins, cyclic alkanes, and B*a*P by mussels (*Mytilus edulis*) following a spill of fuel oil near Le Havre, France. B*a*P was accumulated to a maximum of 56 ppb in fresh tissue after 9 days but was released to essentially background level in 54 days after the spill. Saturated hydrocarbons were accumulated to much higher concentrations than B*a*P by *Mytilus* but were released rapidly to low though still measurable concentrations in 35–54 days.

Recently, Boehm and Quinn (1977) showed that hard-shell clams (*Mercenaria mercenaria*) from chronically oil-polluted stations in Narraganset Bay, Rhode Island, USA, lost only about 30 per cent of their hydrocarbon content in 120 days after transfer to a clean laboratory system. During the depuration period the aromatic fraction dropped slightly from 10·3 per cent of the total hydrocarbon tissue extract to 9·9 per cent after 96 days.

These results, as well as those from several similar studies, seem to indicate that composition of the hydrocarbon mixture and duration of exposure to the pollutant hydrocarbons, as well as the nutritional status of the animals, may all influence patterns of accumulation and release of petroleum hydrocarbons by molluscs.

The studies reported above provide little information on relative bioavailability to aquatic organisms of various oil-derived PAH. When aquatic animals are exposed to a single PAH or a simple mixture of PAH in solution, there is usually a good correlation between octanol/water partition coefficient of each PAH and its bioaccumulation factor (concentration in tissue/concentration in water) (Neely *et al.*, 1974; Neff *et al.*, 1976a; Lu *et al.*, 1977). Therefore, bioaccumulation factors increase as molecular weight of the PAH increases. Based on these considerations, one might predict that the aromatic fraction in tissues of oil-contaminated aquatic animals would be enriched in higher molecular weight PAH as compared to the aromatic fraction of the contaminating oil. This is not the case when short-term exposure to dispersed oil occurs. Oysters (*Crassostrea virginica*) subjected for 8 h to a concentrated oil-in-water dispersion of No. 2 fuel oil (302 ppm total hydrocarbons at 8 h), accumulated a wide spectrum of hydrocarbons (Table 75) (Neff *et al.*, 1976b). By the end of the exposure period, oysters had accumulated a total of 311 ppm total hydrocarbons, including 76·7 ppm PAH. The relative concentrations of different PAH in the oyster tissues were similar to those in the oil, with no evidence of selective accumulation of higher or lower molecular weight PAH. When returned to oil-free seawater, the oysters released n-paraffins rapidly; PAH were released more slowly. All PAH showed roughly similar behavior. Concentrations of different PAH remained essentially constant during the first 120-h post-exposure. However, after 672 h (28 days), the concentration of PAH in oyster tissues reached background levels. When clams (*Rangia cuneata*) were exposed to the same oil dispersion they accumulated only 89·6 ppm total hydrocarbons in 8 h (28·5 per cent of the amount accumulated by oysters in the same time period). The pattern of petroleum hydrocarbon accumulation and release by the clams was qualitatively similar to that of the oysters.

Bieri and Stamoudis (1977) obtained similar results when they exposed oysters (*Crassostrea virginica*) and hard-shell clams (*Mercenaria mercenaria*) to an experimental spill of No. 2 fuel oil. The oysters rapidly accumulated aliphatic and aromatic hydrocarbons after the spill. Between 6–25 h after the spill, most of the n-alkanes had been released from oyster tissues. Branched alkanes and olefins were released more slowly. Patterns of PAH accumulation and release were more complex (Table 76). The concentrations in oyster tissues of alkyl naphthalenes with up to five alkyl carbons, biphenyls with up to two alkyl carbons, and fluorenes with up to one methyl group increased until 25 h after the spill. Maximum tissue concentrations for more highly substituted naphthalenes, biphenyls, and fluorenes as well as for dibenzothiophenes and phenanthrenes were attained 100 h post-spill.

Many of the PAH which reached peak concentration at 25 h were released to low or undetectable levels after 100 h. Higher molecular weight PAH were released to undetectable levels in 242–509 h. All PAH reached maximum concentrations in the water by 6 h then decreased to low or undetectable levels after 25 h.

Bieri and Stamoudis suggested that continued increase of PAH concentrations in the tissues of oysters long after water-accommodated

TABLE 75

ACCUMULATION OF PETROLEUM HYDROCARBONS BY OYSTERS (*Crassostrea virginica*) DURING EXPOSURE TO DISPERSED NO. 2 FUEL OIL IN A FLOW THROUGH SYSTEM AND SUBSEQUENT RELEASE OF HYDROCARBONS WHEN OYSTERS WERE RETURNED TO OIL-FREE SEAWATER. THE COMPOSITION OF THE NO. 2 FUEL OIL IS INCLUDED FOR COMPARISON (FROM ANDERSON *et al.*, 1974c; NEFF *et al.*, 1976b, WITH PERMISSION OF THE PUBLISHER)

| Time (*h*) | Petroleum hydrocarbon concentration (*ppm, μg/g wet weight*)[a] | | | | | | | | | | | | | | |
|---|---|---|---|---|---|---|---|---|---|---|---|---|---|---|---|
| | *n-P* | *N* | *1-MN* | *2-MN* | *DMN* | *TMN* | *B* | *MB* | *F* | *MF* | *P* | *MP* | *DMP* | *DBT* | *Total* |
| *Exposure* | | | | | | | | | | | | | | | |
| 0 | —[b] | 0·2 | 0·1 | 0·3 | 1·0 | 0·8 | — | — | — | — | — | — | — | — | 2·4 |
| 8 | 235 | 14·7 | 8·7 | 15·0 | 21·8 | 9·1 | 0·3 | 0·5 | 1·0 | 1·2 | 1·9 | 1·9 | 0·3 | 0·3 | 311·7 |
| *Depuration* | | | | | | | | | | | | | | | |
| 3 | 156 | 12·0 | 8·4 | 12·0 | 22·7 | 10·8 | 0·3 | 0·4 | 0·7 | 0·7 | 1·3 | 1·3 | 0·2 | 0·3 | 227·1 |
| 6 | 68 | 7·3 | 5·1 | 7·3 | 13·2 | 5·7 | 0·1 | 0·2 | 0·4 | 0·2 | 0·6 | 0·6 | 0·1 | 0·1 | 108·9 |
| 24 | 18 | 6·5 | 5·7 | 7·6 | 14·8 | 9·5 | 0·2 | 0·2 | 0·5 | 0·7 | 1·2 | 1·3 | 0·3 | 0·2 | 66·7 |
| 120 | 10 | 4·7 | 4·7 | 6·8 | 13·4 | 4·9 | 0·1 | 0·1 | 0·2 | 0·1 | 0·4 | 0·4 | 0·2 | 0·1 | 46·1 |
| 672 | — | — | — | 0·1 | 0·5 | 0·9 | — | — | — | — | — | — | — | — | 1·5 |
| No. 2 fuel oil (% composition) | 7·38 | 0·4 | 0·82 | 1·89 | 3·11 | 1·84 | 0·16 | | 0·36 | | 0·53 | | | 0·07 | 16·56 |

[a] n-P, $C_{12}$–$C_{24}$ n-paraffins; N, naphthalene; 1-MN, 1-methylnaphthalene; 2-MN, 2-methylnaphthalene; DMN, dimethylnaphthalenes; TMN, trimethylnaphthalenes; B, biphenyl; MB, methylbiphenyls; F, fluorene; MF, methylfluorenes; P, phenanthrene; MP, methylphenanthrenes; DMP, dimethylphenanthrenes; DBT, dibenzothiophene.
[b] Less than 0·1 ppm.

TABLE 76

CONCENTRATIONS OF PAH IN THE TISSUES OF OYSTERS (*Crassostrea virginica*) EXPOSED TO A SPILL OF 85 liters OF NO. 2 FUEL OIL IN A 24 × 24-m ENCLOSED SHORELINE AREA NEAR YORKTOWN, VIRGINIA, USA. CONCENTRATIONS IN THE OYSTERS IN ppm (μg/g wet weight) AND IN SEAWATER IN ppb (μg/liter) (FROM BIERI AND STAMOUDIS, 1977, WITH PERMISSION OF PERGAMON PRESS)

| Compound | Concentration with respect to time after spill:[a] | | | | | | | |
|---|---|---|---|---|---|---|---|---|
| | 6 *h* | | 25 *h* | | 100 *h* | | 242 *h* | 509 *h* |
| | *Water* | *Oysters* | *Water* | *Oysters* | *Water* | *Oysters* | *Oysters* | *Oysters* |
| 2-Methylnaphthalene | 11 | 0·13 | 0·2 | 0·15 | — | — | — | — |
| 1-Methylnaphthalene | 9 | 0·11 | 0·2 | 0·12 | — | — | — | — |
| Biphenyl and 2,6-dimethylnaphthalene | 8 | 0·20 | 0·2 | 0·28 | — | — | — | — |
| 1,3-Dimethylnaphthalene | 11 | 0·25 | 0·3 | 0·35 | — | — | — | — |
| 1,5- and 2,3-Dimethylnaphthalene | 3 | 0·07 | — | 0·13 | — | — | — | — |
| 3- and 4-Methylbiphenyl and $C_3$-naphthalene | 2 | 0·10 | — | 0·16 | — | — | — | — |
| $C_3$-Naphthalenes | 7 | 0·32 | — | 0·48 | — | 0·25 | 0·03 | — |
| 2,3,5-Trimethylnaphthalene and $C_4$-naphthalene | 2 | 0·13 | — | 0·23 | — | 0·07 | — | — |
| $C_2$-Biphenyl, $C_3$-naphthalene, and $C_4$-naphthalene | 1 | 0·07 | — | 0·19 | — | — | — | — |
| Fluorene, $C_4$-naphthalenes, methylacenaphthenes, and $C_2$-biphenyl | 2 | 0·09 | — | 0·18 | — | — | — | — |
| $C_4$-Naphthalene and $C_2$-biphenyl | 0·4 | 0·04 | — | 0·09 | — | 0·13 | 0·05 | — |
| $C_4$-Naphthalenes and $C_5$-naphthalene | 0·5 | 0·05 | — | 0·13 | — | 0·37 | 0·13 | — |
| $C_5$-Naphthalene, $C_2$-biphenyl, and methylfluorenes | 2 | 0·16 | — | 0·40 | — | 0·47 | — | — |
| $C_5$-Naphthalene, $C_2$-biphenyl, and $C_3$-biphenyl | — | 0·08 | — | 0·14 | — | 0·24 | 0·10 | — |

TABLE 76—*contd.*

| *Compound* | *Concentration with respect to time after spill:*[a] | | | | | | | |
|---|---|---|---|---|---|---|---|---|
| | 6 *h* | | 25 *h* | | 100 *h* | | 242 *h* | 509 *h* |
| | *Water* | *Oysters* | *Water* | *Oysters* | *Water* | *Oysters* | *Oysters* | *Oysters* |
| $C_6$-Naphthalene, $C_2$-biphenyl, $C_3$-biphenyl, methylfluorene, and $C_4$-biphenyl | — | 0·05 | — | 0·11 | — | 0·20 | 0·10 | — |
| Dibenzothiophene, $C_4$-biphenyl, $C_2$-fluorene, and phenanthrene | 2 | 0·20 | — | 0·71 | — | 1·41 | 0·56 | — |
| Methyldibenzothiophene | 0·3 | — | — | 0·18 | — | 0·32 | 0·17 | — |
| Methyldibenzothiophene, $C_3$-fluorene, and 2-methylphenanthrene | 1 | 0·07 | — | 0·33 | — | 0·82 | 0·27 | — |
| 1-Methylphenanthrene, $C_2$-dibenzothiophene, and $C_3$-fluorene | 1 | 0·05 | — | 0·17 | — | 0·34 | 0·17 | — |
| $C_2$-Dibenzothiophene | 0·1 | 0·02 | — | 0·06 | — | 0·18 | 0·13 | — |
| 3,6-Dimethylphenanthrene and $C_2$-phenanthrene | 0·2 | 0·03 | — | 0·09 | — | 0·35 | 0·20 | — |
| $C_2$-phenanthrenes and $C_3$-dibenzothiophene | 0·4 | 0·10 | — | 0·27 | — | 0·77 | 0·48 | — |
| $C_3$-Phenanthrene | — | 0·03 | — | 0·04 | — | 0·21 | 0·15 | — |
| Total unresolved envelope[b] | | 5 | | 9 | | 24 | 12 | 4 |

[a] —, Compound either absent, below background or cannot be identified.
[b] Semiquantitative estimate from unresolved area.

PAH had disappeared indicated that oysters acquired PAH from oil-contaminated organic detritus via the digestive tract. The authors went on to hypothesize that apparent PAH bioaccumulation factors and changes in PAH concentrations in the tissues of animals exposed to an oil spill in their natural environment are determined mainly by residence times of different PAH in the environment. Hard-shell clams accumulated only about one-tenth as much PAH as the oysters did, indicating a substantial interspecific difference in ability of molluscs to accumulate petroleum hydrocarbons.

Bieri *et al.* (1977) used a similar experimental design to investigate accumulation of petroleum hydrocarbons from spills of new and weathered South Louisiana crude oil by the mummichog *Fundulus heteroclitus*. The concentration of all n-alkanes in the fish reached a maximum 76 h after a spill of fresh crude. Naphthalene, monomethylnaphthalenes, and 2,3-dimethylnaphthalene reached maximum concentrations in fish tissues 31 h after the spill, while all other PAH analyzed reached maximum concentrations after 76 h (Table 77). All PAH except naphthalene, 2-methylnaphthalene, and the dibenzothiophene–$C_4$-biphenyl–$C_5$-naphthalene group were still detectable in fish tissues 216 h post-spill. Nearly all PAH reached a maximum concentration in the exposure water 31 h after the spill. Naphthalene, methylnaphthalenes, and some dimethylnaphthalenes disappeared rapidly from the water between 31 and 76-h sampling times.

Mean apparent bioaccumulation factors (concentration in tissues/concentrations in water) were about 1000 at 31 h for PAH and 290 for aliphatics. At 76 h, bioaccumulation factors were 2700 for PAH and 220 for aliphatics. Bioaccumulation factors for alkyl naphthalenes were similar although higher than that for naphthalene, but there was no relation between degree of alkylation and bioaccumulation factor. PAH were accumulated by the fish much more rapidly from the weathered crude oil and all experimental fish died within 120 h. The difference was partially explained by the observation that PAH were accommodated into the water column much more rapidly from weathered than fresh crude oil.

Although maximum concentrations of individual PAH in the water following spills of crude oil (fish experiment) were much lower than maximum PAH concentrations in the water following the spill of No. 2 fuel oil (oyster experiment), fish accumulated substantially higher concentrations of PAH than did the oysters. Mummichogs also tended to retain PAH in their tissues much longer than oysters did. Thus, ability to metabolize PAH (high in fish, low in oysters) may have little effect on the rate of PAH uptake and release in an acute oil-spill situation. The greater

TABLE 77

CONCENTRATIONS OF PAH IN THE TISSUES OF MUMMICHOGS (*Fundulus heteroclitus*) EXPOSED TO A SPILL OF 570 liter OF SOUTH LOUISIANA CRUDE OIL IN A 810-$m^2$ ENCLOSED SHORELINE AREA NEAR YORKTOWN, VIRGINIA, USA. CONCENTRATIONS ARE IN ppm ($\mu g/g$ WET WEIGHT) (FROM BIERI *et al.*, 1977, WITH PERMISSION OF THE AMERICAN PETROLEUM INSTITUTE)

| *Compound(s)* | *Concentration with respect to time after spill:*[a] | | | |
|---|---|---|---|---|
| | *6 h* | *31 h* | *76 h* | *216 h* |
| Naphthalene | 0·25 | 0·68 | 0·14 | — |
| 2-Methylnaphthalene | 0·34 | 1·56 | 1·12 | — |
| 1-Methylnaphthalene | 0·34 | 1·26 | 0·98 | 0·13 |
| Biphenyl and 2,6-dimethylnaphthalene | 0·25 | 0·81 | 0·99 | 0·25 |
| 1,3-Dimethylnaphthalene | 0·36 | 1·18 | 1·54 | 0·40 |
| 1,5-Dimethylnaphthalene | 0·13 | 0·37 | 0·47 | 0·16 |
| 2,3-Dimethylnaphthalene | 0·04 | 0·05 | 0·03 | 0·05 |
| 3- and 4-Methylbiphenyl and $C_3$-naphthalene | 0·10 | 0·27 | 0·56 | 0·14 |
| $C_3$-Naphthalene | 0·09 | 0·20 | 0·42 | 0·12 |
| Methylbiphenyl and $C_3$-naphthalene | 0·07 | 0·17 | 0·44 | 0·09 |
| 2,3,5-Trimethylnaphthalene | 0·07 | 0·13 | 0·25 | 0·08 |
| $C_3$-Naphthalene and $C_2$-biphenyl | 0·03 | 0·09 | 0·22 | 0·03 |
| Fluorene, $C_4$- and $C_5$-naphthalenes. and $C_2$-biphenyl | 0·04 | 0·11 | 0·45 | 0·05 |
| $C_4$-Naphthalene, $C_2$- and $C_3$-biphenyls | 0·03 | 0·04 | 0·11 | 0·03 |
| Methylfluorene, $C_4$-naphthalene, and $C_3$-biphenyl | 0·11 | 0·23 | 0·54 | 0·13 |
| Dibenzothiophene, $C_4$-biphenyl, and $C_5$-naphthalene | 0·01 | 0·02 | — | — |
| Phenanthrene | 0·08 | 0·11 | 0·26 | 0·08 |

[a] —, Peak not detectable on the gas chromatogram.

accumulation of PAH by mummichogs is probably attributable primarily to the fact that they could not escape from the contaminants, while oysters could remain isolated from the contaminated medium for long periods of time through valve closure.

In the experiments described above, animals were exposed to both dispersed oil droplets and water-soluble fractions of oil. It is likely that some PAH measured in the animals were actually present as unassimilated microoil droplets in the gut or adsorbed to the gills or other body surfaces. This may partly account for the lack of differential uptake of various PAH from the oil as opposed to exposure to water-soluble fractions (WSF) of oil alone where this form of uptake is sometimes observed. Rossi and Anderson (1977) reported that mature male polychaete worms (*Neanthes arenaceodentata*) accumulated higher concentrations of methylnaphthalenes than naphthalene and dimethylnaphthalenes during 8-h exposure to a WSF of No. 2 fuel oil, although the WSF contained a higher concentration of naphthalene than of methylnaphthalenes. Gravid female worms accumulated more dimethylnaphthalenes than methylnaphthalenes and naphthalene. All naphthalenes were released at the same rate when the worms were returned to clean seawater. However, females retained the hydrocarbons until they spawned (300–500 h after the beginning of the depuration period). The majority of the hydrocarbons accumulated by gravid female worms were undoubtedly retained in lipid-rich gonads and eggs. Neff *et al.* (1976b) reported 24-h bioaccumulation factors of 2·3 (naphthalene), 8·1–8·5 (methylnaphthalenes), 17·1 and 26·7 (dimethyl- and trimethylnaphthalenes) for clams exposed to a WSF of No. 2 fuel oil. Cox *et al.* (1975) spilled No. 2 fuel oil on the surface of a shrimp mariculture pond and measured concentrations of naphthalenes in the water and tissues of several species of marine animals from the pond. Naphthalenes were greatly enriched in the tissues of crustaceans in comparison to aqueous concentrations (Table 78). Alkyl naphthalenes were more enriched than naphthalenes. There were quantitative differences in the amounts of naphthalenes accumulated by the three species, but the qualitative pattern of naphthalene and alkyl naphthalene uptake was similar. When coho

TABLE 78

CONCENTRATIONS OF NAPHTHALENES IN THE WATER AND SELECTED CRUSTACEANS FROM AN EXPERIMENTAL SHRIMP POND TO WHICH NO. 2 FUEL OIL WAS ADDED. THE WATER SAMPLES WERE COLLECTED 24 h AFTER THE EXPERIMENTAL SPILL AND THE TISSUE SAMPLES WERE COLLECTED 72 h AFTER THE SPILL (FROM COX *et al.*, 1975, WITH PERMISSION OF THE AMERICAN PETROLEUM INSTITUTE)

| *Sample* | *Mean concentration (ppb)* | | |
|---|---|---|---|
| | *Naphthalene* | *Methylnaphthalenes* | *Dimethylnaphthalenes* |
| Water | 2·3 | 15·4 | 15·2 |
| Brown shrimp (*Penaeus aztecus*) | 450 | 3 610 | 14 700 |
| Fiddler crab (*Uca minax*) | 750 | 4 520 | 16 800 |
| Wharf crab (*Sesarma cinereum*) | 930 | 6 050 | 24 700 |

salmon (*Oncorhynchus kisutch*) were exposed to a dilute WSF of Prudhoe Bay crude oil for five weeks, they accumulated the more highly alkylated benzenes and naphthalenes in muscle tissue faster than the less-substituted aromatics (Table 79) (Roubal *et al.*, 1977a). $C_4$- and $C_5$-benzenes and 2-methylnaphthalene had the highest bioaccumulation factors. A similar pattern of differential accumulation of benzenes and

TABLE 79

CONCENTRATIONS OF AROMATIC HYDROCARBONS IN THE WSF OF PRUDHOE BAY CRUDE OIL AND IN THE MUSCLE TISSUE OF YOUNG COHO SALMON (*Oncorhynchus kisutch*). THE RESULTING BIOACCUMULATION FACTORS ARE ALSO INCLUDED. THE FISH WERE EXPOSED TO THE WSF AT 10 °C FOR 5 WEEKS IN A FLOW-THROUGH SYSTEM (FROM ROUBAL *et al.*, 1977a, WITH PERMISSION OF THE AMERICAN PETROLEUM INSTITUTE)

| *Hydrocarbon* | *Concentration (ppb)* | | *Bioaccumulation factor* |
|---|---|---|---|
| | *Water* | *Muscle tissue* | |
| $C_2$-Benzenes | 352 | 490 | 1·39 |
| $C_3$-Benzenes | 40 | 1 500 | 37·5 |
| $C_4$- and $C_5$-Benzenes | 12 | 5 500 | 458·3 |
| Naphthalene | 4 | 240 | 60 |
| 1-Methylnaphthalene | 4 | 400 | 100 |
| 2-Methylnaphthalene | 4 | 560 | 140 |
| $C_2$-Naphthalenes | 10 | 850 | 85·0 |
| $C_3$-Naphthalenes | 6 | 680 | 113·3 |

naphthalenes was observed in the starry flounder *Platichthys stellatus* during two weeks exposure to the WSF of Prudhoe Bay crude oil (Roubal *et al.*, 1978). The flounder accumulated aromatics to higher concentrations in its muscle tissue than did the coho salmon. The coho salmon released aromatics from its muscle tissue more rapidly than did the starry flounder upon return to clean seawater. Therefore, there is differential bio-accumulation of PAH from water-soluble fractions of oil but not from dispersed oil.

### 6.3.1 Distribution of PAH in Tissues of Aquatic Animals

Patterns of accumulation and release of PAH by different body regions of oil-exposed aquatic animals also vary. When shrimp (*Penaeus aztecus*) were exposed to a dilute WSF of No. 2 fuel oil for 20 h, maximum concentrations of naphthalenes were reached in the head region, abdomen, gill, and exoskeleton within the first hour of exposure (Fig. 25) (Neff *et al.*,

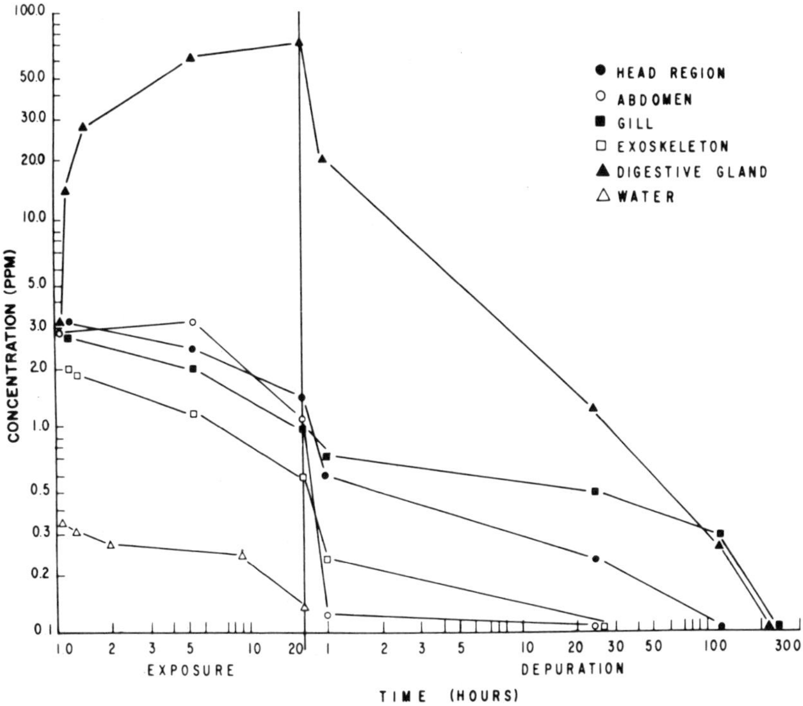


FIG. 25. Accumulation and retention of naphthalenes (naphthalene, methylnaphthalenes, and dimethylnaphthalenes) by different body regions of the juvenile brown shrimp *Penaeus aztecus* exposed to a 20 per cent dilution of the water-soluble fraction of No. 2 fuel oil (from Neff *et al.*, 1976b, with permission of Springer-Verlag).

1976b). The digestive gland continued to accumulate naphthalenes for the full 20-h exposure period and contained more than ten times the concentration of naphthalenes than other tissues analyzed at the end of the exposure period. When shrimp were returned to clean seawater, the abdominal muscle and exoskeleton released naphthalenes very rapidly to undetectable levels after 25 h of depuration. The head region released naphthalenes more slowly. Nearly 250 h were required for complete depuration of naphthalenes from the gill and hepato-pancreas, suggesting that in shrimp the hepato-pancreas is an important site of PAH storage and metabolism and the gills are an important route of PAH excretion.

Naphthalenes also accumulated very rapidly in the Gulf killifish *Fundulus similis* during 2-h exposure to the WSF of No. 2 fuel oil (Fig. 26)

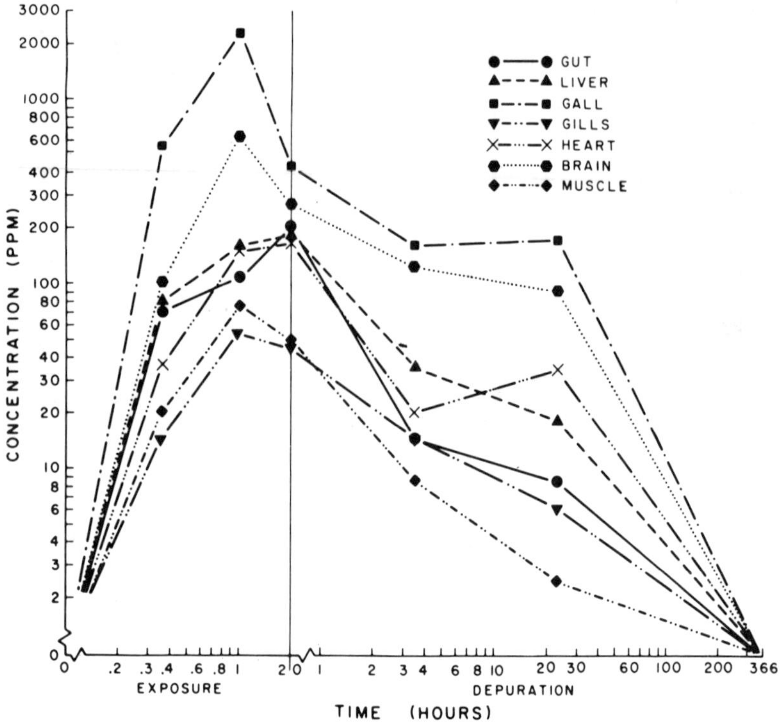


FIG. 26. Distribution of total naphthalenes in the tissues of the Gulf killifish *Fundulus similis* during exposure to the water-soluble fraction of No. 2 fuel oil and at different times following exposure (from Neff *et al.*, 1976b, with permission of Springer-Verlag).

(Neff *et al.*, 1976b). Maximum naphthalene concentrations were reached in most tissues after 1-h exposure. The gall bladder and brain contained the highest concentrations of total naphthalenes (2 300 and 620 ppm, respectively, after 1 h). When fish were returned to clean seawater, all organ systems immediately began to release naphthalenes. The somatic muscles released naphthalenes most rapidly, whereas the gall bladder and brain released naphthalenes much more slowly. Complete depuration of naphthalenes from all tissues required 366 h. Slightly different results were obtained when pink salmon fry (*Oncorhynchus gorbuscha*) were exposed to a WSF of Cook Inlet crude oil for four days (Rice *et al.*, 1977). Maximum concentrations of naphthalenes were reached in the gut, gill, and skeletal muscle (the only tissues analyzed) after 10 h of exposure; the highest concentration of naphthalene was found in the gut at all sampling times, the next highest in skeletal-muscle tissue. Methyl naphthalenes were released more rapidly than dimethylnaphthalenes from the gut during the latter part of the exposure period and after fish were returned to clean seawater.

Mallard ducks (*Anas platyrhynchos*) fed 5 ml of South Louisiana crude oil per day for 14 days demonstrated high concentrations of aromatic hydrocarbons in their tissues (Lawler *et al.*, 1978). The skin and underlying adipose tissue accumulated greater concentrations of aromatic hydrocarbons than did other tissues examined including liver, breast muscle, heart muscle, brain, uropygial gland, and blood. Two- and three-ring aromatics were accumulated to a greater extent than were benzenes. Ringed seals (*Phoca hispida*) rapidly accumulated petroleum hydrocarbons, including aromatics, when exposed to Normal Wells crude oil by immersion or ingestion (Engelhardt *et al.*, 1977). Relatively low levels of hydrocarbons were detected in such tissues as blubber, brain, liver, kidney, muscle, and lung. Somewhat higher levels were found in whole blood. Hydrocarbon concentrations in the bile and urine were high, indicating that these were major routes of hydrocarbon excretion by the seals.

### 6.3.2 Effect of Dissolved Organic Matter on Uptake of PAH

Marine and fresh natural waters nearly always contain significant amounts of dissolved and colloidal organic matter and suspended inorganic and organic particulate matter. These materials may interact with petroleum or PAH entering the water column in such a way as to change bioavailability of these compounds to aquatic organisms. Sandborn and Malins (1977) reported that $^{14}$C-naphthalene was accumulated nearly four times more rapidly than $^{14}$C-naphthalene complexed to bovine serum albumen by stage-V spot shrimp (*Pandalus platyceros*). Dunn and Young (1976) inferred from their data on B*a*P contamination in mussels (*Mytilus edulis*) from California that B*a*P associated with particulate matter of pyrolytic origin was not readily available to marine molluscs. Studies on bioavailability of PAH from food and sediments also seem to indicate that, in most cases, PAH complexed to colloidal organic materials or adsorbed to organic or inorganic particulates are less bioavailable than PAH in solution or fine dispersion in water. Several types of dissolved organic compounds may actually increase the solubility or accommodation of PAH and other petroleum hydrocarbons in water (see Section 4.1.2). Humic substances are complex organic macromolecules of plant origin commonly found in fresh and coastal marine waters in soluble or colloidal form. Because of their detergency, they tend to increase accommodation of poorly soluble organic materials in water. Boehm and Quinn (1976) studied the effect of natural dissolved organic matter (DOM), primarily humic substances, from Narraganset Bay, Rhode Island, on the bioavailability of selected petroleum hydrocarbons and No. 2 fuel oil to thc hard-shell clam *Mercenaria mercenaria*. Removal of DOM from seawater significantly increased uptake of n-hexadecane from water. No significant difference in uptake of phenanthrene from seawater containing DOM compared to seawater from which DOM had been removed was evident. The clams accumulated, on the average, seven times more total hydrocarbons from dispersions of No. 2 fuel oil in DOM-free seawater than from dispersions in seawater containing natural levels of DOM. In the absence of DOM, uptake of saturated hydrocarbons from fuel oil was increased 17-fold while uptake of aromatic hydrocarbons was increased 5-fold. Because of extremely low aqueous solubilities, alkanes above about $C_6$ are present in water primarily as finely dispersed droplets (McAuliffe, 1966). Addition of DOM either increases their solubility or decreases droplet size so that alkanes can pass through the gill filter of *Mercenaria*. Dispersed oil droplets would behave similarly to alkanes. However, phenanthrene is sufficiently soluble at the concentrations used (1 mg/liter) that most would be in solution and therefore its physical state in the water would not be affected by the presence of DOM.

## 6.4 ACCUMULATION OF PAH FROM SEDIMENT

As discussed in previous sections, bottom sediments of lakes, rivers, and coastal marine waters are the ultimate repository for much of the PAH entering the aquatic environment from various sources. Large populations of microbes, plants, and animals live on the surface of or within bottom sediments particularly in estuarine and coastal marine environments. Some benthic animals actually ingest bottom sediment and remove organic materials from it as a source of nutrition (deposit feeders). The question of bioavailability of sediment-adsorbed PAH to these benthic organisms has received relatively little attention. This is unfortunate since assimilation of PAH from sediments by benthic organisms would provide a mechanism by which PAH could be cycled from sediments into the aquatic food chain. In addition, bioaccumulation of PAH from sediment by benthic organisms might pose a health hazard to the predators (including man) of benthic organisms.

Rossi (1977) exposed marine polychaete worms (*Neanthes arenaceodentata*) to sediment contaminated with No. 2 fuel oil. Total concentrations of naphthalenes in the sediment dropped from 9 to 3 ppm during the 38-day exposure period. However, naphthalenes' concentrations in water of the flow-through exposure system never rose above the 0·01 ppm detection limit. Analysis of more than 20 replicate samples of worm tissues, taken periodically throughout the exposure period, showed that the polychaetes contained less than 0·1 ppm total naphthalenes at all sampling times. The worms were apparently unable to accumulate naphthalenes from the sediment although they were observed to ingest sediment and pass it through their digestive tracts. Anderson *et al.* (1977) obtained similar results when they exposed sipunculid worms (*Phascolosoma agassizii*) to sediments contaminated with Prudhoe Bay crude oil. Total hydrocarbon concentration in the sediment ranged from 475 to 765 ppm and total naphthalenes from 2·6 to 3·7 ppm during the two-week exposure period. The highest mean concentrations of total naphthalenes in the worms were 3·8 to 4·8 ppm, reached after 40-h exposure. Worms exposed to contaminated sediment for two weeks contained a mean 1·9 ppm total naphthalenes. When worms were returned to clean sediment, they rapidly released accumulated naphthalenes to undetectable levels in two weeks or less. Sipunculids never contained naphthalene concentrations significantly higher than those in the contaminated sediment, and much of the hydrocarbon present in the worms was probably associated with sediment in the gut. The authors concluded that the sipunculid worms were unable to accumulate naphthalenes from sediment to a significant extent.

Fucik *et al.* (1977) transferred clams (*Rangia cuneata*) to sediments at several stations in the vicinity of an oil-separator platform in Trinity Bay, Texas. The bottom water near the platform contained 0·19 ppb total naphthalenes and bottom sediments contained 27·7 ppm total naphthalenes (see Table 37). Clams accumulated up to 33·6 ppm total naphthalenes in 97 days. There was a good correlation between rates of uptake of naphthalenes by the clams and levels of naphthalenes in sediments at the different stations. Gas chromatograms of clam extracts showed a large unresolved envelope of weathered petroleum similar to that in sediments. The authors concluded that at least part of the naphthalenes accumulated by *R. cuneata* was derived from sediments. The majority of tissue samples contained lower concentrations of naphthalenes than did sediments, indicating that uptake of naphthalenes from sediments by *R. cuneata* was very inefficient.

Accumulation of naphthalenes from oil-contaminated sediment and detritus by the detritivorous clam *Macoma inquinata* was investigated by Roesijadi *et al.* (1978b). Sandy sediments artificially contaminated with Prudhoe Bay crude oil contained initial concentrations of 90·2 to 2750·8 ppm total hydrocarbons and 0·45 to 11·96 ppm total naphthalenes. Total hydrocarbon and naphthalenes concentrations in the sediments decreased by 44·6 and 88·8 per cent, respectively, during the 15-day exposure period. Tissues of sediment-exposed clams contained 0·01 to 0·15 ppm total naphthalenes compared to 0·01 to 0·07 ppm in controls. Thus, the clams were unable to accumulate naphthalenes from the heavily contaminated sediment. Clams exposed to detritus contaminated with Prudhoe Bay crude oil spiked with $^{14}$C-2-methylnaphthalene failed to accumulate significant concentrations of $^{14}$C-2-methylnaphthalene from

TABLE 80

CONCENTRATIONS OF ALIPHATIC AND DIAROMATIC HYDROCARBONS IN THE TISSUES OF THREE BENTHIC INFAUNAL INVERTEBRATES, *Phascolosoma agassizii*, *Macoma inquinata*, AND *Protothaca staminea* FOLLOWING EXPOSURE IN THE FIELD FOR 40 OR 60 DAYS TO SEDIMENT CONTAMINATED WITH PRUDHOE BAY CRUDE OIL AT AN INITIAL CONCENTRATION OF 887·4 μg TOTAL HYDROCARBONS/g SEDIMENT (FROM ROESIJADI *et al.*, 1978a, REPRODUCED BY PERMISSION OF THE MINISTER OF SUPPLY AND SERVICES CANADA)

| *Species and treatment time* | *Hydrocarbon (μg/g wet weight)*[a] | | | | |
|---|---|---|---|---|---|
| | $C_{12}$–$C_{28}$ | *N* | *MN* | *DMN* | *TN* |
| Controls | | | | | |
| *P. agassizii* | <0·10 | <0·005 | <0·005 | <0·01 | <0·02 |
| *M. inquinata* | <0·10 | <0·005 | <0·005 | <0·01 | <0·02 |
| *P. staminea* | <0·10 | <0·005 | <0·005 | <0·01 | <0·02 |
| Exposure for 40 days | | | | | |
| *P. agassizii* | 1·90 | <0·005 | 0·23 | 0·60 | 0·83 |
| *P. agassizii* | 0·73 | <0·005 | 0·01 | 0·15 | 0·16 |
| *M. inquinata* | 0·69 | <0·005 | 0·06 | 0·89 | 0·95 |
| *P. staminea* | <0·10 | <0·005 | <0·005 | <0·01 | <0·02 |
| Exposure for 60 days | | | | | |
| *P. agassizii* | 1·48 | 0·01 | 0·06 | 0·18 | 0·25 |
| *M. inquinata* | 0·54 | 0·02 | 0·27 | 2·39 | 2·68 |
| *M. inquinata* | 3·62 | 0·02 | 0·26 | 1·96 | 2·24 |
| *P. staminea* | 0·10 | <0·005 | 0·02 | 0·16 | 0·18 |

[a] $C_{12}$–$C_{28}$, $C_{12}$–$C_{28}$ n-paraffins; N, naphthalene; MN, methylnaphthalanes; DMN, dimethylnaphthalenes; TN, the sum of naphthalene, methylnaphthalenes and dimethylnaphthalenes.

the detritus. All of the 2-methylnaphthalene in clam tissues could be accounted for via uptake from seawater, presumably of solubilized material or of $^{14}C$-methylnaphthalene adsorbed to very fine suspended matter.

Roesijadi *et al.* (1978a) studied accumulation of Prudhoe Bay crude oil and specific PAH from oil-contaminated sediments by three benthic infaunal invertebrate species, the sipunculid worm *Phascolosoma agassizii* and clams *Macoma inquinata* and *Protothaca staminea*. *P. agassizii* and *M. inquinata* are deposit feeders while *P. staminea* is a suspension feeder. The animals were exposed on the lower shore of Sequim Bay, Washington, in boxes containing beach sand contaminated with Prudhoe Bay crude oil. Total hydrocarbon concentrations in the exposure sediment were 887·4 μg/liter initially and declined to 443·8 and 420·6 μg/g after 40 and 60 days in the field. Hydrocarbon accumulation was generally greater in the two deposit feeding species than in the suspension feeder (Table 80). The amounts of aliphatic and diaromatic hydrocarbons accumulated by all species were quite low even after 60 days of exposure to contaminated sediment. *M. inquinata* accumulated up to 2·24–2·68 ppm total naphthalenes (mostly dimethylnaphthalenes) in 60 days. These results again illustrate the very limited bioavailability of sediment-adsorbed petroleum hydrocarbons even to sediment-ingesting benthic invertebrates.

Clams (*M. inquinata*) were also exposed in the laboratory to sediments containing detritus contaminated with Prudhoe Bay crude oil and spiked with four different $^{14}C$-PAH. One group of clams was exposed directly to the sediment while another group was suspended in the overlying water column. This allowed the investigators to estimate the relative efficiency and magnitude of PAH uptake from detritus and water. After seven days the sediment, water, and clams were analyzed for $^{14}C$ activity. Efficiency of PAH uptake from sediments was much lower than from water (Table 81). Bioaccumulation factors for uptake of the four PAH from contaminated sediments were 0·2 or less indicating no significant bioaccumulation of PAH by this route. Bioaccumulation factors for uptake of the four PAH from seawater were in the 10·3 to 1349 range indicating a significant potential for bioaccumulation, particularly of higher molecular weight PAH, from seawater.

Recently, McCain *et al.* (1978) reported the accumulation of aromatic hydrocarbons from experimentally oiled sediments by the English sole *Parophrys vetulus*. Aromatic hydrocarbons accumulated rapidly in the tissues of the fish and reached maximum concentrations in the three tissues examined (skin, muscle, and liver) after 11 days. At all sampling times, the liver contained higher concentrations of aromatics than did the skin or

TABLE 81

ACCUMULATION OF $^{14}C$-PAH FROM SEDIMENT BY THE DEPOSIT-FEEDING CLAM *Macoma inquinata*. CLAMS WERE EXPOSED FOR 7 DAYS TO SEDIMENT CONTAINING AN INITIAL CONCENTRATION OF 2000 μg/g PRUDHOE BAY CRUDE OIL SPIKED WITH 10 μCi OF THE PAH INDICATED (FROM ROESIJADI *et al.*, 1978a, REPRODUCED BY PERMISSION OF THE MINISTER OF SUPPLY AND SERVICES CANADA)

| *Parameter* | *PAH*[a] | | | |
|---|---|---|---|---|
| | *Phe* | *Chry* | *DMBA* | *BaP* |
| Net uptake from sediment (μg/g) | 0·096 | 0·308 | 0·297 | 0·059 |
| Uptake from seawater (μg/g) | 0·038 | 0·297 | 0·856 | 0·037 |
| Concentration in sediment at 7 days (μg/g) | 0·49 | 8·37 | 4·63 | 0·64 |
| Concentration in seawater at 7 days (μg/ml) | $3{\cdot}7 \times 10^{-3}$ | $4{\cdot}3 \times 10^{-4}$ | $6{\cdot}3 \times 10^{-4}$ | $4{\cdot}3 \times 10^{-5}$ |
| Bioaccumulation factor for uptake from sediment[b] | 0·20 | 0·04 | 0·06 | 0·09 |
| Bioaccumulation factor for uptake from seawater[c] | 10·3 | 694 | 1 349 | 861 |

[a] Phe, Phenanthrene; Chry, chrysene; DMBA, dimethylbenz[*a*]anthracene; B*a*P, benzo[*a*]pyrene.
[b] Calculated as net uptake from sediment/concentration in sediment at 7 days.
[c] Calculated as net uptake from seawater/concentration in seawater at 7 days.

muscle. After 27 days of exposure, only the liver contained measurable concentrations of aromatics and after 51 days the liver contained only 124 ppb 1,2,3,4-tetramethylbenzene and 60 ppb 2-methylnaphthalene. The concentrations of aromatics found in the tissues of the fish throughout the experiment were of the same order of magnitude as aromatic hydrocarbon concentrations in the oiled sediment. However, trimethylnaphthalene, fluorene, and phenanthrene, although present in the oiled sediment, were not detected in the tissues of the fish. These results show that accumulation of aromatic hydrocarbons from oiled sediment by fish is substantially less efficient than uptake from the water.

Thus, accumulation of PAH from sediment, when it occurs at all, may be attributed in large part to uptake of PAH desorbed from sediment particles into the interstitial water. All studies discussed above lead to the general conclusion that sediment-adsorbed PAH are not readily assimilated by benthic invertebrates. However, relatively few species have been studied.

Nothing is known about PAH uptake from sediments by deposit-feeding fish (e.g., mullet) or attached aquatic plants.

## 6.5 EFFECTS OF ENDOGENOUS AND EXOGENOUS FACTORS ON THE ACCUMULATION AND RELEASE OF PAH BY AQUATIC ORGANISMS

Several endogenous biological factors such as size, nutritional status, body composition, age, and sex, as well as exogenous physical factors such as salinity and temperature, may affect patterns of uptake and release of PAH by aquatic organisms. Relatively few investigations have been performed in these areas. Studies of this sort might provide valuable information on mechanisms of PAH accumulation and retention by aquatic organisms.

### 6.5.1 Endogenous Factors

Stegeman and Teal (1973) observed that oysters (*Crassostrea virginica*) containing high concentrations of tissue lipids accumulated petroleum hydrocarbons to higher concentrations than did low fat oysters. Several other investigators have noted a positive correlation between lipid content of aquatic organisms and their ability to accumulate petroleum hydrocarbons, including PAH, from water. Harris *et al.* (1977a) studied accumulation of $^{14}$C-naphthalene by nine species of marine and estuarine copepods. Retention of $^{14}$C-naphthalene following 24-h exposure to this PAH in solution varied nearly 16-fold among the different copepod species. Strong positive correlations were drawn between body weight, ash-free dry weight or total lipid content, and naphthalene retention, but especially between total lipid content and naphthalene retention. The authors went on to study the uptake and retention of $^{14}$C-naphthalene by male and female *Calanus helgolandicus* which had either been starved for five days or fed *Biddulphia* cells before the uptake experiment began. Starved male and female copepods showed significantly lower levels of both total lipid and lipid as percentage of ash-free dry weight than did fed copepods. Starved animals also accumulated significantly less $^{14}$C-naphthalene than fed individuals. Interestingly, both starved and fed male copepods contained nearly three times as much total lipid as starved and fed females, yet males and females accumulated roughly equivalent concentrations of $^{14}$C-naphthalene. It was suggested that lipid composition of males and females may be different and that these compositional differences may affect naphthalene retention by each sex. Although $^{14}$C-naphthalene uptake and retention from solution were positively correlated with total lipid content of the copepods, rate of assimilation of naphthalene from food was unrelated to total lipid content of the animals. Lee (1975) observed that rate of accumulation of $^{3}$H-B*a*P was significantly greater in large than small copepods.

Mature male and gravid female polychaete worms (*Neanthes arenaceodentata*) accumulated naphthalenes at a similar rate from the water-soluble fraction of No. 2 fuel oil (Rossi and Anderson, 1977). However, males rapidly released naphthalenes when returned to oil-free seawater, while gravid females retained them until spawning occurred about 300–500 h after the beginning of the depuration period. Newly released zygotes contained high concentrations of naphthalenes and retained them during early developmental stages to the trocophore stage several days later. The trocophore and later juvenile stages rapidly released naphthalenes. Gravid female *N. arenaceodentata* contain high concentrations of lipids, primarily associated with the ovaries and developing eggs. Naphthalenes apparently accumulated in these gonadal lipid stores and were released in eggs at spawning. Yolk lipid stores are not utilized in this species until larvae reach the trocophore stage. When the lipids were mobilized, naphthalenes were released rapidly.

Thus, hydrocarbons which become associated with stable lipid pools such as depot lipids and gonadal lipid stores may be retained until the animal mobilizes lipids for nutritional purposes, while hydrocarbons which become associated with more labile hydrophobic compartments such as membrane lipids and cellular macromolecules may be released rapidly when ambient levels of hydrocarbons decrease. Such a two-compartment model may partially explain the observation of several investigators that depuration of chronically accumulated hydrocarbons is a two-phase process characterized by an initial rapid release of hydrocarbons followed by a second phase of very gradual release of remaining hydrocarbons.

### 6.5.2 Exogenous Factors

Temperature and salinity of the ambient medium have a profound effect on many physiological functions in aquatic organisms. These factors also affect solubility, adsorption–desorption kinetics, octanol/water partition coefficients, etc. of PAH in water. Therefore, one would expect that salinity and temperature would have a significant effect on accumulation and release of PAH by aquatic organisms. Nevertheless, little research has been done in this area.

Young (1977) acclimated groups of the estuarine grass shrimp

*Palaemonetes pugio* to salinities of 5, 15, and 35‰ S (parts per thousand salinity) or 2, 17, and 32‰ S. Each group was then exposed to either 3 ppm naphthalene or 0·3 ppm phenanthrene at the acclimation salinity. Naphthalene uptake was rapid at all salinities. Tissue naphthalene concentrations reached a maximum in 2 h. The highest level (57 ppm) was observed in animals acclimated and exposed at the intermediate salinity, 15‰. At 5 and 35‰ S, maximum naphthalene uptake was much less with 2-h values of 27·5 and 41 ppm, respectively. Similar results were obtained in

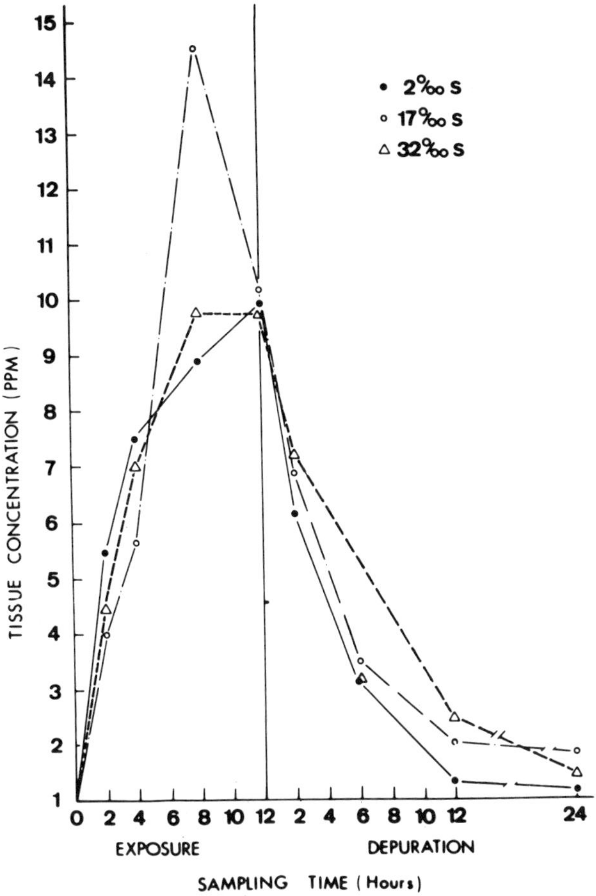


FIG. 27. Accumulation and retention of phenanthrene by grass shrimp (*Palaemonetes pugio*) which were acclimated to, and exposed to, 0·3 ppm phenanthrene at salinities of 2, 17 or 32‰ (from Young, 1977).

the phenanthrene exposure. Phenanthrene uptake was maximal (14·7 ppm) at the intermediate salinity and significantly lower at 2 and 32‰ S (both approximately 9·8 ppm) (Fig. 27). When returned to phenanthrene-free seawater of the acclimation salinity, all three groups of shrimp released phenanthrene rapidly and at approximately the same rate. *P. pugio* is an excellent osmoregulator over its entire normal environmental salinity range. Body fluids are regulated hyperosmotic to the medium at low salinities and hypoosmotic to the medium at high salinities (Roesijadi *et al.*, 1976). At intermediate salinities in the 15–17‰ range, body fluids of the shrimp are isosmotic to the medium. Water turnover rate measurements show that permeability of *P. pugio* is greatest at the isosmotic salinity and is reduced at salinities which are associated with active osmoregulation. These changes in apparent permeability of the shrimp are reflected in changes in rate of uptake but not release of PAH at different salinities.

By comparison, salinity had only a marginally significant effect on rate of naphthalenes uptake from a WSF of Southern Louisiana crude oil by the estuarine marsh clam *Rangia cuneata* (Fucik and Neff, 1977). Naphthalenes uptake was lowest at 30‰ S and variable at lower salinities (Fig. 28). *R. cuneata* is an osmoconformer throughout most of its normal salinity

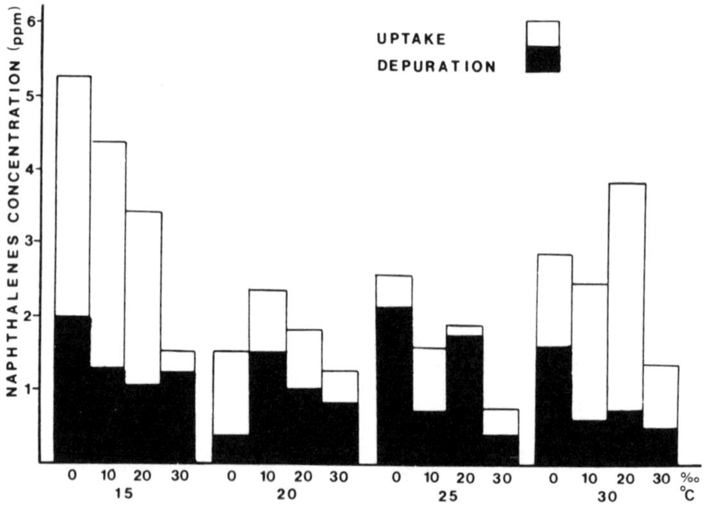


FIG. 28. Uptake and release of total naphthalenes by the clam *Rangia cuneata* exposed to a 25 per cent water-soluble fraction of Southern Louisiana crude oil following acclimation for 14 days to different combinations of salinity and temperature (from Fucik and Neff, 1977, with permission of Pergamon Press).

regime and osmoregulates only at salinities below about 5‰ (Bedford and Anderson, 1972). For another clam, *Protothaca staminea*, which is an osmoconformer throughout its entire environmental salinity regime, salinity did not have a statistically significant effect on uptake of naphthalenes from the WSF of Southern Louisiana crude oil. Therefore, physiological processes underlying osmoregulation by aquatic animals may influence bioavailability to the animals of PAH in solution.

Temperature had a highly significant effect on rate of uptake of naphthalenes from a WSF of Southern Louisiana crude oil by both *Rangia cuneata* (Fig. 28) and *Protothaca staminea* (Fucik and Neff, 1977). Rate of naphthalenes uptake was highest at the lowest temperature used and decreased with increasing temperature. This inverse relationship between temperature and naphthalene uptake rate was not due to the influence of temperature on filtration rate of clams or on residence time of naphthalenes in the water. Filtration rate (the rate at which water was pumped over the gills) of *R. cuneata* increased in a nearly linear fashion with increasing temperature, so that the gills, presumed to be the major sites of PAH uptake by bivalve molluscs (Lee *et al.*, 1972a), were actually exposed to larger volumes of the WSF at higher temperatures where naphthalenes uptake was lowest. Temperatures in the range used in these experiments did not have a statistically significant effect on initial concentration or residence time of naphthalenes in the exposure water. Rate of naphthalenes release from oil-contaminated clams returned to clean seawater was not significantly affected by temperature. The influence of temperature on rate of accumulation of $^{14}C$-naphthalene and $^{14}C$-2-methylnaphthalene from solution by *R. cuneata* was also investigated with similar results (Fucik and Neff, unpublished observations). There was a nearly linear inverse relationship between $^{14}C$ uptake rate and temperature.

Harris *et al.* (1977a) also demonstrated an inverse relationship between temperature and amount of $^{14}C$-naphthalene accumulated by the copepod *Calanus helgolandicus* during exposure to 1 $\mu$g/liter $^{14}C$-naphthalene for 24 h (Fig. 29). Accumulation of $^{14}C$-naphthalene in each copepod decreased by about 39 pg (picograms), or by 3·23 pg/$\mu$g copepod lipid, per 10 °C rise in temperature. The authors suggested that the effect of temperature on $^{14}C$-naphthalene accumulation could be explained if rate of hydrocarbon metabolism increased more rapidly than rate of uptake as temperature increased. This would not explain the effect of temperature on naphthalenes accumulation by molluscs, since they have little, if any, ability to metabolize naphthalenes. A better explanation of this phenomenon is provided by Herbes (1977). He measured the effect of

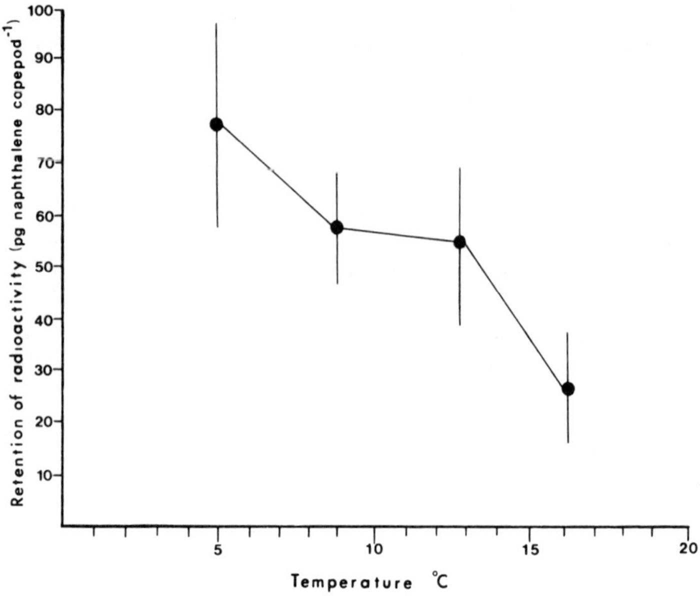


FIG. 29. Effect of temperature on the accumulation of $^{14}$C-naphthalene by the copepod *Calanus helgolandicus* during exposure for 24 h to 1 $\mu$g $^{14}$C-naphthalene/liter seawater (from Harris *et al.*, 1977a, with permission of Pergamon Press).

temperature on the adsorption of anthracene to non-living yeast cells from solution (0·02 $\mu$g/liter). The fraction of anthracene adsorbed by the yeast cells decreased significantly with increasing temperature. Calculated heat of adsorption for this process was 5·2 kcal/mole which is characteristic of simple physical (Van der Waals) adsorption. As temperature rises, the strength of this weak chemical bond decreases, favoring desorption of PAH from the particles. Because partitioning of PAH between soluble and adsorbed phases is determined by relative rates of adsorptive and desorptive reactions, adsorption will be increasingly favored over desorption as temperature decreases. Therefore, temperature exerts its effect on PAH uptake primarily at the initial step of the process—adsorption of PAH from water onto the surface of a biological membrane.

From the very limited research in this area, it is apparent that many biotic and abiotic factors influence rates and patterns of PAH uptake, retention, and release, by aquatic organisms. Since many of these variables are not controlled in laboratory hydrocarbon uptake studies and none of them are controlled in field studies, it is not surprising that published rates of hydrocarbon uptake and release by aquatic organisms are so variable. For instance, several investigators have reported that the majority of hydrocarbons accumulated by marine animals are released rapidly when the animals are returned to clean seawater (Lee *et al.*, 1972a,b; Stegeman and Teal, 1973; Neff *et al.*, 1976b), while others have reported that, once accumulated, hydrocarbons are released very slowly, if at all, even after long periods of depuration (Blumer *et al.*, 1970; Boehm and Quinn, 1977; Melancon and Lech, 1978). These differences may be attributed to differences in such factors as the duration of exposure to hydrocarbons, lipid content, and nutritional status of experimental animals, age and sex of the animals, and salinity, temperature, and other physical environmental factors during the exposure and depuration periods.

*CHAPTER 7*

# *Biological Effects of PAH in the Aquatic Environment*

A great deal of research has been performed in recent years concerning acute and chronic effects of petroleum and specific petroleum hydrocarbons on aquatic organisms and ecosystems. Much of this research has been reviewed in several recent volumes (Carthy and Arthur, 1968; Smith, 1970; Nelson-Smith, 1970, 1973; Baker, 1976; the American Institute of Biological Sciences, 1976; Braunstein *et al.*, 1977; Wolfe, 1977; Malins, 1977a; McIntyre and Whittle, 1977; Vandermeulen, 1978). Much of petroleum's toxicity to aquatic organisms has been attributed to aromatic hydrocarbons contained in it (Boylan and Tripp, 1971; Moore *et al.*, 1973; Moore and Dwyer, 1974; Anderson *et al.*, 1974b,c). However, several other compounds which have been identified in petroleum or water-soluble extracts of petroleum have been shown to be highly toxic to certain aquatic organisms (Larson *et al.*, 1976; Winters *et al.*, 1976, 1977; Donahue *et al.*, 1977; Lee and Nicol, 1978; Wofford and Neff, 1978). These compounds include several heterocyclics and oxygen- or nitrogen-substituted aromatics. Therefore, a discussion of the biological effects of petroleum is not germane to this review. The information reviewed below deals with the effect on aquatic organisms of PAH alone.

## 7.1 MECHANISMS OF PAH TOXICITY

PAH can interact with cells in two ways to cause toxic responses. They may bind reversibly to lipophilic sites in the cell and thereby interfere with several cellular processes. Alternatively, their metabolites, being more hydrophilic, reactive, and electrophilic may bind covalently to many cellular structures causing long-term damage.

Van Overbeek and Blondeau (1954) showed that aromatic hydrocarbons became bound to the plasma membrane of plant cells and increased its permeability to water. Aromatic hydrocarbons also disrupted the orderly arrangement of grana in chloroplasts, interfering with light reactions of photosynthesis. The most toxic aromatics were those in the $C_8$–$C_{14}$ range ($C_2$-benzene to phenanthrene). Acute toxicity to plant cells decreased with increasing carbon number above $C_{14}$. However, the higher molecular weight PAH caused chronic toxicity. This was attributed to their slower rate of penetration through the cell wall. Goldacre (1968) observed that hydrocarbons, including several aromatics, caused swelling and eventual disruption of the plasma membrane of protozoans.

Roubal (1974) and Roubal and Collier (1975) used spin-labeling techniques to study mode of action of petroleum hydrocarbons on marine organisms. *In vivo* and *in vitro* studies with the coho salmon *Oncorhynchus kisutch* revealed that hydrocarbons were bound primarily to membranes in neural tissues. Paraffins penetrated into, and were associated with, the hydrophobic interior of membranes whereas aromatic hydrocarbons were bound to various nonpolar and electron-interactive sites on the membrane surface. It was suggested that a charge-transfer mechanism may direct aromatics away from the membrane interior to the surface. The spin-labeling spectra revealed that binding of aromatic hydrocarbons to the surface of the membrane caused perturbations in the membrane surface organization. Ability of aromatic hydrocarbons to alter membrane surfaces may explain in part why these compounds are toxic to living organisms. Many physiological phenomena are dependent directly on the unique properties of biological membranes. Osmotic and ionic exchange between the cell and the extracellular milieu is mediated by the cell membrane and is essential for maintenance of cellular integrity and for such physiological functions as neurotransmission, muscle contraction, and osmoregulation. Aromatic hydrocarbon-induced changes in the surface organization of membranes may interfere with these essential processes. In addition, many enzymes associated with membranes are weakly bound to membrane lipids. This association between enzyme and membrane lipid is essential for normal enzyme function. if the membrane surface is altered, polar and other interactions between the membrane and enzymes may alter enzyme activity.

Relatively little is known about acute toxicity of the major metabolites of PAH to aquatic organisms. Because of their greater reactivity and solubility, some of them may have a higher acute toxicity than the parent compound. Larson *et al.* (1976, 1977) showed that hydroperoxides produced by photooxidation of aromatic hydrocarbons in oil were highly toxic to yeasts. Some PAH exhibit considerable chronic toxicity, primarily in the form of mutagenesis, carcinogenesis, and teratogenesis. These chronic toxicological effects of PAH are now nearly universally attributed to the covalent binding of certain electrophilic metabolites of PAH to cellular macromolecules (Heidelberger, 1976). There is a good correlation between degree of carcinogenicity of a PAH and the extent of covalent binding of its metabolites to proteins and DNA of mouse skin.

## 7.2 THE ACUTE TOXICITY OF PAH TO AQUATIC ANIMALS

Aromatic hydrocarbons vary substantially in their acute toxicity to aquatic animals. Published values for the acute toxicity of several PAH to invertebrates and fish are summarized in Table 82. Toxicity data for several monocyclic aromatics are included for comparison. In general, acute toxicity of aromatic hydrocarbons to a particular species increases as molecular weight increases. However, the high molecular weight PAH (chrysene, B*a*P, and dibenzanthracene), have low acute toxicity, probably because of their extremely low aqueous solubilities. Within an aromatic series, acute toxicity increases with increasing alkyl substitution on the aromatic nucleus. Thus, in the benzene series acute toxicity increases in the order benzene, toluene, xylenes, trimethylbenzene, tetramethylbenzene. The same trend is seen for naphthalenes and phenanthrenes. There are wide interspecific differences in relative toxicities of different aromatic hydrocarbons. In most cases crustaceans are the most sensitive, polychaete worms intermediate in sensitivity, and fish the most resistant. Within a phyletic group, toxicity of a particular aromatic hydrocarbon to diverse species may vary by as much as two orders of magnitude (i.e., toxicity of toluene to striped bass and mosquito fish and toxicity of naphthalene to spot shrimp larvae and grass shrimp). Some of these differences may be due to differences in bioassay methods. In all but a few cases, the concentrations of aromatic hydrocarbons that are acutely toxic to aquatic animals are several orders of magnitude higher than concentrations found even in the most heavily polluted marine and fresh waters. Aquatic sediments from polluted regions, however, may contain aromatic hydrocarbons at concentrations similar to, or higher than, those which are acutely toxic. The limited bioavailability of sediment-adsorbed aromatic hydrocarbons probably would render them substantially less acutely toxic than aromatic hydrocarbons in solution.

TABLE 82

ACUTE TOXICITY OF AROMATIC HYDROCARBONS TO FRESHWATER AND MARINE ANIMALS

| *Compound* | *Species* | *Concentration (ppm)* | *Effect* ($LC_{50}$)[a] | *Reference* |
|---|---|---|---|---|
| Benzene | *Cancer magister* (stage 1 zoeae, Dungeness crab) | 108 | 96 h | Caldwell *et al.*, 1977 |
| | *Palaemonetes pugio* (grass shrimp) | 27 | 96 h | Neff *et al.*, 1976a |
| | *Crago franciscorum* (bay shrimp) | 20 | 96 h | Benville and Korn, 1977 |
| | *Gambusia affinis* (mosquito fish) | 386 | 96 h | Wallen *et al.*, 1957 |
| | *Marone saxatilis* (striped bass) | 5·8 | 96 h | Benville and Korn, 1977 |
| Toluene | *Cancer magister* (stage 1 zoeae, Dungeness crab) | 28 | 96 h | Caldwell *et al.*, 1977 |
| | *Palaemonetes pugio* (grass shrimp) | 9·5 | 96 h | Neff *et al.*, 1976a |
| | *Crago franciscorum* (bay shrimp) | 4·3 | 96 h | Benville and Korn, 1977 |
| | *Gambusia affinis* (mosquito fish) | 1180 | 96 h | Wallen *et al.*, 1957 |
| | *Marone saxatilis* (striped bass) | 7·3 | 96 h | Benville and Korn, 1977 |
| | *Oncorhynchus gorbuscha* (pink salmon fry) | 5·38 | 24 h | Thomas and Rice, 1978 |
| | *Carassius auratus* (goldfish) | 22·80 | 96 h | Brenniman *et al.*, 1976 |

TABLE 82—*contd.*

| *Compound* | *Species* | *Concentration (ppm)* | *Effect* ($LC_{50}$)[a] | *Reference* |
|---|---|---|---|---|
| Xylenes | *Palaemonetes pugio* (grass shrimp) | 7·4 | 96 h | Neff *et al.*, 1976a |
| | *Carassius auratus* (goldfish) | 16·94 | 96 h | Brenniman *et al.*, 1976 |
| *m*-Xylene | *Cancer magister* (stage 1 zoeae, Dungeness crab) | 12·0 | 96 h | Caldwell *et al.*, 1977 |
| | *Crago franciscorum* (bay shrimp) | 3·7 | 96 h | Benville and Korn, 1977 |
| | *Marone saxatilis* (striped bass) | 9·2 | 96 h | Benville and Korn, 1977 |
| *o*-Xylene | *Cancer magister* (stage 1 zoeae, Dungeness crab) | 6·0 | 96 h | Caldwell *et al.*, 1977 |
| | *Crago franciscorum* (bay shrimp) | 1·3 | 96 h | Benville and Korn, 1977 |
| | *Marone saxatilis* (striped bass) | 11·0 | 96 h | Benville and Korn, 1977 |
| *p*-Xylene | *Crago franciscorum* (bay shrimp) | 2·0 | 96 h | Benville and Korn, 1977 |
| | *Marone saxatilis* (striped bass) | 2·0 | 96 h | Benville and Korn, 1977 |
| Ethylbenzene | *Cancer magister* (stage 1 zoeae, Dungeness crab) | 13 | 96 h | Caldwell *et al.*, 1977 |
| | *Crago franciscorum* (bay shrimp) | 0·49 | 96 h | Benville and Korn, 1977 |

| | | | | |
|---|---|---|---|---|
| | *Marone saxatilis* (striped bass) | 4·3 | 96 h | Benville and Korn, 1977 |
| | *Eurytemora affinis* (estuarine copepod) | 3·8 | 24 h | Ott *et al.*, 1978 |
| 1,2,4-Trimethylbenzene | *Cancer magister* (stage 1 zoeae, Dungeness crab) | 5·1 | 96 h | Caldwell *et al.*, 1977 |
| | *Elasmopus pectenicrus* (marine amphipod) | 4·35 | 96 h | Lee and Nicol, 1978 |
| | *Palaemonetes pugio* (grass shrimp) | 5·4 | 96 h | Neff *et al.*, 1976a |
| 1,3,5,-Trimethylbenzene | *Carassius auratus* (goldfish) | 12·52 | 96 h | Brenniman *et al.*, 1976 |
| 1,2,4,5-Tetramethylbenzene | *Cancer magister* (stage 1 zoeae, Dungeness crab) | 2·1 | 96 h | Caldwell *et al.*, 1977 |
| Naphthalene | *Neanthes arenaceodentata* (marine polychaete) | 3·8 | 96 h | Rossi and Neff, 1978 |
| | *Cancer magister* (stage 1 zoeae, Dungeness crab) | >2·0 | 96 h | Caldwell *et al.*, 1977 |
| | *Cancer magister* (stage 1 zoeae, Dungeness crab) | 0·008–0·012 | 100% mortality in 18–24 h | Sanborn and Malins, 1977 |
| | *Pandalus platyceros* (stages I and V spot shrimp larva) | 0·008–0·012 | 100% mortality in 18–24 h | Sanborn and Malins, 1977 |
| | *Elasmopus pectenicrus* (marine amphipod) | 2·68 | 96 h | Lee and Nicol, 1978 |
| | *Eurytemora affinis* (estuarine copepod) | 0·05 | 30% mortality in 10 days | Berdugo *et al.*, 1977 |
| | *Palaemonetes pugio* (grass shrimp) | 2·4 | 96 h | Neff *et al.*, 1976a |

TABLE 82—*contd.*

| *Compound* | *Species* | *Concentration (ppm)* | *Effect* $(LC_{50})^a$ | *Reference* |
|---|---|---|---|---|
| Naphthalene | *Penaeus aztecus* (brown shrimp) | 2·5 | 24 h | Anderson *et al.*, 1974c |
| | *Gambusia affinis* (mosquito fish) | 150 | 96 h | Wallen *et al.*, 1957 |
| | *Oncorhynchus gorbuscha* (pink salmon fry) | 0·92 | 24 h | Thomas and Rice, 1978 |
| | *Cyprinodon variegatus* (sheep's-head minnow) | 2·4 | 24 h | Anderson *et al.*, 1974c |
| 1-Methylnaphthalene | *Cancer magister* (stage 1 zoeae, Dungeness crab) | 1·9 | 96 h | Caldwell *et al.*, 1977 |
| | *Cyprinodon variegatus* (sheep's-head minnow) | 3·4 | 24 h | Anderson *et al.*, 1974c |
| 2-Methylnaphthalene | *Cancer magister* (stage 1 zoeae, Dungeness crab) | 1·3 | 96 h | Caldwell *et al.*, 1977 |
| | *Eurytemora affinis* (estuarine copepod) | 1·5 | 24 h | Ott *et al.*, 1978 |
| | *Palaemonetes pugio* (grass shrimp) | 1·1 | 96 h | Neff *et al.*, 1976a |
| | *Penaeus aztecus* (brown shrimp) | 0·7 | 24 h | Anderson *et al.*, 1974c |
| | *Cyprinodon variegatus* (sheep's-head minnow) | 2·0 | 24 h | Anderson *et al.*, 1974c |

| | | | | |
|---|---|---|---|---|
| Dimethylnaphthalenes | *Cancer magister* (stage 1 zoeae, Dungeness crab) | 0·60 | 96 h | Caldwell *et al.*, 1977 |
| | *Penaeus aztecus* (brown shrimp) | 0·08 | 24 h | Anderson *et al.*, 1974c |
| | *Cyprinodon variegatus* (sheep's-head minnow) | 5·1 | 24 h | Anderson *et al.*, 1974c |
| 2,6-Dimethylnaphthalene | *Neanthes arenaceodentata* (marine polychaete) | 2·60 | 96 h | Rossi and Neff, 1978 |
| | *Palaemonetes pugio* (grass shrimp) | 0·7 | 96 h | Neff *et al.*, 1976a |
| | *Eurytemora affinis* (estuarine copepod) | 0·85 | 24 h | Ott *et al.*, 1978 |
| 2,3,5-Trimethylnaphthalene | *Eurytemora affinis* (estuarine copepod) | 0·32 | 24 h | Ott *et al.*, 1978 |
| 2,3,6-Trimethylnaphthalene | *Neanthes arenaceodentata* (marine polychaete) | 2·0 | 96 h | Rossi and Neff, 1978 |
| Fluorene | *Neanthes arenaceodentata* (marine polychaete) | 1·0 | 96 h | Rossi and Neff, 1978 |
| | *Palaemonetes pugio* (grass shrimp) | 0·32 | 96 h | Wofford and Neff, 1978 |
| | *Cyprinodon variegatus* (sheep's-head minnow) | 1·68 | 96 h | Wofford and Neff, 1978 |
| Dibenzothiophene | *Palaemonetes pugio* (grass shrimp) | 0·28 | 96 h | Wofford and Neff, 1978 |
| | *Cyprinodon variegatus* (sheep's-head minnow) | 3·18 | 96 h | Wofford and Neff, 1978 |

TABLE 82—*contd.*

| *Compound* | *Species* | *Concentration (ppm)* | *Effect* ($LC_{50}$)[a] | *Reference* |
|---|---|---|---|---|
| Phenanthrene | *Neanthes arenaceodentata* (marine polychaete) | 0·6 | 96 h | Rossi and Neff, 1978 |
| | *Palaemonetes pugio* (grass shrimp) | 0·37 | 24 h | Young, 1977 |
| | *Gambusia affinis* (mosquito fish) | 150 | 96 h | Environmental Protection Agency, 1970 |
| 1-Methylphenanthrene | *Neanthes arenaceodentata* (marine polychaete) | 0·3 | 96 h | Rossi and Neff, 1978 |
| Fluoranthene | *Neanthes arenaceodentata* (marine polychaete) | 0·5 | 96 h | Rossi and Neff, 1978 |
| Benz[*a*]anthracene | *Lepomis macrochirus* (bluegill) | 1·0 | 87% mortality in 6 months | Brown *et al.*, 1977 |
| Chrysene | *Neanthes arenaceodentata* (marine polychaete) | 1·0 | NAT[b] | Rossi and Neff, 1978 |
| Benzo[*a*]pyrene | *Neanthus arenaceodentata* | 1·0 | NAT[b] | Rossi and Neff, 1978 |
| Dibenz[*a*]anthracene | *Neanthus arenaceodentata* | 1·0 | NAT[b] | Rossi and Neff, 1978 |

[a] $LC_{50}$, Median lethal concentration in the time specified, unless stated otherwise.
[b] NAT, not acutely toxic in 96 h.

## 7.3 SUBLETHAL EFFECTS OF PAH ON AQUATIC ORGANISMS

It is obvious that acute toxicity data discussed above do not adequately reflect the potential impact of chronic low-level PAH contamination on aquatic organisms and ecosystems. Chronic exposure to low concentrations of PAH in water, sediment, or food may produce subtle sublethal responses in aquatic organisms. These sublethal effects may be detrimental to the long-term survival of the organism in its normal environment. In addition, chronic exposure to certain PAH may induce cancer in sensitive species.

### 7.3.1 Effects of PAH on Bacteria

Harrison and Raabe (1967) showed that B*a*P in colloidal suspension in the growth medium was highly toxic to *Escherichia coli* when the bacteria were exposed simultaneously to 355-nm wavelength illumination. Neither B*a*P nor illumination alone caused death. Oxygen was required to induce mortality. Significant binding of B*a*P to bacterial cell proteins occurred in the presence of 355-nm illumination and air, but very little binding occurred in the darkness or the absence of oxygen. Thus, lethal properties of B*a*P were probably due primarily to its photooxidation products.

Hass and Applegate (1975) studied the effect of several 3-, 4-, and 5-ring unsubstituted PAH on growth of the bacterium *Escherichia coli*. Ten PAH at concentrations of $10^{-5}$, $10^{-6}$, and $10^{-7}$ M in the growth medium were evaluated. PAH with a highly angular configuration (benz[*a*]anthracene, dibenz[*ah*]anthracene and B*a*P) promoted growth. At most concentrations, anthracene, phenanthrene, chrysene, dibenz[*ac*]anthracene, and pentacene inhibited bacterial growth. The authors concluded that an angular acene configuration was a necessary condition for growth stimulation in *E. coli*.

Slightly different results were obtained by Calder and Lader (1976). They studied the effect of dissolved PAH on growth rate, lag time before initiation of growth, and maximum light absorbance at the end of log-growth phase (an indication of cell numbers) of two marine bacteria, *Serratia marinorubra* and *Vibrio parahaemolyticus*, grown in batch culture. Most of the PAH tested caused an increase in lag time and a decrease in growth rate and maximum absorbance. Pyrene caused an increase in growth rate. Magnitude of growth inhibition was a function of both the concentration of the PAH and its inherent toxic properties. Inherent toxicity of the PAH to bacteria increased inversely with solubility.

Thus, naphthalene at a concentration of 100 μM (12·8 mg/liter) demonstrated a toxic effect similar to B*a*P at a concentration of 0·02 μM (5 μg/liter).

### 7.3.2 Effects of PAH on Aquatic Plants

Graf and Nowak (1966) reported that several PAH at concentrations of 10–20 μg/liter stimulated growth of the freshwater algae *Chlorella vulgaris*, *Scenedesmus obliquus*, and *Ankistrodesmus oraunii*. Degree of growth stimulation by a PAH was directly correlated with its carcinogenicity in animals: B*a*P > benz[*a*]anthracene > benzo[*b*]fluoranthene and indeno-[1,2,3-*cd*]pyrene > benzo[*ghi*]perylene. Low concentrations of several PAH also stimulated growth of sporelings of the marine red alga

TABLE 83

EFFECT OF SELECTED PAH ON THE GROWTH OF SPORELINGS OF THE MARINE RED ALGA *Antithamnion plumula*. SPORELINGS WERE IMMERSED FOR DIFFERENT LENGTHS OF TIME IN SEAWATER CONTAINING FINE SUSPENSIONS OF THE PAH ADDED AS KNOWN CONCENTRATIONS IN METHYL CELLOSOLVE (FROM BONEY AND CORNER, 1962, WITH PERMISSION OF CAMBRIDGE UNIVERSITY PRESS)

| *Hydrocarbon* | *Concentration (mg/liter)* | *Immersion time (h)* | *Growth (cells/10 sporelings/day ± SD)* | *Growth stimulation (%)* |
|---|---|---|---|---|
| Control (methyl cellosolve in seawater) | 0·1 | 96 | 77 ± 4 | — |
| 3-Methylcholanthrene | 0·003 | 96 | 93 ± 3 | 21 |
| | 0·01 | 96 | 145 ± 7 | 88 |
| | 0·1 | 96 | 117 ± 5 | 52 |
| | 1·0 | 1·0 | 118 ± 7 | 53 |
| | 1·0 | 0·5 | 83 ± 1 | 8 |
| | 2·0 | 1·0 | 85 ± 2 | 10 |
| | 2·0 | 0·5 | 114 ± 1 | 48 |
| 9,10-Dimethylbenz[*a*]anthracene | 0·01 | 96 | 180 ± 7 | 133 |
| | 0·1 | 96 | 120 ± 4 | 56 |
| 6-Methylbenz[*a*]anthracene | 0·01 | 96 | 186 ± 4 | 142 |
| | 0·1 | 96 | 173 ± 6 | 125 |
| Benzo[*a*]pyrene | 0·01 | 96 | 115 ± 1 | 49 |
| | 0·1 | 96 | 119 ± 6 | 55 |

*Antithamnion plumula* (Boney and Corner, 1962). The most effective PAH were 6-methylbenz[*a*]anthracene and 9,10-dimethylbenz[*a*]anthracene, both carcinogens (Table 83). Degree of growth stimulation by 3-methylcholanthrene varied with exposure concentration and immersion time. 3-Methylcholanthrene-induced growth stimulation was also demonstrated in two other red algae, *Spermothamnion repens* and *Callithamnion tetricum*. Response of *A. plumula* sporelings to four benz[*a*]anthracene derivatives was dose-dependent. Benz[*a*]anthracene at concentrations of 10 to 300 μg/liter caused a progressive inhibition of growth. A concentration of 10 μg/liter dibenz[*ac*]anthracene and dibenz[*aj*]anthracene stimulated growth while higher concentrations inhibited it. Dibenz[*ah*]anthracene at concentrations of 10 and 30 μg/liter stimulated growth, while higher concentrations inhibited it. Subsequently, Boney (1974) reported that various PAH may stimulate, inhibit, or not affect growth of *A. plumula* sporelings (Table 84). Growth stimulation was

TABLE 84

EFFECTS OF SEVERAL PAH ON THE GROWTH OF SPORELINGS OF THE RED ALGA *Antithamnion plumula*. ALL PAH WERE APPLIED AT A CONCENTRATION OF 0·3 mg/liter (FROM BONEY, 1974, WITH PERMISSION OF PERGAMON PRESS)

| *Compound* | *Growth stimulation (%) compared to controls* |
|---|---|
| Acenaphthene | −3 |
| Anthracene | −20 |
| 2-Methylanthracene | −12 |
| 9-Methylanthracene | +14 |
| Chrysene | +58 |
| Perylene | 0 |
| Pyrene | 0 |
| Triphenylene | −0·2 |

greatest with chrysene and inhibition was most pronounced with anthracene. Boney (1974) showed that the growth-stimulating effect of some PAH was one of an increased rate of cell production by the apical cell cavity. However, the increased rates of cell production induced by B*a*P and other PAH were accompanied by some reduction in cell size.

Naphthalene at concentrations between 3·3 and 30 ppm decreased final cell numbers and duration of the lag period between inoculation and the beginning of cell division in cultures of the green alga *Chorella vulgaris* (Kauss and Hutchinson, 1975). Extent of inhibition was dose-dependent. The algae recovered rapidly and growth rates of naphthalene-exposed cells were similar to that of controls after about one or two days. Similar results were reported by Kauss *et al.* (1973). Naphthalene also inhibited photosynthesis (as measured by $H^{14}CO_3^-$ assimilation rate) in *Chlamydomonas angulosa* (Kauss *et al.*, 1973). When algae were cultured in open flasks, allowing rapid volatilization of naphthalene, photosynthesis was depressed by 89 per cent after 2-h exposure to 30 ppm naphthalene but increased to 59 per cent of the control rate after 20 h. The cells were non-motile during the first 2 h of exposure. When algae were exposed to naphthalene in a closed system which prevented volatilization of naphthalene, inhibition of photosynthesis increased with time from 45 per cent after 30 min to 98 per cent after 2 h at a nominal exposure concentration of 3 ppm naphthalene.

More recently, Vandermeulen and Ahern (1976) reported that exposure to 1 ppm naphthalene resulted in an approximately 40 per cent inhibition of photosynthetic carbon fixation in the unicellular marine algae *Fragilaria* sp., *Monochrysis lutheri*, and *Dunaliella* sp. Exposure to 20 ppm naphthalene resulted in an 80 per cent inhibition in the three species. When cells were resuspended in naphthalene-free medium after preincubation for 4 h in 5 ppm naphthalene, photosynthesis increased immediately and reached control rates after 5 h. Naphthalene exposure had no effect on chlorophyll *a* content of the cells but did result in a statistically significant decrease in cellular ATP concentrations both in the light and in the dark. The authors suggested that the naphthalene-mediated inhibition of photosynthetic carbon fixation may have been a secondary effect of inhibition elsewhere in the cellular machinery, such as partial inhibition of oxidative phosphorylation.

### 7.3.3 Effects of PAH on Aquatic Animals

Ceas (1974) exposed unfertilized and newly fertilized eggs of sea urchins (*Paracentrotus lividus*) to solutions containing B*a*P bound to horse serum proteins. When unfertilized eggs were exposed to 5 μg/ml serum B*a*P for 3 h and then fertilized, they commenced cleavage at the same time as controls; cleavage was very abnormal and stopped at the morula stage. The morulae of exposed embryos contained many dead cells. If the initiation of exposure to serum-B*a*P was delayed until the blastula or gastrula stage, subsequent development of the embryos proceeded normally. The embryonic development of the sea urchin *Psammechinus miliaris* was also unaffected by exposure to 1·4–2·3 μg/ml of a B*a*P–ovalbumin adduct if exposure was initiated after fertilization (Bresch *et al.*, 1972). Slightly different results were reported by DeAngelis and Giordano (1974). Fertilized eggs of *P. lividus* were transferred before the first cleavage to seawater containing $1 \times 10^{-7}$ to $1 \times 10^{-4}$ M B*a*P or 7,12-dimethylbenz[*a*]anthracene. Development to the swimming blastula stage was completely normal in all groups of treated eggs. Continued development was normal in B*a*P-treated embryos and they reached the pluteus stage at the same time as controls. In dimethylbenzanthracene-treated embryos, mesenchyme cells in the blastula were irregular in shape and became disorganized and scattered in the blastocoel. The embryonic skeleton of the pluteus was reduced in size and often abnormal in form. More than 70 per cent of the embryos exposed to $1 \times 10^{-5}$ M dimethylbenzanthrene developed abnormally. When unfertilized eggs and sperm were exposed to B*a*P or dimethylbenzanthracene for 2–5 h before fertilization, subsequent fertilization and development progressed normally. Dimethylbenzanthracene was observed to accumulate to a greater extent than B*a*P in the nucleus of embryos. This may partly explain differences in responses of the embryos to these highly carcinogenic PAH.

Continuous exposure to naphthalene at concentrations of 0·019 to 0·17 ppm had no effect on survival of Dungeness crabs (*Cancer magister*) during complete larval development (Caldwell *et al.*, 1977). However, duration of the first three zoeal stages was increased slightly by chronic exposure to 0·13 ppm naphthalene. The mean delay in development through these stages was 1·5 days. Chronic exposure to naphthalene had no effect on the size of fifth-stage zoeae surviving exposure.

Laughlin and Neff (1978) studied the interactive effects of salinity, temperature, and chronic exposure to naphthalene or phenanthrene on survival and development rate of the larvae of mud crabs (*Rhithropanopeus harrisii*). A phenanthrene concentration of 200 ppb was acutely toxic to the larvae. The first and second zoeal stages were the most sensitive to phenanthrene and the majority of mortalities occurred at, or immediately after, molting. Larvae were substantially more sensitive to phenanthrene at low than at high salinities. Temperatures between 20 and 30 °C had only a slight effect on larval sensitivity to phenanthrene, sensitivity increasing as temperature decreased. Predicted response surfaces for survival (Fig. 30) indicate that larval *R. harrisii* would be severely stressed by combinations of low salinity and high phenanthrene concentrations. This is in marked contrast to the extremely euryplastic response predicted for the controls over the test range.

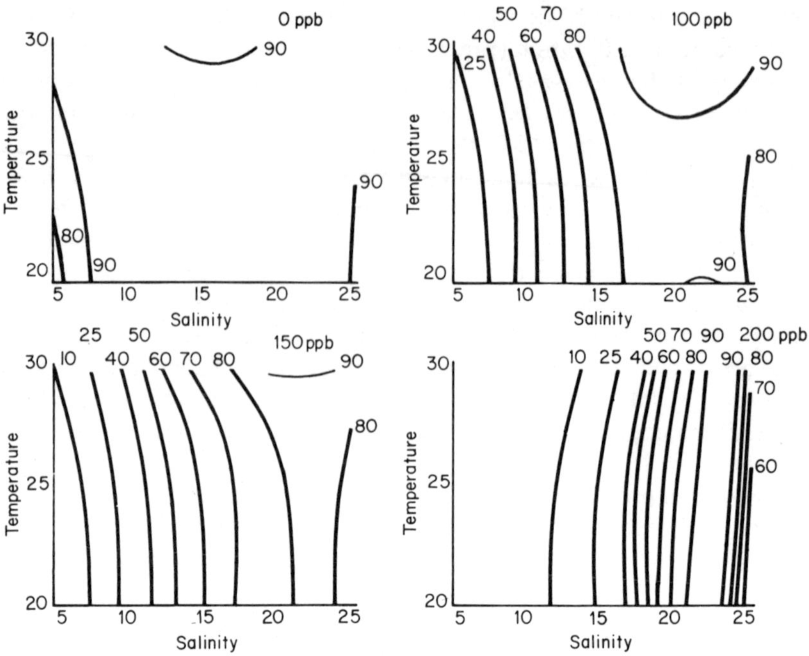


FIG. 30. Predicted response surfaces for percentage survival of larvae of mud crabs (*Rhithropanopeus harrisii*) through complete larval development during continuous exposure to phenanthrene at different combinations of temperature and salinity. Temperatures are in °C and salinities are in ‰ (parts per thousand total dissolved solutes) (from Laughlin and Neff, 1978).

Naphthalene at concentrations of 125 to 500 ppb was not acutely toxic to *R. harrisii* zoeae. The highest mortality occurred in zoeae exposed to 250 or 500 ppb naphthalene at 5‰ S. Duration of larval development to megalops stage was increased by exposure to phenanthrene and decreased by exposure to naphthalene. Temperature had a greater effect than salinity on the magnitude of this response.

Megalops stage larvae of *R. harrisii* which had been exposed continuously during zoeal development to naphthalene or phenanthrene had significantly lower weights than controls (Laughlin, 1977). Phenanthrene had a greater effect than naphthalene on the weight of the megalops. Suboptimal salinities and temperatures increased the effect of hydrocarbons on megalopal weight. Juvenile crabs that were exposed to phenanthrene for 10 days had a statistically significant increase in percentage body water. The results of this study strongly suggest that hydrocarbons interfered with critical osmoregulatory adjustments that accompany molting, resulting in deviations in timing of the molt, increased mortality at time of molt, a decrease in incremental weight, and size increase immediately following molting. These effects were exacerbated by nonoptimal salinities and temperatures.

Therefore, this study demonstrates quite convincingly that variations in physical environmental factors such as salinity and temperature, even when they are well within the normal tolerance range of the species, may substantially modify the nature and magnitude of the organism's response to pollutants. Slightly stressful environmental conditions may significantly increase sensitivity of aquatic organisms to PAH.

Larvae of grass shrimp (*Palaemonetes pugio*) were slightly more resistant than adults to phenanthrene (Young, 1977). Phenanthrene concentrations of 25 to 100 ppb had no effect on duration of the first larval stage, duration of total larval development, or survival of larvae through larval development. Larvae which were exposed continuously from hatching for 30 days to 100 ppb phenanthrene grew significantly more slowly than either controls or larvae exposed to lower concentrations of phenanthrene. Some exhibited growth anomalies. When juveniles were returned to clean seawater their growth was more rapid than that of controls so that after 40 days mean length was not significantly different from that of controls. Thus, grass shrimp showed a substantial ability to recover from chronic exposure to phenanthrene.

Berdugo *et al.* (1977) reported that the feeding rate of copepods (*Eurytemora affinis*) was decreased by 10 and 16 per cent following exposure for 24 h to 1 and 2 ppm naphthalene, respectively. Exposure for 10 days to 10 or 50 ppb naphthalene resulted in an increase in mortality, and decrease in egg production, and feeding rates of the copepods. Only the effect of naphthalene on mortality was statistically significant. Subsequently, Ott *et al.* (1978) reported that chronic exposure of *E. affinis* to approximately 10 $\mu$g/liter (ppb) naphthalene, 2-methylnaphthalene, 2,6-dimethylnaphthalene, or 2,3,5-trimethylnaphthalene for the duration of their adult life (up to 29 days) resulted in statistically significant reductions in the length of life, total number of nauplii produced, and brood size. The rate of egg production and total number of eggs produced per female were both reduced by approximately 50 per cent. The effect of trimethylnaphthalene on reproductive parameters was greater than that of the other naphthalenes.

Donahue *et al.* (1977) studied the effect of several PAH on behavior of nauplii of barnacles (*Balanus amphitrite niveus*). Larvae were placed in vertical glass tubes containing the test solution. The top half of the tube was illuminated while the bottom half was shielded from light. Since the nauplii are positively phototropic they normally tend to aggregate at the illuminated end of the tube. The percentage of larvae which sank into the lower half of the tube was used as an index of toxic effects of PAH. When expressed as a percentage of the saturated solution, methylnaphthalene and dimethylnaphthalenes were most toxic to the larvae (Table 85). However,

TABLE 85

THE EFFECT OF SELECTED PAH ON NAUPLII OF BARNACLES (*Balanus amphitrite niveus*). CONCENTRATIONS ARE GIVEN AS A PERCENTAGE OF SATURATED SOLUTION IN SEAWATER OF 30‰ S (FROM DONAHUE *et al.*, 1977, WITH PERMISSION OF APPLIED SCIENCE PUBLISHERS)

| *Compound* | *Value of the 1-h $D_{50}$*[a] | *Activity of nauplii*[b] *A* | *B* | *C* |
|---|---|---|---|---|
| Naphthalene | 13 | 20 | 25 | |
| 1-Methylnaphthalene | 6·4 | 5 | 10 | 50 |
| Dimethylnaphthalenes | 6 | 6 | 10 | 50 |
| Biphenyl | 27 | 40 | 50 | |
| Fluorene | 44 | 60 | 85 | |
| Phenanthrene | 70 | 100 | | |
| Anthracene | no effect | 100 | | |

[a] Percentage of saturated solution at which half the larvae were sent to the bottom in one hour.
[b] Percentage of saturated solution at which larvae recovered from the bottom were: A, all swimming; B, barely swimming or few swimming; C, mostly dead.

all the PAH except anthracene decreased the positive phototropic response of the larvae. The lack of effect by anthracene is probably due to its extremely low aqueous solubility. Since solubility of PAH decreases as molecular weight increases, differences in relative toxicities of the various PAH are not as great as indicated. For instance, a 6 per cent solution of dimethylnaphthalene in seawater (the most toxic PAH tested) contains approximately 0·14 ppm dimethylnaphthalene while a 70 per cent solution of phenanthrene in seawater (one of the least toxic PAH to the larvae) contains approximately 0·42 ppm phenanthrene. PAH concentrations required to cause death or significant inhibition of activity were substantially higher than those required to inhibit phototaxis.

When hydrocorals (*Millepora* sp.) from Bermuda were exposed to phenanthrene at concentrations of 25 to 500 ppb, rates of $^{45}Ca$ deposition in the skeleton and of photosynthetic $H^{14}CO_3^-$ fixation by the symbiotic zooxanthallae were decreased (Fucik and Neff, 1978) and the response was dose-dependent. Since calcification by the coral is partly dependent on photosynthesis by its symbiotic algae, depressed calcification rates were probably due to inhibitory effects of phenanthrene on photosynthesis by the algae.

In other experiments, Neff and Fucik (1978) studied the effects of naphthalene, phenanthrene, and dibenzothiophene on the rate of $^{45}Ca$ deposition in spines of the coral-reef sea urchin *Diadema antillarum* from Bermuda. Exposure to 500 ppb solutions of the three compounds caused a 20 to 70 per cent decrease in rate of $^{45}Ca$ deposition in the spines. There was little difference in response of the sea urchins to each of the three compounds.

Haranghy (1956) studied effects of B*a*P on the filtration rate of freshwater mussels (*Driessena polymorpha*, *Anodonta cygnea*, *Unio pictorum*, and *U. tumidus*). Injections of 0·2 to 1·0 μg B*a*P into the visceral sac of the molluscs resulted in a significant decrease in filtration rate. It was suggested that B*a*P inhibited ciliary activity or possibly the entire metabolism of the mussels.

Oxygen consumption (respiratory rate) is a good index of the integrated metabolic response of an organism to many endogenous and exogenous variables. As such, it has been used often as an index of sublethal pollutant stress. Laughlin (1977) studied interactive effects of temperature, salinity, salinity shock (a sudden change in salinity immediately before respirometry), and chronic exposure to naphthalene or phenanthrene on the respiratory rates of early and late zoeae and megalops of mud crabs (*Rhithropanopeus harrisii*). At most combinations of constant temperature and salinity, respiratory rates of the crabs were increased by exposure to phenanthrene or naphthalene. Phenanthrene had a greater effect than naphthalene on respiratory rates. The effect of the hydrocarbons was greater at non-optimal combinations of temperature and salinity, particularly combinations of high temperature and low salinity. Early zoeae were more sensitive than late zoeae or megalops. Salinity shock greatly increased respiratory response of the crabs upon exposure to hydrocarbons, indicating that hydrocarbon exposure increased the metabolic cost of adjustment to new environmental salinities. Thus, the net effect of phenanthrene and naphthalene on early life stages of *R. harrisii* is an increase in metabolic cost of maintenance and thereby a shunting of limited energy reserves away from biomass accretion. Nonoptimal salinity–temperature regimes, because they also increase metabolic cost of maintenance, exacerbate effects of the hydrocarbons.

Thomas and Rice (1978) studied the effect of naphthalene on oxygen consumption and opercular breathing rate of pink salmon fry (*Oncorhynchus gorbuscha*). During exposure to 0·76 ppm naphthalene, oxygen consumption rose sharply to 400 per cent of the control rate after 8 h. Also, a 55 per cent increase in opercular breathing rate was evident after 4 h. Continued exposure resulted in a decrease in both consumption and breathing rate toward control values. There was a direct linear relationship between breathing rate and naphthalene concentration and the effect of naphthalene on breathing rate of the fish was greater at 4 than at 12 °C.

A wide variety of histopathological and physiological responses to naphthalene exposure was described in the mummichog *Fundulus heteroclitus* (DiMichele and Taylor, 1978). Gill hyperplasia was the most consistent pathological condition noted and occurred even in fish exposed to as little as 2 $\mu$g/liter (ppb) naphthalene for 15 days. Hemorrhages of the gill filaments were also noted in fish exposed to naphthalene concentrations of 2 mg/liter (ppm) and higher. At all but the lowest naphthalene concentration, there were signs of necrosis of neurosensory cells in epithelia of the lips, mouth, and pharynx, and in the olfactory organ and lateral-line system. In those fish exposed to higher concentrations of naphthalene (2–30 ppm) focal necrosis, hyperplasia, and ischemia were observed in several organs including the pancreas, liver, and brain. These latter histopathalogies were attributed to blood stasis. In extensively damaged areas of the pancreas, liver, and brain, small blood vessels were engorged with red blood cells, blocking circulation to the affected areas. It was hypothesized that blood stasis was caused by a naphthalene-induced increase in rigidity of the erythrocyte membrane which prevented free passage of erythrocytes through finer blood vessels and capillaries. Hematocrit values of naphthalene-exposed fish were not markedly different from those of controls. However, fish exposed to 0·002–0·2 ppm naphthalene had significantly elevated titers of blood cortisol, glucose, and protein.

Results of this investigation indicate that toxicity of naphthalene in fish is due primarily to its effect on blood components in acutely lethal situations. At sublethal concentrations of naphthalene, there is evidence of neurosensory damage and metabolic stress (adrenocortical stress syndrome) which may affect the animal's overall ability to survive in its natural environment.

The investigations discussed above show that potentially detrimental sublethal responses to PAH can occur at ambient concentrations substantially lower than those which cause acute toxicity. In most cases, PAH concentrations required to elicit significant sublethal responses were higher than those normally encountered in all but heavily polluted aquatic environments. Environmentally realistic PAH concentrations (1–50 ppb) did cause significant sublethal responses in some sensitive organisms.

## 7.4 PAH-INDUCED CANCER IN AQUATIC ORGANISMS

It has been recognized for many years that some PAH can cause cancer in laboratory mammals and possibly man, and correlations have been made between occupational or other exposure to PAH and the incidence of human cancer (International Agency for Research on Cancer, 1973; National Academy of Sciences, 1972; Bridboard *et al.*, 1976). Not all PAH are carcinogenic in mammals, and among those which have been shown to cause cancer, carcinogenicity varies widely. The relative carcinogenicity of several PAH are listed in Table 86; 4-, 5-, and 6-ring PAH are more carcinogenic than either smaller or larger ring systems, and highly angular configurations are more carcinogenic than either linear or highly condensed ring systems. Alkylation may substantially modify carcinogencity of a PAH. Position of alkyl substituents is extremely important. The degree of carcinogenicity of a PAH is related to structure and reactivity of its major metabolites. Newman (1976) compared the carcinogenic activity of 12 monomethylbenz[*a*]anthracenes. 7-Methylbenz[[*a*]anthracene was most active, 6-, 8-, and 12-methyl isomers were slightly active, and the remaining monomethylbenz[*a*]anthracenes were inactive. The author concluded that the 7-position in benz[*a*]anthracene is the position at which detoxification occurs. To obtain significant carcinogenic activity, the 7-position must be blocked. The 5-position is that at which metabolism occurs to produce cancer. Blockage of this position destroys carcinogenic activity. Similar results were obtained with methylchrysenes (Hecht *et al.*, 1976). 5-Methylchrysene demonstrated very high carcinogenic activity, equal to or greater than that of B*a*P. 2-Methylchrysene had moderate activity; other methylchrysenes were inactive. Thus, alkylation affects carcinogenicity of PAH by altering the position of initial enzymatic attack on the PAH molecule by the MFO–cytochrome P-450 system.

Different species of organisms vary substantially in sensitivity to PAH-induced carcinogenesis. This may result from interspecific differences in the levels of MFO–cytochrome P-450 activity, stereochemistry of the reactions

TABLE 86

RELATIVE CARCINOGENICITY OF PAH TO LABORATORY MAMMALS (REPRODUCED FROM *Particulate Polycyclic Organic Matter* (1972), PP. 6–11, WITH PERMISSION OF THE NATIONAL ACADEMY OF SCIENCES, WASHINGTON, D.C.)

| *Compound* | *Carcinogenicity*[a] | *Compound* | *Carcinogenicity*[a] |
|---|---|---|---|
| Anthracene | — | Aceanthrylene | — |
| Phenanthrene | — | Benz[*j*]aceanthrylene = cholanthrene | ++ |
| Benz[*a*]anthracene | + | 3-Methylcholanthrene | ++++ |
| 7,12-Dimethylbenz[*a*]-anthracene | ++++ | Naphthacene | — |
| Dibenz[*aj*]anthracene | + | Pyrene | — |
| Dibenz[*ah*]anthracene | +++ | Benzo[*a*]pyrene | +++ |
| Dibenz[*ac*]anthracene | + | Benzo[*e*]pyrene | — |
| Benzo[*a*]phenanthrene | +++ | Dibenzo[*al*]pyrene | ± |
| Fluorene | — | Dibenzo[*ah*]pyrene | +++ |
| Benzo[*a*]fluorene | — | Dibenzo[*ai*]pyrene | +++ |
| Benzo[*b*]fluorene | — | Dibenzo[*cd,jk*]pyrene | — |
| Benzo[*c*]fluorene | — | Indeno[1,2,3-*cd*]pyrene | + |
| Dibenzo[*ag*]fluorene | + | Chrysene | ± |
| Dibenzo[*ah*]fluorene | ± | Dibenzo[*b,def*]chrysene | ++ |
| Dibenzo[*ac*]fluorene | ± | Dibenzo[*def,p*]chrysene | + |
| Fluoranthene | — | Dibenzo[*def,mno*]chrysene = anthanthrene | — |
| Benzo[*b*]fluoranthene | ++ | Perylene | — |
| Benzo[*j*]fluoranthene | ++ | Benzo[*ghi*]perylene | — |
| Benzo[*k*]fluoranthene | — | Coronene | — |
| Benzo[*mno*]fluoranthene | — | | |

[a] —, not carcinogenic; ± uncertain or weakly carcinogenic; +, carcinogenic; ++, +++, ++++, strongly carcinogenic.

catalyzed by enzymes from different species, and rate at which the active metabolites are converted to less-active products. As discussed in Section 5.4, many aquatic organisms possess the requisite enzyme systems for metabolic activation of PAH. However, it is quite uncertain in most cases whether these enzymes produce the same metabolites as those produced by mammalian enzymes. The limited data available (see Table 56) indicate that some aquatic animals do produce the active metabolites necessary for carcinogenesis (e.g., the 5,6-dihydrodiol of benz[*a*]anthracene and the 7,8-dihydrodiol of B*a*P).

Stegeman (1977) provided evidence that the MFO–cytochrome P-450 system of fish is able to produce carcinogenic or mutagenic metabolites, at least *in vitro*. He added bacteria *Salmonella typhimurium* strain TA-98 to a

TABLE 87

BIOACTIVATION OF BENZO[*a*]PYRENE TO TOXIC AND MUTAGENIC DERIVATIVES BY SCUP (*Stenostomus versicolor*) LIVER MIXED-FUNCTION OXYGENASES. SCUP LIVER MICROSOMES WERE INCUBATED WITH *Salmonella typhimurium* STRAIN TA-98 AND APPROPRIATE COFACTORS (FROM STEGEMAN, 1977, WITH PERMISSION OF *Oceanus* MAGAZINE)

| *Incubation conditions* | *Bacterial survivors* ($\times 10^7$) | *Mutant fraction* ($\times 10^{-8}$)[a] |
|---|---|---|
| Complete (12·5 μg B*a*P/ml) | 53 | 47·2 |
| Minus B*a*P | 1 088 | 0·55 |
| Minus NADPH | 945 | 0·21 |

[a] Mutant fraction refers to the number of $His^+$ revertants per number of survivors.

reaction mixture containing B*a*P and MFO enzymes from scup (*Stenostomus versicolor*) liver. After a suitable incubation period, there appeared a reduced number of bacterial survivors and an increased number of bacterial mutants (Table 87), indicating that toxic and mutagenic metabolites had been produced by the fish liver enzymes. Similar results were obtained by Payne *et al.* (1978) with hepatic microsomes of the rainbow trout *Salmo gairdneri* and a PAH-enriched fraction of used crankcase oil. Thus, both marine and freshwater fish are able to produce mutagenic, and by inference carcinogenic, metabolites from PAH.

### 7.4.1 Laboratory Studies

Many, but not all, carcinogenic chemicals are also mutagenic or teratogenic. That is, they can cause abnormal growth in organisms exposed to the chemical or birth defects in the offspring of exposed parents. Thus, carcinogenicity of a chemical is often inferred from its mutagenicity or teratogenicity. Some of the sublethal effects of PAH discussed above are indicative of mutagenic or teratogenic effects of PAH (e.g., growth stimulation in algae and abnormal development of sea urchin embryos).

Unfortunately, with the exception of amphibians, there has been relatively little laboratory research on the carcinogenicity, mutagenicity and teratogenicity of PAH in aquatic organisms. Exposure of colonial calcareous sponges *Leucosolenia complicata* and *L. variabilis* to 5 g/liter

B*a*P resulted in choanocyte damage and abnormal growth of the oscular tube (Korotkova and Tokin, 1968). The solitary sponge *Sycon raphanus* was unaffected.

Foster (1969) exposed adult planarians (*Dugesia dorotocephala*) to either 3-methylcholanthrene or B*a*P during regeneration of excised body parts. Nine per cent of the planaria exposed to 3-methylcholanthrene and 7 per cent of those exposed to B*a*P developed lethal papilliform tumors and other malformations. He suggested that the malformations and possibly also the tumors were derived from stimulated regenerative cells—totipotent neoblasts. Offspring of surviving PAH-exposed adults developed abnormal growths similar to those of the exposed adults. 3-Methylcholanthrene was more carcinogenic and teratogenic than B*a*P to the offspring (Table 88). It

TABLE 88

LETHAL TUMORS AND DEVELOPMENTAL ANOMALIES IN THE OFFSPRING OF PLANARIA (*Dugesia dorotocephala*) TREATED WITH 3-METHYLCHOLANTHRENE OR BENZO[*a*]-PYRENE (FROM FOSTER, 1969, WITH PERMISSION OF THE NATIONAL CANCER INSTITUTE)

| *Treatment of parents* | *Number of offspring* | *Tumors in offspring* | *Malformations in offspring* |
|---|---|---|---|
| 3-Methylcholanthrene (regenerated) | 40 | 12 (papilliform tumors) | 6 eyespots poorly developed; 4 enlarged heads and eyes; 3 fused eyespots |
| Benzo[*a*]pyrene (regenerated) | 75 | 2 (nodular tumors) | 3 enlarged heads and eyes; 2 small heads and eyes; 2 fused eyespots |
| Acetone control (regenerated) | 65 | 1 (nodular tumor) | variation in pigmentation; 2 enlarged heads and eyes |
| Untreated controls (not regenerated) | 77 | none | 1 small head and eyes; 1 fused eyespot |

is tempting to speculate, based on these results, that either planaria possess the MFO–cytochrome P-450 system or that PAH which accumulated in their transparent tissues were photooxidized to mutagenic products.

Recently Khudolei and Sirenko (1977) induced the formation of neoplasms in the digestive gland and hematopoietic system of freshwater mussels (*Unio pictorum*) by subjecting the bivalves to 200–400 ppm diethyl- and dimethylnitrosamines in water. However, there are no published reports of cancer induction in bivalve molluscs by exposure to, or injection of, PAH. Such an investigation would be very informative for several reasons. The majority of bivalve molluscs studied to date lack, or have very low, MFO activity and therefore should not be able to activate PAH to carcinogenic metabolites. Yet there are many reports of natural populations of bivalves with high incidences of neoplastic disease (Section 7.4.2). Some of these populations are from oil-polluted habitats. If bivalves are not sensitive to PAH-induced carcinogenesis, another causative agent for these outbreaks of neoplasia must be sought. If PAH can cause cancer in bivalves, a closer look at how this is accomplished may lead to new insights into the mechanisms of PAH-induced carcinogenesis.

Bourne and Jones (1973) exposed cell cultures derived from fibroblastic gonad cells of the rainbow trout *Salmo gairdneri* (RTG-2) and epithelioid cells of the fathead minnow *Pimephales promelas* (FHM) to 7,12-dimethylbenz[*a*]anthracene. The PAH contaminant inhibited mitosis and induced an increase in the number of multinucleate cells. RTG-2 cells were more sensitive than FHM cells to the inhibition of mitosis, although FHM cells showed a much greater incidence of multinucleate cells than did RTG-2 cells. In FHM cultures, many cells were extremely abnormal with a reduction in cytoplasm, clumped chromation, and basophilic cytoplasm. These results suggest that the fish cells were converting 7,12-dimethylbenz[*a*]anthracene to reactive metabolites which bonded chemically to nuclear proteins and DNA.

When zebra fish (*Brachydanio rerio*) embryos were exposed to 0·56 ppm 7,12-dimethylbenz[*a*]anthracene they developed tail necrosis, tumor-like growths, an enlarged pericardium, and shortened body (Jones and Huffman, 1957). Epithelial cells and nuclei of PAH-exposed embryos were enlarged and granular. The growth of some parts of the embryos was accelerated. Ermer (1970) painted 0·5 mg of 3-methylcholanthrene on the skin of three freshwater fish, three-spined stickleback (*Gasterosteus aculetus*), bitterling (*Rhodeus amarus*), and carp (*Cyprinus carpio*), twice a week for 3–6 months. This treatment produced epitheliomas (skin cancer) in *G. aculetus* and *R. amarus* but not in *C. carpio*. Similar results were produced when B*a*P was used. Epitheliomas were induced in the first two but not the third fish. He also injected *G. aculetus* with 0·1 ml of 1 per cent B*a*P in glycerol 10 times (10 mg B*a*P) and maintained the fish for 4 months. B*a*P produced injection-site necrosis but no neoplasms.

Neukomn (1974) induced cancerous lesions in newts (*Triturus cristatus*) by subcutaneous injection of several PAH into regenerative areas of the epidermis. Epithelial hyperplasia was followed by infiltration of underlying tissues. Neoplastic infiltration occurred within 4–5 days after injection, with activity continuing for 12–20 days and culminating in a diffuse tumor.

The PAH tested varied in their carcinogenicity to the newt in the following order of decreasing potency: dibenz[*ah*]anthracene, 3-methylcholanthrene, B*a*P, and 9,10-dimethylbenz[*a*]thracene. Chrysene and benz[*a*]anthracene were weakly active and pyrene was inactive.

Seilern-Aspang and Kratochwil (1962, 1963) induced epitheliomas in newts (*Triturus cristatus*) by subcutaneous injection of B*a*P. The cancer often showed regression and differentation into non-malignant tissue when a regeneration process was initiated in the animals by amputating a limb or the tail. Pizzarello and Wolsky (1966) showed that if regeneration was initiated before injection of 3-methylcholanthrene or dibenz[*ah*]anthracene into *T. viridescans* the cancer did not form. However, the presence of carcinogenic PAH in the animals retarded the process of regeneration. These results suggest that regeneration and carcinogenesis are antagonistic processes in the newts. Regeneration inhibits malignant growth, but the presence of carcinogenic tendencies in the animal during regeneration retards reconstruction of tissues lost by amputation.

Crystals of 3-methylcholanthrene, 7,12-dimethylbenz[*a*]anthracene, or B*a*P were implanted subcutaneously in the tail of the toad *Bufo arenarum* tadpoles (De Lustig and Matos, 1971). After a few days, papillomas and lymphomas were detected in 90 per cent of the treated animals. Tumor-like cells invaded normal structures in the tail, displacing the newly formed tissues. Several of the PAH-treated tadpoles developed an accessory tail or notochord. Fluorene and paraffin were completely inactive in inducing cancer or accessory body structures. Subsequently, Matos and De Lustig (1973) showed that if the tail of the tadpole was amputated at the center of the treated area 9 days after implantation of PAH crystals, in order to promote the formation of a regenerative blastema, the PAH-induced cancerous nodules were encapsulated and eventually destroyed. Incidence of PAH-induced teratogenic effects, including supernumerary fins and accessory notochords, was increased by a regenerative field produced by tail amputation. 7,12-Dimethylbenz[*a*]anthracene was substantially more teratogenic than either 3-methylcholanthrene or B*a*P. In a similar investigation, Ruben and Balls (1964) induced lymphosarcomas in the African clawed toad *Xenopus laevis* by implantation of 3-methylcholanthrene into the forelimbs or abdomen. The presence of a regenerating limb system in the carcinogenic environment did not diminish carcinogenic activity. Accessory limb structures were obtained near the sites of crystal implantation in some cases.

These studies show that carcinogenic PAH can produce cancer-like growths and cause teratogenesis and mutagenesis in some aquatic invertebrates and vertebrates. The number of species examined to date is very low, however, and there are no reports of induction of cancer by exposure of aquatic animals to environmentally realistic levels of carcinogenic PAH in the water, food, or sediments.

### 7.4.2 Field Studies

Published literature on the incidence of cancer or cancer-like growths in tissues of natural populations of marine and freshwater invertebrates and fish is growing rapidly. Much of the recent literature is reviewed in several papers from two recent symposium volumes (Dawe *et al.*, 1976; Kraybill *et al.*, 1977). In many cases, organisms from polluted environments have a higher incidence of tumors and hyperplastic diseases than those from uncontaminated environments. The causative agent (or agents) is unknown in the vast majority of cases. A great many different inorganic and organic chemicals can induce cancer in sensitive species and polluted aquatic habitats nearly always contain a wide variety of potential carcinogens (Bergel, 1974). Fish are known to be susceptible to tumor induction by chemical agents, ionizing radiation, physical factors, and viruses (Stich and Acton, 1976). In addition, fish neoplasms can be genetically induced. To date, carcinogenic PAH have not been unequivocally identified as the causative agent for an increased incidence of cancer in any natural population of aquatic organisms.

There are several reports of increased incidence of cancerous growths in aquatic animals in the vicinity of an oil spill (Hodgins *et al.*, 1977). Yevich and Barszcz (1977) reported a high incidence of gonodal neoplasms in soft-shell clams (*Mya arenaria*) from Long Cove, Searsport, Maine, and of hematopoietic neoplasms in the same species from Harpswell, Neck, Maine. Although both sites had been contaminated with refined oil, the authors concluded that they could not say that oil was in any way a causative factor in inducing the neoplasms.

Powell *et al.* (1970) induced hyperplasia of ovicells in the estuarine bryozoan *Schizoporella unicornis* by placing normal colonies in close proximity to coal-tar derivatives in an estuary. The authors attributed uncontrolled growth of the reproductive structures to stimulation by several PAH known to be present in coal tar. However, Straughan and Lawrence (1975) found no ovicell hyperplasia in bryozoans from surface, subsurface, and benthic kelp fronds in the vicinity of Coal Point, California, an area of chronic submarine oil seepage.

Brown *et al.* (1973, 1977) reported a significantly higher frequency of tumors in 2121 freshwater fish from the polluted Fox River, Illinois, USA than in 4639 fish from relatively unpolluted Lake of the Woods, Ontario, Canada. In fish from the Fox River, incidence of neoplasia ranged from 1·17 to 12·2 per cent in different species with a mean of 4·38 per cent. Fish from the unpolluted habitat had a mean incidence of neoplasia of 1·03 per cent with a range of 0 to 2·56 per cent in different species. The brown bullhead *Ictalurus nebulosus* was the most seriously infected fish in the Fox River sample. Several organic and inorganic pollutants were detected at higher concentrations in the Fox River than in the Lake of the Woods. Concentrations of benzene, toluene, naphthalene, and benzanthracene in the Fox River in 1976 were 0·2, 0·1, 0·2, and 0·05 ppm, respectively: These chemicals were not detected in Canadian water. Several chlorinated hydrocarbons and heavy metals were also present at higher concentrations in the Fox River than in the Lake of the Woods. Some of the tumors were apparently caused by viruses but the etiology of other tumors is still unknown.

A stronger case can be made for implicating PAH as the causative agent of a high incidence of neoplasia in tiger salamanders (*Ambystoma tigrinum*) from a 13-hectare sewage effluent lagoon at Reese Air Force Base, Texas (Rose, 1976, 1977). Here, the incidence of neoplastic and non-neoplastic lesions in neotenic salamanders reached a maximum of 53 per cent in 1975. Water and sediment of the lagoon were analyzed for organochlorine and organophosphate pesticides, nitrosamines, several heavy metals, and PAH. All but the PAH were below the limits of detection or were present at normal expected concentrations. Lagoon water contained 0·085 μg/liter B*a*P, and sludge on the floor of the lagoon contained high concentrations of several PAH, especially perylene (Table 89). The sewage effluent lagoon is quite eutrophic and there are frequent large blooms of phytoplankton, *Daphnia*, and copepods. Dead planktonic organisms accumulate in large mats on the anoxic bottom sludge. These are ideal conditions for indirect biosynthesis of PAH by reduction of extended quinones of biological origin (Section 2.1.1) or direct biosynthesis by anaerobic bacteria in the sludge (Section 2.1.2). The high concentration of perylene supports the former hypothesis. Other sources of this unusual PAH assemblage are also possible.

Salamanders from the lagoon had elevated hepatic AHH activity. AHH activity in laboratory-reared salamanders could be induced to a level similar to that in lagoon animals by exposing them to 3-methylcholanthrene in water. Other inducers of AHH activity—organochlorine and organophosphate insecticides—were absent from lagoon water. All these observations provide strong circumstantial evidence that PAH are critical to the high incidence of neoplasia in salamanders from the lagoon.

TABLE 89

PAH ISOLATED FROM SLUDGE IN A SEWAGE EFFLUENT LAGOON CONTAINING A POPULATION OF TIGER SALAMANDER NEOTENES (*Ambystoma tigrinum*) WITH A HIGH INCIDENCE OF NEOPLASIA (FROM ROSE, 1977, WITH PERMISSION OF THE NEW YORK ACADEMY OF SCIENCES)

| *Compound* | *Concentration* (*μg/kg, ppb*) |
|---|---|
| Perylene | 300·0 |
| Pyrene | 5·8 |
| Fluoranthene | 5·7 |
| Alkyl pyrenes | 4·9 |
| Benz[*a*]anthracene | 1·4 |
| Chrysene | 1·3 |
| Triphenylene | 0·5 |
| Benzo[*a*]pyrene | 0·5 |
| Benzo[*e*]pyrene | 0·2 |
| Anthanthrene | 0·2 |

A cancer-like hyperplasia was reported in the marine alga *Porphyra tenera* cultivated in coastal waters near industrial wastewater outfalls from the city of Ohmuta, Japan (Ishio *et al.*, 1971; 1972a,b). The cancer could not be induced by exposing algae to diluted wastewater. However, exposure of the leaves of the alga to bottom sediments from the outfall region for 80 to 320 min resulted in the production of cancerous growths within 36 days. The sediment was separated into several fractions and the greatest carcinogenic potency was found in the neutral fraction. Two carcinogenic compounds were isolated from this fraction, one was confirmed to be benzanthrone and the other which had the empirical formula $C_{25}H_{14}$ was tentatively identified as 12-hydrodibenzo[*cd,ghi*]perylene.

Obviously, a great deal more research is required on identification of the causative agents of neoplasia outbreaks in aquatic organisms. This type of research requires close collaboration of histopathologists, cancer epidemiologists, and environmental chemists.

### 7.4.3 Aquatic PAH and Human Cancer Risk

It has been estimated that from 50 to 90 per cent of all human cancers are causatively related to environmental factors, mainly chemical carcinogens (Maugh, 1974; Wynder, 1976). The contribution of carcinogenic PAH in air, water, and food to human cancer is completely unknown. Near ubiquity of these compounds in the human environment indicates that PAH could be important causative agents of human cancer.

There are many potential sources of PAH for humans, for example drinking water, smoked, roasted or charcoal-broiled foods, vegetables, vegetable fats and oils, and air pollutants (including cigarette smoke). It has been estimated that the amount of carcinogenic PAH consumed by man in drinking water is typically only about 0·1 per cent of the amount accumulated from food (Andelman and Snodgrass, 1972). Andelman and Suess (1970) made calculations based on four samples of drinking water and showed annual human consumption of PAH to be about 6, 9, 22, and 70 μg for the populations served by these water supplies. The 1970 World Health Organization European Standards for Drinking Water (WHO, 1970) recommended that the concentration of total PAH in drinking water not exceed 0·2 μg/liter. Based on a daily human consumption of 2·5 liters, human consumption of drinking water with this concentration of PAH would result in the ingestion of 182·5 μg PAH per year. By comparison, smoke from 100 cigarettes (five packs) contains about 264 μg total PAH and 2·4 μg B*a*P (Severson *et al.*, 1976).

Blumer (1972) suggested that petroleum pollution of the seas could pose a cancer risk to man through contamination of fisheries resources with carcinogenic PAH and of recreational beaches with tar. Sullivan (1974) reviewed reports on B*a*P content of petroleum, rate of petroleum spillage in the sea, and contamination of marine organisms with PAH. He concluded that the amount of B*a*P in marine foods is higher (presumably owing to oil pollution) than in non-marine foods and may pose a cancer hazard to human consumers of fishery products. More recent studies on B*a*P and total PAH contamination of marine organisms indicate that in most cases these organisms contain low or undetectable levels of B*a*P and other PAH (See Section 4.3). B*a*P concentrations in marine animals were usually in the 0–20 μg/kg range except in animals collected from severely polluted locations or from the immediate vicinity of creosoted pilings. B*a*P concentrations as high as 1–5 mg/kg have been reported in some marine animals and repeated consumption of such heavily contaminated animals would pose a cancer risk to humans. Heavily contaminated animals are usually unpalatable because of their oily smell and taste. The only report of B*a*P accumulation by a marine animal following an oil spill is that of Bories *et al.* (1976). Eleven days after a spill, mussels (*Mytilus edulis*) contained approximately 55 μg B*a*P/kg. In comparison, a charcoal-broiled steak may contain up to 50 μg B*a*P/kg (Lijinsky, 1967), vegetable oils may contain up to 36 μg B*a*P/kg (Swallow, 1976), and smoked meats may contain up to 15 μg B*a*P/kg and 141 μg total PAH/kg (Panalaks, 1976). Typically, food plants contain 0·08 to 30 μg B*a*P/kg (Carrier, 1977). Leafy vegetables and grains generally contain higher B*a*P concentrations than root vegetables. Thus, aquatic organisms do not ordinarily contain significantly higher concentrations of B*a*P and other PAH than other human foods. However, consumption of aquatic organisms from regions severely contaminated with petroleum or industrial pollution should certainly be avoided.

*CHAPTER 8*

# *Summary—Cycle of PAH in the Aquatic Environment*

Polycyclic aromatic hydrocarbons are formed by a variety of processes, which include indirect and direct biosynthesis, diagenesis of organic material to produce fossil fuels, and incomplete combustion of organic matter. Indirect biosynthesis of PAH occurs when extended quinones and related polycyclic materials (mostly plant and animal pigments) are exposed to reducing conditions in anoxic sediments. The resulting PAH assemblage is relatively simple compositionally and tends to accumulate in anoxic sediments where it was formed. Direct biosynthesis of PAH by aerobic and anaerobic bacteria, fungi, and plants is more controversial. Evidence both for and against this process has been published. In most cases where PAH biosynthesis has been reported, accumulation of PAH in the organisms could also be attributed to uptake of PAH from exogenous sources. PAH biosynthesis may occur to a limited extent under special environmental conditions when certain bacterial growth substrates are present. The resulting PAH assemblages would have a relatively simple composition.

Fossil fuels including peat, coal, and petroleum are relatively rich in PAH. PAH assemblages in crude and refined petroleum are extremely complex. There are many homologous series of PAH and in each the concentration of alkyl homologs is greater than that of the unalkylated parent compound.

Pyrolysis of organic matter at temperatures between 400 and 2000 °C results in generation of a wide variety of PAH. Reducing conditions in the pyrolytic environment favor PAH production. PAH assemblages manufactured by pyrolysis are extremely complex and are characterized by a predominance of the unalkylated parent PAH rather than alkylated homologs in each homologous series.

There are many routes by which PAH from diverse sources reach the aquatic environment. PAH formed by indirect and direct biosynthesis are apparently quite static and tend to remain in ecosystems where they are formed. PAH biosynthesized by soil microorganisms may be leached out of the soil and may enter fresh and marine waters by surface runoff from land. PAH indigenous to fossil fuels may enter the aquatic environment in coal dust, in leachate of peat bogs, or in petroleum spillage into water bodies. The last-mentioned is the most important quantitatively. Pyrosynthesized PAH are released into the atmosphere adsorbed to soot. Airborne particulate PAH are subject to photooxidation and chemical oxidation but residence time of particulate PAH in the atmosphere is long enough to allow dispersal over long distances. Airborne PAH enter the aquatic environment through dry fallout, rainfall, and vapor-phase deposition onto water surfaces. PAH deposited on land may reach the aquatic environment in surface runoff. Industrial and domestic wastewaters are often rich in PAH. Sewage treatment processes remove some PAH but the remainder are released to the aquatic environment.

It is estimated that nearly 700 metric tons of B*a*P and 230 000 metric tons of total PAH enter the aquatic environment each year from various sources. The main routes of entry of B*a*P are surface runoff from land and fallout or rainout of airborne B*a*P. Main routes of entry of total PAH to the aquatic environment are petroleum spillage and fallout or rainfall. The majority of PAH entering the aquatic environment remain close to sites of deposition, so that lakes, rivers, estuaries, and coastal marine environments near centers of human population are primary repositories of aquatic PAH.

Because of low aqueous solubilities, PAH entering the aquatic environment rapidly become adsorbed to organic and inorganic particulate materials or are accumulated in the tissues of aquatic organisms. PAH in water are deposited rapidly in bottom sediments where they may accumulate to high concentrations. Natural waters generally contain very low concentrations of PAH. Aquatic organisms often contain tissue PAH concentrations orders of magnitude higher than aqueous concentrations, but equal to or less than those in bottom sediments.

Several processes interact to reduce concentrations of PAH in water and sediments. These include evaporation, photooxidation and metabolic degradation by aquatic bacteria, fungi, and animals. Low molecular weight PAH (naphthalenes–phenanthrenes) are removed from the water column primarily by evaporation, microbial oxidation, and sedimentation. Higher molecular weight PAH, including B*a*P, are removed primarily by photooxidation and sedimentation. It is estimated that as much as 50 per cent B*a*P and other high molecular weight PAH entering the water column are eventually deposited in bottom sediments. The rich bacterial and fungal flora at the water–sediment interface readily degrades PAH. If PAH are buried in anoxic sediments they are extremely resistant to degradation and persist indefinitely. Benthic meio- and macro-fauna may participate directly in degradation of sedimentary PAH. The major role of these organisms in sedimentary PAH degradation stems from the fact that they mix bottom sediments and return buried PAH to the superficies where such compounds can be degraded by surface microflora. Although bottom sediments are major sinks of PAH, bottom-water currents and storms as well as human and benthic animal activities can resuspend PAH-contaminated sediments allowing them to be transported to new locations or accumulated in the aquatic food chain. Desorption and solubilization of PAH from surface sediments may occur to a limited extent, especially if the sediments are heavily contaminated with PAH (e.g., after an oil spill). Available evidence indicates that recycling of PAH from sediments back into the water column and biota is very slow.

Aquatic organisms are able to accumulate PAH from water, food, and sediment. In most cases, accumulation from water is more efficient than from food or sediment. Sediment-adsorbed PAH have only a very limited bioavailability to aquatic organisms. Benthic infaunal animals rarely contain higher concentrations of PAH than the sediments in which they live.

Most aquatic animals have the ability to degrade PAH to more polar metabolites and excrete them rapidly. Even species lacking PAH-metabolizing abilities are able to release accumulated PAH rapidly when they are returned to a PAH-free environment. Thus, food-chain biomagnification of PAH occurs only to a very limited extent, if at all.

PAH are acutely toxic to aquatic animals at concentrations of about 0·2–10 ppm. Deleterious sublethal responses in aquatic organisms are sometimes observed at concentrations in the 5–100 ppb PAH range. Many of these PAH-induced responses involve interference of PAH with key membrane-mediated physiological and biochemical processes.

Metabolic intermediates of some PAH are highly carcinogenic, mutagenic, and/or teratogenic. PAH-induced cancer-like growths or developmental anomalies have been described in several aquatic animals and plants. Attempts to identify PAH as causative agents in high incidences of cancer in natural populations of aquatic organisms have met with only limited success. The importance of PAH to carcinogenesis in aquatic organisms or in man is unknown. PAH-contaminated water and fishery products represent minor sources of PAH toxicity for man.

# *Bibliography*


Acheson, M. A., R. M. Harrison, R. Perry, and R. A. Wellings (1976) Factors affecting the extraction and analysis of polynuclear aromatic hydrocarbons in water. *Water Res.* **10:** 207–212.

Ahearn, D. G., S. P. Meyers, and P. G. Standard (1971) The role of yeasts in the decomposition of oils in marine environments. *Dev. Ind. Microbiol.* **12:** 126–134.

Ahokas, J. T., N. T. Karki, A. Oikari, and A. Soivio (1976a) Mixed function monooxygenase of fish as an indicator of pollution of aquatic environment by industrial effluent. *Bull. Environ. Contam. Toxicol.* **16:** 270–274.

Ahokas, J. T., O. Pelkonen, and N. T. Karki (1975) Metabolism of polycyclic hydrocarbons by a highly active aryl hydroxylase system in the liver of a trout species. *Biochem. Biophys. Res. Commun.* **63:** 635–641.

Ahokas, J. T., O. Pelkonen, and N. T. Karki (1976b) Cytochrome P-450 and drug-induced spectral interactions in the hepatic microsomes of trout *Salmo trutta lacustris. Acta Pharmacol. Toxicol.* **38:** 440–449.

Ahokas, J. T., O. Pelkonen, and N. T. Karki (1977) Xenobiotic metabolism in fish. *Xenobiotica* **7:** 104–105.

Aizenshtat, Z. (1973) Perylene and its geochemical significance. *Geochim. Cosmochim. Acta* **37:** 559–567.

American Institute of Biological Sciences (1976) *Sources, Effects and Sinks of Hydrocarbons in the Aquatic Environment.* Washington, D.C.: AIBS. 578 pp.

Andelman, J. B., and J. E. Snodgrass (1972) Incidence and significance of polynuclear aromatic hydrocarbons in the water environment. *CRC Crit. Rev. Environ. Contr.* **4**(1): 69–83.

Andelman, J. B., and M. J. Suess (1970) Polynuclear aromatic hydrocarbons in the water environment. *Bull. Wld. Hlth. Org.* **43:** 479–508.

Anderson, J. W., R. C. Clark, and J. J. Stegeman (1974a) Petroleum hydrocarbons. pp. 36–75 In: *Marine Bioassays Workshop Proceedings.* Sponsored by API, EPA and Marine Technol. Soc. Washington, D.C.: Marine Technology Society.

Anderson, J. W., L. J. Moore, J. W. Blaylock, D. L. Woodruff, and S. L. Kiesser (1977) Bioavailability of sediment-sorbed naphthalenes to the sipunclid worm, *Phascolosoma agassizii.* pp. 276–285 In: D. A. Wolfe (ed.) *Fate and Effects of Petroleum Hydrocarbons in Marine Ecosystems and Organisms.* New York: Pergamon Press.

Anderson, J. W., J. M. Neff, B. A. Cox, H. E. Tatem, and G. M. Hightower (1974b) Characteristics of dispersions and water-soluble extracts of crude and refined oils and their toxicity to estuarine crustaceans and fish. *Mar. Biol.* **27:** 75–88.

Anderson, J. W., J. M. Neff, B. A. Cox, H. E. Tatem, and G. M. Hightower (1974c) Effects of oil on estuarine animals: toxicity, uptake and depuration, respiration. pp. 285–310 In: F. J. Vernberg and W. B. Vernberg (eds.) *Pollution and Physiology of Marine Organisms.* New York: Academic Press.

Anderson, R. S. (1978) Benzo[*a*]pyrene metabolism in the American oyster *Crassostrea virginica.* Environmental Protection Agency. Ecological Research Series. EPA-600/3-78-009. 18 pp.

Armstrong, H. W., K. Fucik, J. W. Anderson, and J. M. Neff (1977) Effects of oil-field brine effluent on benthic organisms in Trinity Bay, Texas. Washington, D.C.: American Petroleum Institute. API Publication No. 4291. 82 pp.

Badger, G. M. (1962) Mode of formation of carcinogens in the human environment. *Natl. Cancer Inst. Monogr.* **9:** 1–16.

Badger, G. M., J. K. Donnelly, and T. M. Spotswood (1966) Formation of aromatic hydrocarbons at high temperatures. XXVII. The pyrolysis of isoprene. *Austr. J. Chem.* **19:** 1023–1043.

Badger, G. M., R. W. L. Kimber, and J. Novotny (1964) The formation of aromatic hydrocarbons at high temperatures. XXI. Pyrolysis of n-butylbenzene over a range of temperatures from 300 to 900 °C at 50 °C intervals. *Austr. J. Chem.* **17:** 778–786.

Badger, G. M., R. W. L. Kimber, and T. M. Spotswood (1960) Mode of formation of BP in human environment. *Nature (Lond.)* **187:** 663–665.

Baker, J. M. (ed.) (1976) *Marine Ecology and Oil Pollution.* New York: John Wiley and Sons. 566 pp.

Barnsley, E. A. (1975) The bacterial degradation of fluoranthene and benzo[*a*]pyrene. *Can. J. Microbiol.* **21:** 1004–1008.

Barbier, M., D. Joly, A. Saliot, and D. Tourres (1973) Hydrocarbons from sea water. *Deep-Sea Res.* **20:** 305–314.

Basu, D. K., and J. Saxena (1978) Polynuclear aromatic hydrocarbons in selected U.S. drinking waters and their raw water sources. *Environ. Sci. Technol.* **12:** 795–798.

Bedford, W. B., and J. W. Anderson (1972) The physiological response of the estuarine clam *Rangia cuneata* (Grey) to salinity. I. Osmoregulation. *Physiol. Zool.* **45:** 255–260.

Bend, J. R., M. O. James, and P. M. Dansette (1977) *In vitro* metabolism of xenobiotics in some marine animals. pp. 505–521 In: H. F. Kraybill, C. J. Dawe, J. C. Harshbarger, and R. G. Tardiff (eds.) *Aquatic Pollutants and Biologic Effects with Emphasis on Neoplasia.* New York: Ann. N. Y. Acad. Sci. Vol. 298.

Bentley, P., F. Oesch, and H. Glatt (1977) Dual role of epoxide hydratase in both activation and inactivation of benzo[*a*]pyrene. *Arch. Toxicol.* **39:** 65–75.

Benville, P. E., Jr., and S. Korn (1977) The acute toxicity of six monocyclic aromatic crude oil components to striped bass (*Marone saxatilis*) and bay shrimp (*Crago franciscorum*). *Calif. Fish & Game* **63:** 204–209.

Berdugo, V., R. P. Harris, and S. C. O'Hara (1977) The effect of petroleum hydrocarbons on reproduction of an estuarine planktonic copepod in laboratory cultures. *Mar. Pollut. Bull.* **8:** 138–143.

Bergel, F. (1974) Carcinogenic hazards in natural and manmade environments. *Proc. Roy. Soc. (Lond.) Ser. B.* **185:** 165–181.

Bidleman, T. F., C. P. Rice, and E. C. Olney (1976) High molecular weight hydrocarbons in the air and sea: rates and mechanisms of air/sea transfer. pp. 323–352 In: H. L. Windom and R. A. Duce (eds.) *Marine Pollutant Transfer*. Lexington, Massachusetts: D. C. Heath and Co.

Bieri, R. H., and V. C. Stamoudis (1977) The fate of petroleum hydrocarbons from a No. 2 fuel oil spill in a seminatural estuarine environment. pp. 332–344 In: D. A. Wolfe (ed.) *Fate and Effects of Petroleum Hydrocarbons in Marine Ecosystems and Organisms*. New York: Pergamon Press.

Bieri, R. H., V. C. Stamoudis, and M. K. Cueman (1977) Chemical investigations of two experimental oil spills in an estuarine ecosystem. pp. 511–516. *Proc.* 1977 *Oil Spill Conference (Prevention, Behavior, Control, Cleanup)*. Washington, D.C.: American Petroleum Institute.

Blum, S. C., and R. E. Swarbrick (1975) Hydroponic growth of crops in presence of $^{14}$C-benzo[*a*]pyrene. In: *Symposium on Chemistry, Occurrence, and Measurement of Polynuclear Aromatic Hydrocarbons.* Preprints. Div. Petrol. Chem., Inc. A.C.S. **20:** 869–875.

Blumer, M. (1972) Oil contamination and the living resources of the sea. pp. 476–481 In: M. Ruivo (ed.) *Marine Pollution and Sea Life*. London: Fishing News (Books).

Blumer, M. (1975) Curtisite, idrialite and pendletonite, polycyclic aromatic hydrocarbon minerals: their composition and origin. *Chem. Geol.* **16:** 245–256.

Blumer, M. (1976) Polycyclic aromatic compounds in nature. *Sci. Amer.* **234**(3): 34–45.

Blumer, M., W. Blumer, and T. Reich (1977) Polycyclic aromatic hydrocarbons in soils of a mountain valley: correlation with highway traffic and cancer incidence. *Environ. Sci. Technol.* **11:** 1082–1084.

Blumer, M., G. Souza, and J. Sass (1970) Hydrocarbon pollution of edible shellfish by an oil spill. *Mar. Biol.* **5:** 195–202.

Blumer, M., and W. W. Youngblood (1975) Polycyclic aromatic hydrocarbons in soils and recent sediments. *Science* **188:** 53–55.

Boehm, P. D., and J. G. Quinn (1973) Solubilization of hydrocarbons by the dissolved organic matter in sea water. *Geochim. Cosmochim. Acta* **37:** 2459–2477.

Boehm, P. D., and J. G. Quinn (1976) The effect of dissolved organic matter in sea water on the uptake of mixed individual hydrocarbons and No. 2 fuel oil by a marine filter-feeding bivalve (*Mercenaria mercenaria*). *Estuar. Cstl. Mar. Sci.* **4:** 93–105.

Boehm, P. D., and J. G. Quinn (1977) The persistence of chronically accumulated hydrocarbons in the hard-shell clam *Mercenaria mercenaria. Mar. Biol.* **44:** 227–233.

Bohm-Gossl, T., and R. Kruger (1965) Solubilization of 3,4-benzopyrene by alkylbenzene sulfonic acids. *Kolloid-Z.Z. Polymere* **206:** 65–70 (German).

Bohon, R. L., and W. F. Claussen (1951) The solubility of aromatic hydrocarbons in water. *J. Amer. Chem. Soc.* **73:** 1571–1578.

Boney, A. D. (1974) Aromatic hydrocarbons and the growth of marine algae. *Mar. Pollut. Bull.* **5:** 185–186.

Boney, A. D., and E. D. S. Corner (1962) On the effects of some carcinogenic hydrocarbons on the growth of sporelings of marine red algae. *J. Mar. Biol. Assn. U.K.* **43:** 579–585.

Bories, G., J. Tulliez, J.-C. Peltier, and R. Fleckinger (1976) Change in the concentration of aliphatic and naphthenic hydrocarbons, as well as 3,4-benzopyrene, in mussels from a coastal zone polluted by a spill of fuel oil. *C. R. Acad. Sci. (Paris) Ser. D* **282:** 1641–1644 (French).

Borneff, J., and R. Fischer (1962) Carcinogenic substances in water and soil. *Arch. Hyg. (Berlin)* **146:** 334–345 (German).

Borneff, J., and H. Kunte (1964) Carcinogenic substances in water and soil. Part XVI: Evidence of polycyclic aromatic hydrocarbons in water samples through direct extraction. *Arch. Hyg. (Berlin)* **148:** 585–597 (German).

Borneff, J., and H. Kunte (1965) Carcinogenic substances in water and soil. Part XVII: Concerning the origin and estimation of the polycyclic aromatic hydrocarbons in water. *Arch. Hyg. (Berlin)* **149:** 226–243 (German).

Borneff, J., and H. Kunte (1967) Carcinogenic substances in water and soil. Part XIX: The effect of sewage purification on polycyclic aromatic hydrocarbons. *Arch. Hyg. (Berlin)* **151:** 202–210 (German).

Borneff, J., and H. Kunte (1969) Carcinogenic substances in water and soil. Part XXVI. Routine method for analysis of polycyclic aromatic hydrocarbons in water. *Arch. Hyg. (Berlin)* **153:** 220–229 (German).

Borneff, J., F. Selenka, H. Kunte, and A. Maximos (1968a) Experimental studies on the formation of polycyclic aromatic hydrocarbons in plants. *Environ. Res.* **2:** 22–29.

Borneff, J., F. Selenka, H. Kunte, and A. Maximos (1968b) The synthesis of 3,4-benzopyrene and other polycyclic aromatic hydrocarbons in plants. *Arch. Hyg. (Berlin)* **152:** 279–282 (German).

Boucart, J., and L. Mallet (1965) Marine pollution of the shores of the central region of the Tyrrhenian Sea (Bay of Naples) by benzo-3,4-pyrene-type polycyclic aromatic hydrocarbons. *C. R. Acad. Sci. (Paris) Ser. D* **260:** 3729–3734.

Bourne, E. W., and R. W. Jones (1973) Effects of 7,12-dimethyl-benz[*a*]anthracene (DMBA) in fish cells *in vitro*. *Trans. Amer. Microsc. Soc.* **92:** 140–142.

Boylan, D. B., and B. W. Tripp (1971) Determination of hydrocarbons in seawater extracts of crude oil and crude oil fractions. *Nature (Lond.)* **230:** 44–47.

Braunstein, H. M., E. D. Copenhaver, and H. A. Pfuderer (eds.) (1977) *Environmental, Health, and Control Aspects of Coal Conversion: an Information Overview*. Oak Ridge, Tennessee: Oak Ridge National Laboratory. ORNL/EIS-94-5. 1600 pp.

Bravo, H., S. Salazar, A. V. Botello, and E. F. Mandelli (1978) Polyaromatic hydrocarbons in oysters from coastal lagoons along the eastern coast of the Gulf of Mexico, Mexico. *Bull. Environ. Contam. Toxicol.* **19:** 171–176.

Brenniman, G., R. Hartung, and W. J. Weber, Jr. (1976) A continuous flow bioassay method to evaluate the effects of outboard motor exhausts and selected aromatic toxicants on fish. *Water Res.* **10:** 165–169.

Bresch, H., R. Spielhoff, U. Mohr, and H. Barkemeyer (1972) Use of sea urchin egg for quick screen testing of the biological activities of substances. I. Influence of fractions of a tobacco smoke condensate on early development. *Proc. Soc. Exper. Biol. Med.* **141:** 747–752.

Bridboard, K., J. F. Finklea, J. K. Wagoner, J. B. Moran, and P. Caplan (1976) Human exposure to polynuclear aromatic hydrocarbons. pp. 319–324 In: R. Freudenthal and P. W. Jones (eds.) *Carcinogenesis—A Comprehensive Survey. Vol.* 1. *Polynuclear Aromatic Hydrocarbons. Chemistry, Metabolism, and Carcinogenesis.* New York: Raven Press.

Brisou, J. (1969) Biosynthesis of 3,4-benzopyrene and anaerobiosis. *C. R. Soc. Biol. (Paris)* **163:** 772–774 (French).

Brisou, J. (1972) Lipids, sterols, terpenes and bacterial biosynthesis of 3,4-benzopyrene. pp. 181–183 In: L. Mallet (ed.) *Pollution des Mileaux Vitaux par les Hydrocarbures Polybenzeniques du Type Benzo*-3,4-*pyrene*. Paris: Librairie Maloine (French).

Broddin, G., L. van Vaeck, and K. van Cauwenberghe (1977) On the size distribution of polycyclic aromatic hydrocarbon-containing particles from a coke oven emission source. *Atmos. Environ.* **11:** 1061–1064.

Brown, E. R., J. J. Hazdra, L. Keith, I. Greenspan, J. B. G. Kwapinski, and P. Beamer (1973) Frequency of fish tumours found in a polluted watershed as compared to nonpolluted Canadian waters. *Cancer Res.* **33:** 189–198.

Brown, E. R., T. Sinclair, L. Keith, P. Beamer, J. J. Hazdra, V. Nair, and O. Callaghan (1977) Chemical pollutants in relation to diseases in fish. pp. 535–546 In: H. F. Kraybill, C. J. Dawe, J. C. Harshbarger, and R. G. Tardiff (eds.) *Aquatic Pollutants and Biologic Effects with Emphasis on Neoplasia.* New York: Ann. N.Y. Acad. Sci. Vol. 298.

Brown, R. A., and H. L. Huffman, Jr. (1976) Hydrocarbons in open ocean waters. *Science* **191:** 847–849.

Brunnemann, K. D., and D. Hoffmann (1976) Analysis of polynuclear aromatic hydrocarbons in the respiratory environment. pp. 283–298 In: R. Freudenthal and P. W. Jones (eds.) *Carcinogenesis—A Comprehensive Survey. Vol.* 1. *Polynuclear Aromatic Hydrocarbons. Chemistry, Metabolism, and Carcinogenesis.* New York: Raven Press.

Burns, K. A. (1976a) Hydrocarbon metabolism in the intertidal fiddler crab *Uca pugnax. Mar. Biol.* **36:** 5–11.

Burns, K. A. (1976b) Microsomal mixed function oxidases in an estuarine fish, *Fundulus heteroclitus*, and their induction as a result of environmental contamination. *Comp. Biochem. Physiol.* **53B:** 443–446.

Burwood, R., and G. C. Speers (1974) Photo-oxidation as a factor in the environmental dispersal of crude oil. *Estuar. Cstl. Mar. Sci.* **2:** 117–135.

Busbee, D. L., D. Colvin, J. Muijsson, F. Rose, and E. Cantrell (1975) Induction of aryl hydrocarbon hydroxylase in *Ambystoma tigrinum. Comp. Biochem. Physiol.* **50C:** 33–36.

Busbee, D. L., D. Colvin, J. Muijsson, F. Rose, and E. Cantrell (1978) Metabolism of benzo[*a*]pyrene in animals with high aryl hydrocarbon hydroxylase levels and high rates of spontaneous cancer. *Cancer Let.* **4:** 61–67.

Cahnmann, H. J., and M. Kuratsune (1957) Determination of polycyclic aromatic hydrocarbons in oysters collected in polluted water. *Anal. Chem.* **29:** 1312–1317.

Calder, J. A., and J. H. Lader (1976) Effect of dissolved aromatic hydrocarbons on the growth of marine bacteria in batch culture. *Appl. Environ. Microbiol.* **32:** 95–101.

Caldwell, R. S., E. M. Caldarone, and M. H. Mallon (1977) Effects of a seawater-soluble fraction of Cook Inlet crude oil and its major aromatic components on larval stages of the Dungeness crab, *Cancer magister* Dana. pp. 210–220 In: D. A. Wolfe (ed.) *Fate and Effects of Petroleum Hydrocarbons in Marine Ecosystems and Organisms*. New York: Pergamon Press.

Carrier, R. F. (1977) Plant interactions. In: H. M. Braunstein, E. D. Copenhaver, and H. A. Pfuderer (eds.) *Environmental, Health, and Control Aspects of Coal Conversion: an Information Overview*. Vol. 2. Oak Ridge, Tennessee: Oak Ridge National Laboratory. ORNL/EIS-95. 118 pp.

Carthy, J. D., and D. R. Arthur (eds.) (1968) *The Biological Effects of Oil Pollution on Littoral Communities*. London: Field Studies Council. 198 pp.

Ceas, M. P. (1974) Effects of 3-4 benzopyrene on sea urchin egg development. *Acta Embryol. Exper.* **3:** 267–272.

Cerniglia, C. E., and D. T. Gibson (1978) Metabolism of naphthalene by cell extracts of *Cunninghamella elegans*. *Arch. Biochem. Biophys.* **186:** 121–127.

Clark, H. F., and L. Diamond (1971) Comparative studies on the interaction of benzpyrene with cells derived from poikilothermic and homeothermic vertebrates. II. Effect of temperature on benzpyrene metabolism and cell multiplication. *J. Cell. Physiol.* **77:** 385–392.

Coleman, H. J., J. E. Dooley, D. E. Hirsch, and C. J. Thompson (1973) Compositional studies of a high-boiling 370–535 °C distillate from Prudhoe Bay, Alaska, crude oil. *Anal. Chem.* **45:** 1724–1737.

Colvin, D. (1974) Induction of aryl hydrocarbon hydroxylase in *Ambystoma tigrinum*. M.S. Thesis. North Texas State Univ., Denton, Texas (cited by F. L. Rose, 1977, op. cit.)

Commins, B. T. (1969) Formation of polycyclic aromatic hydrocarbons during pyrolysis and combustion of hydrocarbons. *Atmos. Environ.* **3:** 565–572.

Corner, E. D. S. (1975) The fate of fossil fuel hydrocarbons in marine animals. *Proc. Roy. Soc. (Lond.) Ser. B.* **189:** 391–413.

Corner, E. D. S., R. P. Harris, C. C. Kilvington, and S. C. M. O'Hara (1976a) Petroleum compounds in the marine food web: short-term experiments on the fate of naphthalene in *Calanus*. *J. Mar. Biol. Assn. U.K.* **56:** 121–123.

Corner, E. D. S., R. P. Harris, K. J. Whittle, and P. R. Mackie (1976b) Hydrocarbons in marine zooplankton and fish. pp. 71–106 In: A. P. M. Lockwood (ed.) *Effects of Pollutants on Aquatic Organisms*. Cambridge, England: Cambridge University Press.

Corner, E. D. S., C. C. Kilvington, and S. C. M. O'Hara (1973) Qualitative studies on the metabolism of naphthalene in *Maia squinado* (Herbst). *J. Mar. Biol. Assn. U.K.* **53:** 819–832.

Cox, B. A., J. W. Anderson, and J. C. Parker (1975) An experimental oil spill: the distribution of aromatic hydrocarbons in the water, sediment, and animal tissues within a shrimp pond. pp. 607–612. *Proc.* 1975 *Conference on Prevention and Control of Oil Pollution*. Washington, D.C.: American Petroleum Institute.

Dailey, N. S. (1977) Process effluents: quantities and control technologies. In: H. M. Braunstein, E. D. Copenhaver, and H. A. Pfuderer (eds.) *Environmental, Health, and Control Aspects of Coal Conversion: an Information Overview*. Oak Ridge, Tennessee: Oak Ridge National Laboratory. ORNL/EIS-94. 157 pp.

Davies, I. W., R. M. Harrison, R. Perry, D. Ratnayaka, and R. A. Wellings (1976)

Municipal incinerator as source of polynuclear aromatic hydrocarbons in environment. *Environ. Sci. Technol.* **10:** 451–453.

Davis, W. W., M. E. Krahl, and G. H. A. Clawes (1942) Solubility of carcinogenic and related hydrocarbons in water. *J. Amer. Chem. Soc.* **64:** 108–110.

Dawe, C. J., D. G. Scarpelli, and S. R. Wellings (eds.) (1976) Tumors in Aquatic Animals. In: *Progress in Experimental Tumor Research.* Vol. 20. Basel, Switzerland: S. Karger. 438 pp.

de Angelis, E., and G. G. Giordano (1974) Sea urchin egg development under the action of benzo[*a*]pyrene and 7,12-dimethylbenz[*a*]anthracene. *Cancer Res.* **34:** 1275–1280.

Dean-Raymond, D., and R. Bartha (1975). Biodegradation of some polynuclear aromatic petroleum components by marine bacteria. *Dev. Ind. Microbiol.* **16:** 97–110.

de Lustig, E. S., and E. L. Matos (1971) Teratogenic effects induced in tail of *Bufo arenarum* tadpoles following treatment with carcinogens. *Experientia* **27:** 555–556.

DePierre, J. N., and L. Ernster (1978) Metabolism of polycyclic hydrocarbons and its relationship to cancer. *Biochem. Biophys. Acta* **473:** 149–186.

Depuis, A. (1960) The problem, origin of cancer. *Tech. Eau* **14:** 25–26 (French).

Dewaide, J. H. (1970) Species differences in hepatic drug oxidation in mammals and fishes in relation to thermal acclimation. *Comp. Gen. Pharmacol.* **1:** 375–384.

Dewaide, J. H., and P.Th. Henderson (1970) Seasonal variation of hepatic drug metabolism in the roach, *Leuciscus rutilus* L. *Comp. Biochem. Physiol.* **32:** 489–498.

DiMichele, L., and M. H. Taylor (1978) Histopathological and physiological responses of *Fundulus heteroclitus* L. to naphthalene exposure. *J. Fish. Res. Bd. Canada* **35:** 1060–1066.

DiSalvo, L. H., H. E. Guard, and L. Hunter (1975) Tissue hydrocarbon burden of mussels as potential monitor of environmental hydrocarbon insult. *Environ. Sci. Technol.* **9:** 247–251.

Dixit, D., and J. W. Anderson (1977) Distribution of naphthalenes within exposed *Fundulus similis* and correlations with stress behavior. pp. 633–636. *Proc.* 1977 *Oil Spill Conference* (*Prevention, Behavior, Control, Cleanup*). Washington, D.C.: American Petroleum Institute.

Donahue, W. H., R. T. Wang, M. Welch, and J. A. C. Nicol (1977) Effects of water-soluble components of petroleum oils and aromatic hydrocarbons on barnacle larvae. *Environ. Pollut.* **13:** 187–202.

Dunn, B. P., and H. F. Stich (1975) The use of mussels in estimating benzo[*a*]pyrene contamination of the marine environment. *Proc. Soc. Exper. Biol. Med.* **150:** 49–51.

Dunn, B. P., and H. F. Stich (1976a) Monitoring procedures for chemical carcinogens in coastal waters. *J. Fish. Res. Bd. Canada* **33:** 2040–2046.

Dunn, B. P., and H. F. Stich (1976b) Release of the carcinogen benzo[*a*]pyrene from environmentally contaminated mussels. *Bull. Environ. Contam. Toxicol.* **14:** 398–401.

Dunn, B. P., and D. R. Young (1976) Baseline levels of benzo[*a*]pyrene in southern California mussels. *Mar. Pollut. Bull.* **7:** 231–234.

Duppel, W., J. M. Lebeault, and M. J. Coon (1973) Properties of a yeast cytochrome P-450-containing enzyme system which catalyzes the hydroxylation of fatty acids, alkanes and drugs. *Eur. J. Biochem.* **36:** 583–592.

Eganhouse, R. P., and J. A. Calder (1976) The solubility of medium molecular weight aromatic hydrocarbons and the effects of hydrocarbon co-solutes and salinity. *Geochim. Cosmochim. Acta* **40:** 555–561.

Eglinton, G., B. R. T. Simoneit, and J. A. Zoro (1975) The recognition of organic pollutants in aquatic sediments. *Proc. Roy. Soc. (Lond.) Ser. B.* **189:** 415–442.

Ehrhardt, M. (1972) Petroleum hydrocarbons in oysters from Galveston Bay. *Environ. Pollut.* **3:** 257–271.

Eisenbrand, J. (1971) On the water solubility of 3,4-benzopyrene and other aromatic hydrocarbons and its increase by solubilizers. *Deut. Lebensmitt-Rundsch.* **67:** 435–444 (German).

Ekwall, P., and L. Sjoblöm (1952) Butyric acid and lactic acid in aqueous solutions as solubilizers for carcinogenic hydrocarbons. *Acta Chem. Scand.* **6:** 96–100.

Elsworthy, P. H., A. T. Florence, and C. B. Macfarlane (1968) *Solubilization by Surface-Active Agents.* London: Chapman and Hall, Ltd.

Engelhardt, F. R., J. R. Geraci, and T. G. Smith (1977) Uptake and clearance of petroleum hydrocarbons in the ringed seal, *Phoca hispida. J. Fish. Res. Bd. Canada* **34:** 1143–1147.

Environmental Protection Agency (1970) *Water Quality Criteria Data Book, Organic Chemical Pollution of Fresh Water.* Vol. 1. EPA Water Pollution Control Research Series 18010 DPV. Washington, D.C.: U.S. Govt. Print. Office.

Environmental Protection Agency (1975) *Scientific and Technical Assessment Report on Particulate Polycyclic Organic Matter (PPOM).* EPA-60016-75-001.

Ermer, M. (1970) Studies with carcinogens in short-lived fish species. *Zool. Anz.* **184:** 175–193 (German).

Ershova, K. P. (1968) Carcinogenic pollution of water by effluents from petroleum refineries. *Gig. Sanit.* **33:** 102 (Russian) (cited by Andelman and Snodgrass, 1972. op. cit.).

Farrington, J. W., N. W. Frew, P. M. Gschwend, and B. W. Tripp (1977) Hydrocarbons in cores of northwestern Atlantic coastal and continental margin sediments. *Estuar. Cstl. Mar. Sci.* **5:** 793–808.

Farrington, J. W., and P. A. Meyers (1975) Hydrocarbons in the marine environment. pp. 109–136 In: G. Eglinton (ed.) *Environmental Chemistry* Vol. 1. London: The Chemical Society.

Fazio, T. (1971) Analysis of oyster samples for polycyclic hydrocarbons. pp. 238–243. *Proc. 7th Nat. Shellfish Sanit. Workshop.* Washington, D.C.: FDA, Division of Shellfish Sanitation.

Fedorenko, Z. P. (1964) The effect of biochemical treatment of waste water of a by-product coke plant on the 3,4-benzpyrene content. *Hyg. Sanit.* **29**(3): 19–21 (Russian).

Ferris, J. P., M. J. Fasco, F. L. Stylianopoulou, D. M. Jerina, J. W. Daly, and A. M. Jeffrey (1973) Monooxygenase activity in *Cunninghamella bainierii.* Evidence for a fungal system similar to liver microsomes. *Arch. Biochem. Biophys.* **156:** 97–103.

Ferris, J. P., L. H. MacDonald, M. A. Patrie, and M. A. Martin (1976) Aryl hydrocarbon hydroxylase activity in the fungus *Cunninghamella bainierii*. Evidence of the presence of cytochrome P-450. *Arch. Biochem. Biophys.* **175:** 443–452.

Flower, A. R. (1978) World oil production. *Sci. Amer.* **238:** 42–49.

Foster, J. A. (1969) Malformations and lethal growths in planaria treated with carcinogens. *Natl. Cancer Inst. Monogr.* **31:** 683–691.

Freudenthal, R., and P. W. Jones (eds.) (1976) *Carcinogenesis—A Comprehensive Survey. Vol. 1. Polynuclear Aromatic Hydrocarbons. Chemistry, Metabolism, and Carcinogenesis.* New York: Raven Press. 450 pp.

Fucik, K. W., H. W. Armstrong, and J. M. Neff (1977) The uptake of naphthalenes by the clam, *Rangia cuneata*, in the vicinity of an oil-separator platform in Trinity Bay, Texas. pp. 637–640. *Proc.* 1977 *Oil Spill Conference* (*Prevention, Behavior, Control, Cleanup*). Washington, D.C.: American Petroleum Institute.

Fucik, K. W., and J. M. Neff (1977) Effects of temperature and salinity on naphthalenes uptake in the temperate clam *Rangia cuneata* and the boreal clam *Protothaca staminea*. pp. 305–312 In: D. A. Wolfe (ed.) *Fate and Effects of Petroleum Hydrocarbons in Marine Ecosystems and Organisms*. New York: Pergamon Press.

Fucik, K. W., and J. M. Neff (1978) Effects of petroleum and aromatic hydrocarbons on $^{45}$calcium uptake and zooxanthellae photosynthesis in corals (unpublished manuscript).

Gerarde, H. W., and D. F. Gerarde (1962) The ubiquitous hydrocarbons. *Assoc. Food Drug Offic.* U.S. **25–26:** 1–47.

Gibson, D. T. (1976) Microbial degradation of carcinogenic hydrocarbons and related compounds. pp. 224–238 In: *Sources, Effects and Sinks of Hydrocarbons in the Aquatic Environment*. Washington, D.C.: American Institute of Biological Sciences.

Gibson, D. T. (1977) Biodegredation of aromatic petroleum hydrocarbons. pp. 36–46 In: D. A. Wolfe (ed.) *Fate and Effects of Petroleum Hydrocarbons in Marine Ecosystems and Organisms*. New York: Pergamon Press.

Gibson, D. T., V. Mahdevan, D. M. Jerina, H. Yagi, and H. J. C. Yeh (1975) Oxidation of the carcinogens benzo(*a*)pyrene and benzo(*a*)anthracene to dihydrodiols by a bacterium. *Science* **189:** 295–297.

Giger, W., and M. Blumer (1974) Polycyclic aromatic hydrocarbons in the environment: isolation and characterization by chromatography, visible, ultraviolet and mass spectrometry. *Anal. Chem.* **46:** 1663–1671.

Giger, W., and C. Schaffner (1978) Determination of polycyclic aromatic hydrocarbons in the environment by glass capillary gas chromatography. *Anal. Chem.* **50:** 243–249.

Gilchrist, C. A., A. Lynes, G. Steel, and B. T. Whitham (1972) The determination of polycyclic aromatic hydrocarbons in mineral oils by thin-layer chromatography and mass spectrometry. *Analyst* **97:** 880–888.

Goldacre, R. J. (1968) The effects of detergents and oils on the cell membrane. pp. 131–139 In: J. D. Carthy and D. R. Arthur (eds.) *The Biological Effects of Oil Pollution on Littoral Communities*. London: Field Studies Council.

Gordon, R. J. (1976) Distribution of airborne polycyclic aromatic hydrocarbons throughout Los Angeles. *Environ. Sci. Technol.* **10:** 370–373.

Graf, W., and H. Diehl (1966) Concerning the naturally caused normal level of carcinogenic polycyclic aromatics and its cause. *Arch. Hyg.* (*Berlin*) **150:** 49–59 (German).

Graf, W., and W. Nowak (1966) Promotion of growth in lower and higher plants by carcinogenic polycyclic aromatics. *Arch. Hyg.* (*Berlin*) **150:** 513–528 (German).

Graf, W., and C. Winter (1968) 3,4-Benzopyrene in crude oil. *Arch. Hyg.* (*Berlin*) **152:** 289–293 (German).

Green, F. A., Jr., and J. M. Neff (1977) Toxicity, accumulation and release of three polychlorinated naphthalenes (Halowax 1000, 1013, and 1099) in postlarval and adult grass shrimp, *Palmaemonetes pugio*. *Bull. Environ. Contam. Toxicol.* **17:** 399–407.

Greffard, J., and J. Meury (1967) Carcinogenic hydrocarbon pollution in Toulon harbour. *Cah. Oceanogr.* **19:** 457–468 (French).

Grimmer, G., and H. Bohnke (1975) Profile analysis of polycyclic aromatic hydrocarbons and metal content in sediment layers of a lake. *Cancer Let.* **1:** 75–84.

Grimmer, G., H. Bohnke, and H. Borwitzky (1978) Profile-analysis of polycyclic aromatic hydrocarbons in sewage sludge by gas chromatography. *Fresenius Z. Anal. Chem.* **289:** 91–95 (German).

Grimmer, G., and D. Duvel (1970) Endogenous formation of polycyclic hydrocarbons in higher plants. 8. Carcinogenic hydrocarbons in the environment of humans. *Z. Naturforsch.* **25b:** 1171–1175 (German).

Gross, G. P. (1974) Gasoline composition and vehicle exhaust gas polynuclear aromatic content. Fourth Annual (Final) Report CRC-ARRAC Project No. CAPE-6-68, Coordinating Research Council, Inc., New York.

Gruger, E. H., Jr., M. M. Wekell, P. T. Numoto, and D. R. Craddock (1977a) Induction of hepatic aryl hydrocarbon hydroxylase in salmon exposed to petroleum dissolved in seawater and to petroleum and polychlorinated biphenyls, separate and together, in food. *Bull. Environ. Contam. Toxicol.* **17:** 512–520.

Gruger, E. H., Jr., M. M. Wekell, and P. A. Robisch (1977b) Effects of chlorinated biphenyls and petroleum hydrocarbons on the activity of hepatic aryl hydrocarbon hydroxylase of coho salmon (*Oncorhynchus kisutch*) and chinook salmon (*O. tshawyscha*). pp. 323–331 In: D. A. Wolfe (ed.) *Fate and Effects of Petroleum Hydrocarbons in Marine Ecosystems and Organisms*. New York: Pergamon Press.

Gubergrits, M., L. Paalme, and U. Kirso (1972) Degredation of benzo[*a*]pyrene and phenol by physicochemical agents during self purification of reservoirs. pp. 49–53. *Vop. Profil. Zagryazneniya Vnesh. Sredy, Chastnosti Vadoėmov. Kantcerogen Veschestvami*. (Russian) (cited by Radding *et al.*, 1976. op. cit.).

Guddal, E. (1959) Isolation of polycyclic aromatic hydrocarbons from the roots of *Chrysanthemum vulgare* Bernh. *Acta Chem. Scand.* **13:** 834–835.

Guerin, M. R. (1977) Energy sources of polycyclic aromatic hydrocarbons. Oak Ridge, Tennessee: Oak Ridge National Laboratory. Conf-770130-2. 78 pp.

Guillory, J. B., and C. F. Cook (1973) Energy transfer processes involving ultraviolet stabilizers. Quenching of singlet oxygen. *J. Polymer Sci., Polymer Chem. Ed.* **11:** 1927.

Gump, B. H., H. S. Hertz, W. E. May, and S. N. Chesler (1977) The hydrocarbon burden in the marine environment surrounding a refinery tanker jetty. *J. Environ. Sci. Health* **A12:** 105–113.

Hancock, J. L., H. G. Applegate, and J. D. Dodd (1970) Polynuclear aromatic hydrocarbons on leaves. *Atmosph. Environ.* **4**(4): 363–370.

Hangebrauck, R. P., D. J. von Lehmden, and J. E. Meeker (1964) Emissions of polynuclear hydrocarbons and other pollutants from heat generation and incineration processes. *J. Air Pollut. Contr. Assn.* **14:** 267–278.

Haranghy, L. (1956) Effects of 3,4-benzpyrene on freshwater mussels. *Acta Biol. Acad. Sci. Hungary.* **7:** 101–108 (German).

Harris, R. P., V. Berdugo, E. D. S. Corner, C. C. Kilvington, and S. C. M. O'Hara (1977a) Factors affecting the retention of a petroleum hydrocarbon by marine planktonic copepods. pp. 286–304 In: D. A. Wolfe (ed.) *Fate and Effects of Petroleum Hydrocarbons in Marine Ecosystems and Organisms.* New York: Pergamon Press.

Harris, R. P., V. Berdugo, S. C. M. O'Hara, and E. D. S. Corner (1977b) Accumulation of $^{14}$C-1-naphthalene by an oceanic and an estuarine copepod during long-term exposure to low-level concentrations. *Mar. Biol.* **42:** 187–195.

Harrison, A. P., and V. E. Raabe (1967) Factors influencing the photodynamic action of benzopyrene on *E. coli. J. Bact.* **93:** 618–626.

Harrison, R. M., R. Perry, and R. A. Wellings (1975) Polynuclear aromatic hydrocarbons in raw, potable and waste waters. *Water Res.* **9:** 331–346.

Harrison, R. M., R. Perry, and R. A. Wellings (1976a) Effect of water chlorination upon levels of some polynuclear aromatic hydrocarbons in water. *Environ. Sci. Technol.* **10:** 1151–1156.

Harrison, R. M., R. Perry, and R. A. Wellings (1976b) Chemical kinetics of chlorination of some polynuclear aromatic hydrocarbons under conditions of water treatment processes. *Environ. Sci. Technol.* **10:** 1156–1160.

Hase, A., and R. A. Hites (1976) On the origin of polycyclic aromatic hydrocarbons in recent sediments: biosynthesis by anaerobic bacteria. *Geochim. Cosmochim. Acta* **40:** 1141–1143.

Hase, A., and R. A. Hites (1977) On the origin of polycyclic aromatic hydrocarbons in the aqueous environment. pp. 205–214 In: L. H. Keith (ed.) *Identification of Organic Pollutants in Water.* Ann Arbor, Michigan: Ann Arbor Science Publishers, Inc.

Hase, A., P. H. Lin, and R. A. Hites (1976) Analysis of complex polycyclic aromatic hydrocarbon mixtures by computerized GC–MS. pp. 435–442 In: R. Freudenthal and P. W. Jones (eds.) *Carcinogenesis—A Comprehensive Survey. Vol.* 1. *Polynuclear Aromatic Hydrocarbons. Chemistry, Metabolism, and Carcinogenesis.* New York: Raven Press.

Hass, B. S., and H. G. Applegate (1975) The effects of unsubstituted polycyclic aromatic hydrocarbons on the growth of *Escherichia coli. Chem.-Biol. Interact.* **10:** 265–268.

Hayatsu, R., R. G. Scott, L. P. Moore, and M. H. Studier (1975) Aromatic units in coal. *Nature* (*Lond.*) **257:** 378–380.

Hecht, S. S., M. Loy, and D. Hoffmann (1976) On the structure and carcinogenicity of the methylchrysenes. pp. 325–340 In: R. Freudenthal and P. W. Jones (eds.) *Carcinogenesis—A Comprehensive Survey. Vol.* 1. *Polynuclear Aromatic Hydrocarbons. Chemistry, Metabolism, and Carcinogenicity.* New York: Raven Press.

Heidelberger, C. (1976) Studies on the mechanisms of carcinogenesis by polycyclic aromatic hydrocarbons and their derivatives. pp. 1–8 In: R. Freudenthal and

P. W. Jones (eds.) *Carcinogenesis—A Comprehensive Survey. Vol. 1. Polynuclear Aromatic Hydrocarbons. Chemistry, Metabolism, and Carcinogenesis.* New York: Raven Press.

Hellmann, von H. (1974) Occurrence and origin of so-called carcinogens and other polycyclic hydrocarbons in water. *Deut. Gewasserkund. Mitteil.* **18:** 155–157 (German).

Herbes, S. E. (1977) Partitioning of polycyclic aromatic hydrocarbons between dissolved and particulate phases in natural waters. *Water Res.* **11:** 493–496.

Herbes, S. E., and G. F. Risi (1978) Metabolic alteration and excretion of anthracene by *Daphnia pulex*. *Bull. Environ. Contam. Toxicol.* **19:** 147–155.

Hites, R. A. (1976) Sources of polycyclic aromatic hydrocarbons in the aquatic environment. pp. 325–332 In: *Sources, Effects and Sinks of Hydrocarbons in the Aquatic Environment.* Washington, D.C.: American Institute of Biological Sciences.

Hites, R. A., and W. G. Biemann (1975) Identification of specific organic compounds in a highly anoxic sediment by GC/MS and HRMS. *Adv. Chem. Ser.* **147:** 188–201.

Hites, R. A., R. E. Laflame, and J. W. Farrington (1977) Sedimentary polycyclic aromatic hydrocarbons: the historical record. *Science* **198:** 829–831.

Hodgins, H. O., B. B. McCain, and J. W. Hawkes (1977) Marine fish and invertebrate diseases, host disease resistance, and pathological effects of petroleum. pp. 95–174 In: D. C. Malins (ed.) *Effects of Petroleum on Arctic and Subarctic Marine Environments and Organisms.* Vol. II. New York: Academic Press.

Hoffmann, D., and E. L. Wynder (1963) Studies on gasoline engine exhaust. *J. Air Pollut. Contr. Assn.* **13:** 322–327.

Huberman, E., L. Sachs, S. K. Yang, and H. V. Gelboin (1976) Identification of mutagenic metabolites of benzo[*a*]pyrene in mammalian cells. *Proc. Natl. Acad. Sci.* **73:** 607–612.

Huffstetler, J. K., and L. W. Rickert (1977) Conversion processes. In: H. M. Braunstein, E. D. Copenhaver, and H. A. Pfuderer (eds.) *Environmental, Health, and Control Aspects of Coal Conversion: an Information Overview.* Vol. 1. Oak Ridge, Tennessee: Oak Ridge National Laboratory ORNL/EIS-94. 125 pp.

Il'nitskii, A. P., and K. P. Ershova (1970) Factors influencing the solubility of benzo[*a*]pyrene in water. *Vopr. Onkol.* **16:** 86 (Russian) (cited by Andelman and Snodgrass, 1972. op. cit.).

Il'nitskii, A. P., K. P. Ershova, A. Y. Khesina, L. G. Rozhkova, V. G. Klubkov, and A. A. Korolev (1971a) Stability of carcinogens in water and efficacy of methods of decontamination. *Hyg. Sanit.* **36:** 9–13 (Russian).

Il'nitskii, A. P., V. G. Klubkov, and L. M. Shabad (1972) Navigation as one of the sources of pollution of water reservoirs by carcinogenic hydrocarbons. *Vopr. Onkol.* **18:** 49 (Russian) (cited by Andelman and Snodgrass, 1972. op. cit.).

Il'nitskii, A. P., A. Y. Khesina, S. N. Cherkinskii, and L. M. Shabad (1968) Effect of ozonation upon aromatic hydrocarbons, including carcinogens. *Hyg. Sanit.* **33:** 323–327 (Russian).

Il'nitskii, A. P., and L. G. Rozhnova (1970) Pollution of reservoirs by carcinogenic hydrocarbons. *Vopr. Onkol.* **16:** 78 (Russian) (cited by Andelman and Snodgrass, 1972. op. cit.).

Il'nitskii, A. P., L. G. Rozhnova, and T. W. Drozdova (1971b) On the content of carcinogenic hydrocarbons in surface waters. *Hyg. Sanit.* **36:** 316 (Russian) (cited by Andelman and Snodgrass, 1972. op. cit.).

Inomata, M., and C. Nagata (1972) Photoinduced phenoxy radical of 3,4-benzopyrene. *Gann* **63:** 119–130.

International Agency for Research on Cancer (1973) *Monograph on the Evaluation of Carcinogenic Risk of the Chemical to Man: Certain Polycyclic Aromatic Hydrocarbons and Heterocyclic Compounds.* Vol. 3. Geneva, Switzerland: World Health Organization.

Ishio, S., K. Kawabe, and T. Tomiyama (1972a) Algal cancer and its causes I. Carcinogenic potencies of water and suspended solids discharged to the River Ohmuta. *Bull. Jap. Soc. Sci. Fish.* **38:** 17–24 (Japanese).

Ishio, S., H. Nagagawa, and T. Tomiyama (1972b) Algal cancer and its causes II. Separation of carcinogenic compounds from sea bottom mud polluted by wastes of the coal chemical industry. *Bull. Jap. Soc. Sci. Fish.* **38:** 571–576 (Japanese).

Ishio, S., T. Yano, and R. Nakagana (1971) Algal cancer and causal substances in wastes from the coal chemical industry. pp. 1–8 In: *Proc. 5th International Water Pollution Research Conference.* London: Pergamon Press.

Jerina, D. M., and J. W. Daly (1974) Arene oxides: a new aspect of drug metabolism. *Science* **185:** 573–578.

Jerina, D. M., J. W. Daly, A. M. Jeffrey, and D. T. Gibson (1971) *Cis*-1,2-dihydroxy-1,2-dihydronaphthalene: A bacterial metabolite from naphthalene. *Arch. Biochem. Biophys.* **142:** 394–396.

Jewell, D. M., R. G. Ruberto, and B. E. Davis (1972) Systematic approach to the study of aromatic hydrocarbons in heavy distillates and residues by elution adsorption chromatography. *Anal. Chem.* **44:** 2318–2321.

Johnston, R. (1976) Mechanisms and problems of marine pollution in relation to commercial fisheries. pp. 3–158 In: R. Johnston (ed.) *Marine Pollution.* London: Academic Press.

Jolley, R. L., G. Jones, W. W. Pitt, and J. E. Thompson (1975) Chlorination of organics in cooling waters and process effluents. pp. 115–152 In: R. L. Jolley (ed.) *Proc. Conf. on Environmental Impact of Water Chlorination.* Oak Ridge, Tennessee.

Jones, A. R., M. R. Guerin, and B. R. Clark (1977) Preparative-scale liquid chromatographic fractionation of crude oils derived from coal and shale. *Anal. Chem.* **49:** 1766–1770.

Jones, R. W., and M. N. Huffman (1957) Fish embryos as bio-assay material in testing chemicals for effects on cell division and differentiation. *Trans. Amer. Microsc. Soc.* **76:** 177–183.

Jonsson, R. (1968) Separation and identification of some naturally occurring alkylphenanthrenes. *Talanta* **15:** 425–431.

Katz, M., and R. C. Pierce (1976) Quantitative distribution of polynuclear aromatic hydrocarbons in relation to particle size of urban particulates. pp. 413–429 In: R. Freudenthal and P. W. Jones (eds.) *Carcinogenesis—A Comprehensive Survey. Vol.* 1. *Polynuclear Aromatic Hydrocarbons. Chemistry, Metabolism, and Carcinogenesis.* New York: Raven Press.

Kauss, P. B., and T. C. Hutchinson (1975) The effects of water-soluble petroleum components on the growth of *Chlorella vulgaris* Beijerinck. *Environ. Pollut.* **9:** 157–174.

Kauss, P., T. C. Hutchinson, C. Soto, J. Hellebust, and M. Griffiths (1973) The toxicity of crude oil and its components to freshwater algae. pp. 703–714. *Proc.* 1973 *Conference on Prevention and Control of Oil Spills.* Washington, D.C.: American Petroleum Institute.

Keegan, R. E. (1971) The trace fluorometric determination of polynuclear aromatic hydrocarbons in natural water. Doctoral dissertation, Graduate School of Chemistry, University of New Hampshire, Durham, New Hampshire (cited by Andelman and Snodgrass, 1972. op. cit.).

Khan, M. A. Q., R. H. Stanton, and G. Reddy (1974) Detoxication of foreign compounds by invertebrates. pp. 177–201 In: M. A. Q. Khan and J. P. Bederka, Jr. (eds.) *Survival in Toxic Environments.* New York: Academic Press.

Khudolei, V. V., and O. A. Sirenko (1977) Tumor development in the bivalve mollusk *Unio pictorum* induced by N-nitroso compounds. *Bull. Exper. Biol. Med.* **83:** 684–686.

Klevens, H. B. (1950) Solubilization of polycyclic hydrocarbons. *J. Phys. Colloid Chem.* **54:** 283–298.

Knorr, M., and D. Schenk (1968) The question of the synthesis of polycyclic aromatic hydrocarbons by bacteria. *Arch. Hyg.* (*Berlin*) **152:** 63–65 (German).

Koe, B. K., and L. Zechmeister (1952) The isolation of carcinogenic and other polycyclic aromatic hydrocarbons from barnacles. Part II: The goose barnacle (*Mitella polymerus*). *Arch. Biochem. Biophys.* **41:** 396–403.

Koons, C. B., and P. H. Monaghan (1976) Input of hydrocarbons from seeps and recent biogenic sources. pp. 84–107 In: *Sources, Effects and Sinks of Hydrocarbons in the Aquatic Environment.* Washington, D.C.: American Institute of Biological Sciences.

Korotkova, G. P., and B. P. Tokin (1968) Stimulation of the process of somatic embryogenesis in some porifera and coelenterata. I. Effect of carcinogenic agents on some porifera. *Acta Biol. Hungary* **19:** 465–474.

Kraybill, H. F., C. J. Dawe, J. C. Harshbarger, and R. G. Tardiff (eds.) (1977) *Aquatic Pollutants and Biologic Effects with Emphasis on Neoplasia.* New York: Ann. N.Y. Acad. Sci. Vol. 298. 604 pp.

Kreiger, R. I., and P. W. Lee (1973) Properties of the aldrin epoxidase system in the gut and fat body of a caddis-fly larva. *J. Econ. Entomol.* **66:** 1–6.

Kuhnhold, W. W., and F. Busch (1978) On the uptake of three different types of hydrocarbons by salmon eggs (*Salmo salar* L.). *Meeresforsch.* **26:** 50–59.

Kupfer, D., and W. H. Bugler (1976) Interactions of chlorinated hydrocarbons with steroid hormones. Pharmacological basis of pesticide toxicity. *Fed. Proc.* **35:** 2603–2608.

Kuratsune, M., and T. Hirohata (1962) Decomposition of polycyclic aromatic hydrocarbons under laboratory illuminations. *Natl. Cancer Inst. Monogr.* **9:** 117–133.

Kurelec, B., S. Britvic, M. Rijavec, W. E. G. Muller, and R. K. Zahn (1977) Benzo[*a*]pyrene monooxygenase induction in marine fish-molecular response to oil pollution. *Mar. Biol.* **44:** 211–216.

Lalou, C. (1963) Contribution to the study of pollution of marine sediments by benzo-3,4-pyrene. Commission Internationale pour l'Exploration Scientifique de la Mer Méditerranée, *Rapport et Procès-Verbeux*, **17:** 711–718 (French).

Lao, R. C., R. S. Thomas, and J. L. Monkman (1975) Computerized gas chromatographic–mass spectrometric analysis of polycyclic aromatic hydrocarbons in environmental samples. *J. Chromatog.* **112:** 681–700.

Larson, R. A., D. W. Blankenship, and L. L. Hunt (1976) Toxic hydroperoxides: photochemical formation from petroleum constituents. pp. 298–308 In: *Sources, Effects and Sinks of Hydrocarbons in the Aquatic Environment.* Washington, D.C.: American Institute of Biological Sciences.

Larson, R. A., and L. L. Hunt (1978) Photooxidation of a refined petroleum oil: inhibition by β-carotene and role of singlet oxygen. *Photochem. Photobiol.* **28:** 553–556.

Larson, R. A., L. L. Hunt, and D. W. Blankenship (1977) Formation of toxic products from a #2 fuel oil by photooxidation. *Environ. Sci. Technol.* **11:** 492–496.

Laughlin, R. B., Jr. (1977) Interactive effects of physical factors and hydrocarbon exposure on the survival, development rate and growth of the mud crab *Rhithropanopeus harrisii* and the horseshoe crab *Limulus polyphemus.* Ph.D. dissertation. The Graduate College, Texas A. & M. University, College Station, Texas. 206 pp.

Laughlin, R. B., Jr., and J. M. Neff (1978) The interactive effects of polynuclear aromatic hydrocarbons, salinity, and temperature on the survival and development rate of the larvae of the mud crab, *Rhithropanopeus harrisii. Mar. Biol.* (in press).

Lawler, G. C., W.-A. Loong, and J. L. Laseter (1978) Accumulation of aromatic hydrocarbons in tissues of petroleum-exposed mallard ducks (*Anas platyrhynchos*). *Environ. Sci. Technol.* **12:** 51–54.

Lee, R. F. (1975) Fate of petroleum hydrocarbons in marine zooplankton. pp. 549–554. *Proc.* 1975 *Conference on Prevention and Control of Oil Pollution.* Washington, D.C.: American Petroleum Institute.

Lee, R. F. (1976) Metabolism of petroleum hydrocarbons in marine sediments. pp. 333–344 In: *Sources, Effects and Sinks of Hydrocarbons in the Aquatic Environment.* Washington, D.C.: American Institute of Biological Sciences.

Lee, R. F. (1977) Fate of petroleum components in estuarine waters of the southeastern United States. pp. 611–616. *Proc.* 1977 *Oil Spill Conference (Prevention, Behavior, Control, Cleanup)* Washington, D.C.: American Petroleum Institute.

Lee, R. F., E. Furlong, and S. Singer (1977) Metabolism of hydrocarbons in marine invertebrates: aryl hydrocarbon hydroxylase from tissues of the blue crab, *Callinectes sapidus* and the polychaete worm, *Nereis* sp. pp. 111–124 In: C. S. Giam (ed.) *Pollutant Effects on Marine Organisms.* Lexington, Massachusetts: D. C. Heath and Co.

Lee, R. F., W. S. Gardner, J. W. Anderson, J. W. Blaylock, and J. Barwell-Clarke (1978a) Fate of polycyclic aromatic hydrocarbons in controlled ecosystem enclosures. *Environ. Sci. Technol.* **12:** 832–838.

Lee, R. F., C. Ryan, and M. L. Neuhauser (1976) Fate of petroleum hydrocarbons taken up from food and water by the blue crab *Callinectes sapidus. Mar. Biol.* **37:** 363–370.

Lee, R. F., R. Sauerheber, and A. A. Benson (1972a) Petroleum hydrocarbons: uptake and discharge by the marine mussel *Mytilus edulis. Science* **177:** 344–346.

Lee, R. F., R. Sauerheber, and G. H. Dobbs (1972b) Uptake, metabolism and discharge of polycyclic aromatic hydrocarbons by marine fish. *Mar. Biol.* **17:** 201–208.

Lee, R. F., S. C. Singer, K. R. Tenore, W. S. Gardner, and R. M. Philpot (1978b) Detoxification system in polychaete worms: importance in the degredation of sediment hydrocarbons. In: F. J. Vernberg, W. B. Vernberg, and A. Calabrese (eds.) *Marine Pollution: Functional Processes.* New York: Academic Press (in press).

Lee, R. F., and M. Takahashi (1977) The fate and effect of petroleum hydrocarbons in controlled ecosystem enclosures. pp. 150–156 In: A. D. McIntyre and K. J. Whittle (eds.) *Petroleum Hydrocarbons in the Marine Environment.* Charlottenlund Slot, Denmark: Cons. Intern. Explor. Mer. Vol. 171.

Lee, W. Y., and J. A. C. Nicol (1978) Individual and combined toxicity of some petroleum hydrocarbons to the marine amphipod, *Elasmopus pectenicrus. Mar. Biol.* **48:** 215–222.

Lehr, R. E., and D. M. Jerina (1977) Metabolic activations of polycyclic hydrocarbons. Structure–activity relationships. *Arch. Toxicol.* **39:** 1–6.

Leo, A., C. Hansch, and D. Elkins (1971) Partition coefficients and their uses. *Chem. Revs.* **71:** 525–616.

Lewis, W. M. (1975) Polynuclear aromatic hydrocarbons in water. *Water Treat. Exam.* **24:** 243–277.

Lijinsky, W. (1967) Detection of carcinogenic chemicals in the environment. *Cancer Bull.* **19:** 63–64.

Lijinsky, W., I. Domsky, G. Mason, H. Y. Ramahi, and T. Safavi (1963) The chromatographic determination of trace amounts of polynuclear hydrocarbons in petrolatum, mineral oil, and coal-tar. *Anal. Chem.* **35:** 952–956.

Lima-Zanghi, C. (1968) Fatty acid balance of marine plankton and pollution by benzo-3,4-pyrene. *Cah. Oceanogr.* **3:** 203–216 (French).

Lu, P.-Y, R. L. Metcalf, N. Plummer, and D. Mandel (1977) The environmental fate of three carcinogens: benzo[*a*]pyrene, benzidine, and vinyl chloride evaluated in laboratory model ecosystems. *Arch. Environ. Contam. Toxicol.* **6:** 129–142.

Lunde, G., and A. Bjorseth (1977) Polycyclic aromatic hydrocarbons in long-range transported aerosols. *Nature* (*Lond.*) **268:** 518–519.

MacKay, D., and A. W. Wolkoff (1973) Rate of evaporation of low-solubility contaminants from water bodies to atmosphere. *Environ. Sci. Technol.* **7:** 611–614.

Malaney, G. W., A. A. Lutin, J. J. Cibulka, and L. H. Hickerson (1967) Resistance of carcinogenic organic compounds to oxidation by activated sludge. *J. Water Pollut. Contr. Fed.* **39:** 2020–2029.

Malins, D. C. (ed.) (1977a) *Effects of Petroleum on Arctic and Subarctic Marine Environments and Organisms. Vol. I. Nature and Fate of Petroleum.* 321 pp. *Vol. II. Biological Effects.* New York: Academic Press. 500 pp.

Malins, D. C. (1977b) Metabolism of aromatic hydrocarbons in marine organisms. pp. 482–496 In: H. F. Kraybill, C. J. Dawe, J. C. Harshbarger, and R. G. Tardiff (eds.) *Aquatic Pollutants and Biological Effects with Emphasis on Neoplasia.* New York: Ann. N.Y. Acad. Sci. Vol. 298.

Mallet, L. (1961) Investigation of benzo-3,4-pyrene-type polycyclic aromatic hydrocarbons in the fauna of marine environments (the Channel, Atlantic and Mediterranean). *C. R. Acad. Sci.* (*Paris*) *Ser. D* **253:** 168–170 (French).

Mallet, L. (1965) Presence of 3,4-benzpyrene-type hydrocarbons in the sediments underlying the Seine bed downstream from Paris. *Bull. Acad. Nat. Med.* (*Paris*) **149:** 656–660 (French).

Mallet, L. (1966) Investigation on the presence of 3,4-benzpyrene-type polycyclic aromatic hydrocarbons in the deep sediments underlying the Seine bed downstream from Paris. *Gaz. Hop.* (*Paris*) **138:** 69–70 (French).

Mallet, L., and M. Heros (1960) Presence of BP in urine from man. *C. R. Acad. Sci.* (*Paris*) *Ser. D* **250:** 943–945 (French).

Mallet, L., M. Heros, and J. Brisou (1972a) Biosynthesis and bioregression of carcinogenic polybenzene hydrocarbons of the benzo-3,4-pyrene type, taken as controls at the expense of lipids. pp. 177–180 In: L. Mallet (ed.) *Pollution des Mileaux Vitaux par les Hydrocarbures Polybenzeniques du Type Benzo*-3,4-*pyrene*. Paris: Librairie Maloine (French).

Mallet, L., and R. Lami (1964) Investigation on pollution of plankton by benzo-3,4-pyrene-type polycyclic aromatic hydrocarbons in the Rance estuary. *C. R. Soc. Biol.* (*Paris*) **158** 2261–2262 (French).

Mallet, L., C. Lima-Zanghi, and J. Brisou (1967) Investigations of the possibilities of biosynthesis of polybenzene hydrocarbons of the benzo-3,4-pyrene type by a *Clostridium putride* in the presence of marine plankton lipids. *C. R. Acad. Sci.* (*Paris*) *Ser. D* **264:** 1534–1537 (French).

Mallet, L., V. Perdriau, and S. Perdriau (1963) Extent of pollution by polycyclic aromatic hydrocarbons of the benzo-3,4-pyrene type in the North Sea and the glacial Arctic Ocean. *Bull. Acad. Nat. Med.* (*Paris*) **147:** 320–325 (French).

Mallet, L., and M.-L. Priou (1967) Retention of benzo-3,4-pyrene type polycyclic aromatic hydrocarbons by the marine sediments, fauna and flora of the Bay of Saint-Malo. *C. R. Acad. Sci.* (*Paris*) *Ser. D* **264:** 969–971 (French).

Mallet, L., M. L. Priou, and M. Leon (1972b) Biosynthesis and biodegradation of the polycyclic aromatic hydrocarbon benzo-3,4-pyrene in the sediments of the Bay of Saint-Malo. pp. 159–163 In: L. Mallet (ed.), *Pollution des Mileaux Vitaux par les Hydrocarbures Polybenzeniques du Type Benzo*-3,4-*pyrene*. Paris: Librairie Maloine (French).

Mallet, L., and J. Sardou (1964) Investigation on the presence of benzopyrene-type polycyclic aromatic hydrocarbons in the plankton environment of the region of the Bay of Villefranche (Alpes-Maritimes region). *C. R. Acad. Sci.* (*Paris*) *Ser. D* **258:** 5264–5267 (French).

Mallet, L., and J. Sardou (1965) Investigation on the presence of the benzo-3,4-pyrene-type polycyclic aromatic hydrocarbons in the plankton environment of the region of the Bay of Villefranche. pp. 331–334 In: *Pollutions Marines par les Microorganismes et les Produits Petroliers*, Monaco Symposium, April 1964, Monaco: Commission Internationale pour l'Exploration Scientifique de la Mer Méditerranée (French).

Mallet, L., and M. Tissier (1969) Experimental biosynthesis of polybenzene hydrocarbons of the benzo-3,4-pyrene type at the expense of forest soils. *C. R. Soc. Biol.* (*Paris*) **163:** 63–65 (French).

Marty, J.-C., A. Saliot, and M. J. Tissier (1978) Inventory, distribution and origin of aliphatic and polycyclic aromatic hydrocarbons in seawater, the surface microlayer, and marine aerosols in the eastern tropical Atlantic Ocean. *C. R. Acad. Sci.* (*Paris*) *Ser. D* **286:** 833–836 (French).

Masuda, Y., and M. Kuratsune (1966) Photochemical oxidation of benzopyrene. *Int. J. Air Wat. Pollut.* **10:** 805–811.

Matos, E. L., and E. S. de Lustig (1973) Teratogenic effects of carcinogen implantation in a regenerative field in *Bufo arenarium* tadpoles. *Teratology* **8:** 167–174.

Maugh, T. H. (1974) Chemical carcinogens: a long-neglected field blossoms. *Science* **183:** 940–944.

May, W. E., S. P. Wasik, and D. H. Freeman (1978a) Determination of the aqueous solubility of polynuclear aromatic hydrocarbons by a coupled column liquid chromatographic technique. *Anal. Chem.* **50:** 175–179.

May, W. E., S. P. Wasik, and D. H. Freeman (1978b) Determination of the solubility behavior of some polycyclic aromatic hydrocarbons in water. *Anal. Chem.* **50:** 997–1000.

McAuliffe, C. (1966) Solubility in water of paraffin, cycloparaffin, olefin, acetylene, cycloolefin, and aromatic hydrocarbons. *J. Physical Chem.* **70:** 1267–1275.

McCain, B. B., H. O. Hodgins, W. D. Gronlund, J. W. Hawkes, D. W. Brown, M. S. Myers, and J. H. Vandermeulen (1978) Bioavailability of crude oil from experimentally oiled sediments to English sole (*Parophrys vetulus*), and pathological consequences. *J. Fish. Res. Bd. Canada* **35:** 657–664.

McGinnes, P. R., and V. L. Snoeyink (1974) Determination of the fate of polynuclear aromatic hydrocarbons in natural water systems. *Water Resources Council Rept.* 80. PB232168/5GA, 62 pp.

McIntyre, A. D., and K. J. Whittle (eds.) (1977) *Petroleum Hydrocarbons in the Marine Environment.* Charlottenlund Slot, Denmark: Cons. Intern. Explor. Mer. Vol. 171. 230 pp.

McKay, J. F., and D. R. Latham (1973) Polyaromatic hydrocarbons in high-boiling petroleum distillates. Isolation by gel permeation chromatography and identification by fluorescence spectrometry. *Anal. Chem.* **45:** 1050–1055.

Meinschein, W. G. (1959) Origin of petroleum. *Bull. Amer. Assoc. Petrol. Geol.* **42:** 925–943.

Melancon, M. J., Jr., and J. J. Lech (1978) Distribution and elimination of naphthalene and 2-methylnaphthalene in rainbow trout during short- and long-term exposures. *Arch. Environ. Contam. Toxicol.* **7:** 207–220.

Meyers, P. A., and J. G. Quinn (1973) Association of hydrocarbons and mineral particles in saline solutions. *Nature* (*Lond.*) **244:** 23–24.

Miller, D. L., J. P. Corliss, R. N. Farragut, and H. C. Thompson, Jr. (1978) Accumulation and elimination of the polynuclear aromatic hydrocarbon chrysene by mangrove snapper (*Lutjanus griseus*) and pink shrimp (*Penaeus duorarum*). Unpublished manuscript from the Southeast Fisheries Center, Miami Lab., NMFS, NOAA, Miami, Florida. 18 pp.

Mix, M. C., R. T. Riley, K. I. King, S. R. Trenholm, and R. L. Schaffer (1977) Chemical carcinogens in the marine environment. Benzo[*a*]pyrene in economically-important bivalve mollusks from Oregon estuaries. pp. 421–431 In: D. A. Wolfe (ed.) *Fate and Effects of Petroleum Hydrocarbons in Marine Ecosystems and Organisms.* New York: Pergamon Press.

Moore, S. F., and R. L. Dwyer (1974) Effects of oil on marine organisms: a critical assessment of published data. *Water Res.* **8:** 819–827.

Moore, S. F., R. L. Dwyer, and A. M. Katz (1973) A preliminary assessment of the environmental vulnerability of Machias Bay, Maine to oil supertankers. Cambridge, Massachusetts Institute of Technology, R. M. Parsons Laboratory for Water Resources and Hydrodynamics R73-15. 162 pp.

Müller, G., G. Grimmer, and H. Bohnke (1977) Sedimentary record of heavy metals and polycyclic aromatic hydrocarbons in Lake Constance. *Naturwissen.* **64:** 422–431.

Nagata, S., and G. Kondo (1977) Photo-oxidation of crude oils. pp. 617–620. *Proc.* 1977 *Oil Spill Conference (Prevention, Behavior, Control, Cleanup).* Washington, D.C.: American Petroleum Institute.

National Academy of Sciences (1972) *Particulate Polycyclic Organic Matter.* Washington, D.C.: NAS: 361 pp.

National Academy of Sciences (1975) *Petroleum in the Marine Environment.* Washington, D.C.: NAS: 107 pp.

Neely, W. B., D. R. Branson, and G. E. Blau (1974) Partition coefficient to measure bioconcentration potential of organic chemicals in fish. *Environ. Sci. Technol.* **8:** 1113–1115.

Neff, J. M., and J. W. Anderson (1975) Accumulation, release, and distribution of benzo[*a*]pyrene-$C^{14}$ in the clam *Rangia cuneata.* pp. 469–472. *Proc.* 1975 *Conference on Prevention and Control of Oil Pollution.* Washington, D.C.: American Petroleum Institute.

Neff, J. M., J. W. Anderson, B. A. Cox, R. B. Laughlin, Jr., S. S. Rossi, and H. E. Tatem (1976a) Effects of petroleum on survival, respiration and growth of marine animals. pp. 515–540 In: *Sources, Effects and Sinks of Hydrocarbons in the Aquatic Environment.* Washington, D.C.: American Institute of Biological Sciences.

Neff, J. M., B. A. Cox, D. Dixit, and J. W. Anderson (1976b) Accumulation and release of petroleum-derived aromatic hydrocarbons by four species of marine animals. *Mar. Biol.* **38:** 279–289.

Neff, J. M., and K. W. Fucik (1978) Effects of No. 2 fuel oil and aromatic hydrocarbons on $^{45}Ca$ deposition in the regenerating spines of the sea urchin *Diadema antillarum* (unpublished manuscript).

Nelson-Smith, A. (1970) The problem of oil pollution of the sea. *Adv. Mar. Biol.* **8:** 215–306.

Nelson-Smith, A. (1973) *Oil Pollution and Marine Ecology.* New York: Plenum Press. 260 pp.

Neukomn, S. (1974) The newt test for studying certain categories of carcinogenic substances. pp. 228–235 In: *Experimental Model Systems in Toxicology and their Significance to Man. Proc. Europ. Soc. Study Drug Tox., Internat. Conf.* Vol. 15.

Newman, M. S. (1976) Carcinogenic activity of benz[*a*]anthracenes. pp. 203–208 In: R. Freudenthal and P. W. Jones (eds.) *Carcinogenesis—A Comprehensive Survey. Vol.* 1. *Polynuclear Aromatic Hydrocarbons. Chemistry, Metabolism, and Carcinogenesis.* New York: Raven Press.

Niaussat, P. (1970) Pollution by *in situ* biosynthesis of carcinogenic hydrocarbons of a lagoon ecosystem: reproduction *in vitro* of this phenomenon. *Rev. Int. Oceanog. Med.* **17:** 87–98 (French).

Niaussat, P., and C. Auger (1970) Distribution of benzo[*a*]pyrene and perylene in various organisms of the Clipperton Lagoon ecosystem. *C. R. Acad. Sci. (Paris) Ser. D.* **270:** 2702–2705 (French).

Niaussat, P., C. Auger, and L. Mallet (1970) Relative quantitative appearance of carcinogenic hydrocarbons in pure cultures of *Bacillus badius* as a result of the presence in the medium of certain chemical compounds. *C. R. Acad. Sci.* (*Paris*) *Ser. D.* **270:** 1042–1045 (French).

Niaussat, P., L. Mallet, and J. Ottenwaelder (1969) Presence of benzo[*a*]pyrene in various marine phytoplankton species cultivated *in vitro*. Possible role of associated bacteria. *C. R. Acad. Sci.* (*Paris*) *Ser. D.* **268:** 1109–1112 (French).

Niaussat, P. M., J. Trichet, M. Heros, N. T. Luong, and J. P. Ehrhardt (1975) Presence of benzo-3,4-pyrene in the water and organic sediments of brackish water of Polynesian atolls. Study of certain associated biotic and abiotic factors. *C. R. Acad. Sci.* (*Paris*) *Ser. D.* **281:** 1031–1034 (French).

Olsen, D., and J. L. Haynes (1969) Preliminary air pollution survey of organic carcinogens. A literature review. APTD 69–43. National Air Pollution Control Administration, Raleigh, N.C.

Orr, W. L., and J. R. Grady (1967) Perylene in basin sediments off southern California. *Geochim. Cosmochim. Acta* **31:** 1201–1209.

Ott, F. S., R. P. Harris, and S. C. M. O'Hara (1978) Acute and sublethal toxicity of naphthalene and three methylated derivatives to the estuarine copepod, *Eurytemora affinis*. *Mar. Environ. Res.* **1:** 49–58.

Oyler, A. R., D. L. Bodenner, K. J. Welch, R. J. Liukkonen, R. M. Carlson, H. L. Kopperman, and R. Caple (1978) Determination of aqueous chlorination reaction products of polynuclear aromatic hydrocarbons by reversed phase high performance liquid chromatography–gas chromatography. *Anal. Chem.* **50:** 837–842.

Palmork, K. H., S. Wilhelmsen, and T. Neppelberg (1973) The contribution from various industries of polynuclear aromatic hydrocarbons to the marine environment. International Council for Exploration of the Sea, Fisheries Improvement Commission. Report No. 1973/E:33. 20 pp.

Panalaks, T. (1976) Determination and identification of polycyclic aromatic hydrocarbons in smoked and charcoal-broiled food products by high pressure liquid chromatography and gas chromatography. *J. Environ. Sci. Health* **B11:** 299–315.

Pancirov, R. J. (1974) Compositional data on API reference oils used in biological studies: a No. 2 fuel oil, a Bunker C, Kuwait crude oil, and South Louisiana crude oil. Washington, D.C. American Petroleum Institute.

Pancirov, R. J., and R. A. Brown (1975) Analytical methods for polynuclear aromatic hydrocarbons in crude oils, heating oils, and marine tissues. pp. 103–113. *Proc.* 1975 *Conference on Prevention and Control of Oil Pollution*, Washington, D.C.: American Petroleum Institute.

Pancirov, R. J., and R. A. Brown (1977) Polynuclear aromatic hydrocarbons in marine tissues. *Environ. Sci. Technol.* **11:** 989–992.

Patterson, A. M., L. T. Capell, and D. F. Walker (1960) *The Ring Index. A List of Ring Systems Used in Organic Chemistry* (*2nd ed.*). Washington, D.C.: American Chemical Society. 1425 pp.

Payer, H. D., C. J. Soeder, H. Kunte, P. Karuwanna, R. Nonhof, and W. Graf (1975) Accumulation of polycyclic aromatic hydrocarbons in cultivated microalgae. *Naturwissen.* **62:** 536–538.

Payne, J. F. (1976) Field evaluations of benzopyrene hydroxylase induction as a monitor for marine petroleum pollution. *Science* **191:** 945–946.

Payne, J. F. (1977) Mixed function oxidases in marine organisms in relation to petroleum hydrocarbon metabolism and detection. *Mar. Pollut. Bull.* **8:** 112–114.

Payne, J. F., I. Martins, and A. Rahimtula (1978) Crankcase oils: are they a major mutagenic burden in the aquatic environment? *Science* **200:** 329–330.

Payne, J. F., and W. R. Penrose (1975) Induction of aryl hydrocarbon (benzo[*a*]pyrene) hydroxylase in fish by petroleum. *Bull. Environ. Contam. Toxicol.* **14:** 112–116.

Peakall, D. B. (1975) PCBs and their environmental effects. pp. 469–508. *CRC Critical Reviews in Environmental Control* Sept. 1975.

Pedersen, M. G., W. K. Hershberger, and M. R. Juchau (1974) Metabolism of 3,4-benzopyrene in rainbow trout (*Salmo gairdneri*). *Bull. Environ. Contam. Toxicol.* **12:** 481–486.

Pedersen, M. G., W. K. Hershberger, P. K. Zachariah, and M. R. Juchau (1976) Hepatic biotransformation of environmental xenobiotics in six strains of rainbow trout (*Salmo gairdneri*). *J. Fish. Res. Bd. Canada* **33:** 666–675.

Perdriau, J. (1964) Marine pollution by carcinogenic benzo-3,4-pyrene-type hydrocarbons—biological incidences. Part II. *Cah. Oceanogr.* **16:** 205–229 (French).

Philpot, R. M., M. O. James, and J. R. Bend (1976) Metabolism of benzo[*a*]pyrene and other xenobiotics by microsomal mixed function oxidases in marine species. pp. 184–199 In: *Sources, Effects and Sinks of Hydrocarbons in the Aquatic Environment*. Washington, D.C.: American Institute of Biological Sciences.

Piccinetti, C. (1967) Distribution of the carcinogenic hydrocarbon benzo-3,4 pyrene in the upper and middle Adriatic Sea. *Arch. Oceanogr. Limnol., Suppl.* **15:** 169–183 (Italian).

Pierce, R. C., and M. Katz (1975) Dependency of polynuclear aromatic hydrocarbon content on size distribution of atmospheric aerosols. *Environ. Sci. Technol.*, **9:** 347–353.

Pizzarello, D. J., and A. Wolsky (1966) Carcinogenesis and regeneration in newts. *Experientia* **22:** 387–388.

Pohl, R. J., J. R. Bend, A. M. Guarino, and J. R. Fouts (1974) Hepatic microsomal mixed function oxidase activity of several marine species from coastal Maine. *Drug Metab. Dispos.* **2:** 545–555.

Powell, N. A., C. S. Sayce, and D. F. Tufts (1970) Hyperplasia in an estuarine bryozoan attributable to coal tar derivatives. *J. Fish Res. Bd. Canada* **27:** 2095–2096.

Pupp, C., R. C. Lao, J. J. Murray, and R. F. Pottie (1974) Equilibrium vapour concentrations of some polycyclic aromatic hydrocarbons, arsenic trioxide ($As_4O_6$) and selenium dioxide and the collection efficiencies of these air pollutants. *Atmos. Environ.* **8:** 915–925.

Radding, S. B., T. Mill, C. W. Gould, D. H. Liu, H. L. Johnson, D. C. Bomberger, and C. V. Fojo (1976) The environmental fate of selected polynuclear aromatic hydrocarbons. EPA560/5-75-009. US Environmental Protection Agency, Washington, D.C. 122 pp.

Rice, S. D., R. E. Thomas, and J. W. Short (1977) Effect of petroleum hydrocarbons on breathing and coughing rates and hydrocarbon uptake-depuration in pink salmon fry. pp. 259–278. In: F. J. Vernberg, A. Calabrese, F. P. Thurberg and W. B. Vernberg (eds.) *Physiological Responses of Marine Biota to Pollutants*. New York: Academic Press.

Riseborough, R. W., B. W. de Lappe, and W. Walker II (1976) Transfer of higher molecular weight chlorinated hydrocarbons to the marine environment. pp. 261–322 In: H. L. Windom and R. A. Duce (eds.) *Marine Pollutant Transfer*. Lexington, Massachusetts: D. C. Heath and Co.

Robinson, E. (1976) Pollutant sources of particulate materials in marine atmospheres. *SCOR–NAS Workshop on Tropospheric Transport of Pollutants to the Ocean*, background paper. Cited by: C. Patterson, D. Settle, B. Schaule and M. Burnett, 1976. Transport of pollutant lead to the oceans and within ocean ecosystems. pp. 23–36 In: H. L. Windom and R. A. Duce (eds.) *Marine Pollutant Transfer*. Lexington, Massachusetts: D. C. Heath and Co.

Roesijadi, G., J. W. Anderson, and J. W. Blaylock (1978a) Uptake of hydrocarbons from marine sediments contaminated with Prudhoe Bay crude oil: influence of feeding type of test species on availability of polycyclic aromatic hydrocarbons. *J. Fish. Res. Bd. Canada* **35:** 608–614.

Roesijadi, G., J. W. Anderson, S. R. Petrocelli, and C. S. Giam (1976) Osmoregulation of the grass shrimp *Palaemonetes pugio* exposed to polychlorinated biphenyls (PCBs). I. Effect on chloride and osmotic concentrations and chloride- and water-exchange kinetics. *Mar. Biol.* **38:** 343–355.

Roesijadi, G., D. L. Woodruff, and J. W. Anderson (1978b) Bioavailability of naphthalenes from marine sediments artificially contaminated with Prudhoe Bay crude oil. *Environ. Pollut.* **15:** 223–229.

Rose, F. L. (1976) Tumorous growths of the tiger salamander, *Ambystoma tigrinum*, asociated with treated sewage effluent. pp. 251–262 In: C. J. Dawe, D. G. Scarpelli, and S. R. Wellings (eds.) *Tumors in Aquatic Animals. Prog. Exper. Tumor Res.* Vol. 20.

Rose, F. L. (1977) Tissue lesions of tiger salamanders (*Ambystoma tigrinum*): relationship to sewage effluents. pp. 270–279 In: H. F. Kraybill, C. J. Dawe, J. C. Harshbarger, and R. G. Tardiff (eds.) *Aquatic Pollutants and Biologic Effects with Emphasis on Neoplasia*. New York: Ann. N.Y. Acad. Sci. Vol. 298.

Rossi, S. S. (1977) Bioavailability of petroleum hydrocarbons from water, sediments and detritus to the marine annelid, *Neanthes arenaceodentata*. pp. 621–626. *Proc.* 1977 *Oil Spill Conference (Prevention, Behaviour, Control, Cleanup)*. Washington, D.C.: American Petroleum Institute.

Rossi, S. S., and J. W. Anderson, (1977) Accumulation and release of fuel oil-derived diaromatic hydrocarbons by the polychaete *Neanthes arenaceodentata*. *Mar. Biol.* **39:** 51–55.

Rossi, S. S., and J. M. Neff (1978) Toxicity of polynuclear aromatic hydrocarbons to the marine polychaete, *Neanthes arenaceodentata*. *Mar. Pollut. Bull.* **9:** 220–223.

Roubal, W. T. (1974) Spin-labeling of living tissue—a method for investigating pollutant–host interaction. pp. 367–380 In: F. J. Vernberg and W. B. Vernberg (eds.) *Pollution and Physiology of Marine Organisms*. New York: Academic Press.

Roubal, W. T., D. H. Bovee, T. K. Collier, and S. I. Stranahan (1977a) Flow-through system for chronic exposure of aquatic organisms to seawater-soluble hydrocarbons from crude oil: construction and applications. pp. 551–556. *Proc.* 1977 *Oil Spill Conference (Prevention, Behaviour, Control, Cleanup)*. Washington, D.C.: American Petroleum Institute.

Roubal, W. T., and T. K. Collier (1975) Spin-labeling techniques for studying mode of action of petroleum hydrocarbons on marine organisms. *Fish. Bull.* **73:** 299–305.

Roubal, W. T., T. K. Collier, and D. C. Malins (1977b) Accumulation and metabolism of carbon-14 labeled benzene, naphthalene, and anthracene by young coho salmon (*Oncorhynchus kisutch*). *Arch. Environ. Contam. Toxicol.* **5:** 513–529.

Roubal, W. T., S. I. Stranahan, and D. C. Malins (1978) The accumulation of low molecular weight aromatic hydrocarbons of crude oil by coho salmon (*Oncorhynchus kisutch*) and starry flounder (*Platichthys stellatus*). *Arch Environ. Contam. Toxicol.* **7:** 237–244.

Ruben, L. N., and M. Balls (1964) The implantation of methylcholanthrene crystals into regenerating and non-regenerating forelimbs of *Xenopus laevis*. *J. Morph.* **115:** 239–254.

Saliot, A. (1969) Contribution to the study of organic compounds dissolved in sea water. Thèse 3-eme cycle. Fac. Sci. Paris. 106 pp. (French) (cited by Barbier *et al.*, 1973. op. cit.).

Samoilovich, L. N., and Y. R. Red'kin (1968) 3,4-Benzopyrene pollution of the River Sunzha caused by the petrochemical industry in Grozny. *Gig. Sanit.* **33:** 6 (Russian) (cited by Andelman and Snodgrass, 1972. op. cit.).

Sanborn, H. R., and D. C. Malins (1977) Toxicity and metabolism of naphthalene: a study with marine larval invertebrates. *Proc. Soc. Exper. Biol. Med.* **154:** 151–155.

Sawicki, E., J. E. Meeker, and M. J. Morgan (1965) The quantitative composition of air pollution source effluents in terms of aza heterocyclic compounds and polynuclear aromatic hydrocarbons. *Int. J. Air Water Pollut.* **9:** 291–298.

Scaccini-Cicatelli, M. (1965) Storage of benzo-3,4-pyrene in the tissues of *Tubifex* worms. *Boll. Pesca Piscic. Idrobiol.* **20:** 245–250 (Italian).

Scaccini-Cicatelli, M. (1966) On the process of accumulation of benzo-3,4-pyrene in the tissues of *Tubifex*. *Boll. Soc. Ital. Biol. Sperim.* **42:** 957–959 (Italian).

Schamp, N., and F. van Wassenhove (1972) Determination of benzo[*a*]pyrene in bitumen and plants *J. Chromatog.* **69:** 421–425.

Schmeltz, I., and D. Hoffmann (1976) Formation of polynuclear aromatic hydrocarbons from combustion of organic matter. pp. 225–240. In: R. Freudenthal and P. W. Jones (eds.) *Carcinogenesis—A Comprehensive Survey. Vol. 1. Polynuclear Aromatic Hydrocarbons. Chemistry, Metabolism, and Carcinogenesis*. New York: Raven Press.

Schwarz, F. P., and S. P. Wasik (1976) Fluorescence measurement of benzene, naphthalene, anthracene, pyrene, fluoranthene, and benzo[*a*]pyrene in water. *Anal. Chem.* **48:** 425–428.

Seilern-Aspang, F., and K. Kratochwil (1962) Induction and differentation of an epithelial tumor in the newt (*Triturus cristatus*). *J. Embryol. Exper. Morph.* **10:** 337–356.

Seilern-Aspang, F., and K. Kratochwil (1963) Spontaneous healing of an infiltrating and metastasizing epithelial tumor of *Triturus cristatus*. *Arch. Geschwulstforsch.* **21:** 292–300.

Severson, R. F., M. E. Snook, H. C. Higman, O. T. Chortyk, and F. J. Akin (1976) Isolation, identification, and quantification of polynuclear aromatic hydrocarbons in tobacco smoke. pp. 253–270 In: R. Freudenthal and P. W. Jones (eds.) *Carcinogenesis—A Comprehensive Survey. Vol.* 1. *Polynuclear Aromatic Hydrocarbons. Chemistry, Metabolism, and Carcinogenesis.* New York: Raven Press.

Sharp, J. R., K. W. Fucik, and J. M. Neff (1978) Physiological basis of differential sensitivity of fish embryonic stages to oil pollution. In: F. J. Vernberg, W. B. Vernberg, and A. Calabrese (eds.) *Marine Pollution: Functional Processes.* New York: Academic Press.

Shimkin, M. B., B. K. Koe, and L. Zechmeister (1951) An instance of the occurrence of carcinogenic substances in certain barnacles. *Science* **113:** 650–651.

Sims, P., L. L. Grover, A. Swaisland, K. Pal, and A. Hewer, (1974) Metabolic activation of benzo[*a*]pyrene proceeds by a diol-epoxide. *Nature* (*London*) **252:** 326–328.

Singer, S. C., and R. F. Lee (1977) Mixed function oxygenase activity in blue crab, *Callinectes sapidus:* tissue distribution and correlation with changes during molting and development. *Biol. Bull.* **153:** 377–386.

Sisler, F. D., and C. E. ZoBell (1947) Microbial utilization of carcinogenic hydrocarbons. *Science* **106:** 521–522.

Smith, J. E. (ed.) (1970) *Torrey Canyon Pollution and Marine Life.* Cambridge, England: Cambridge University Press. 196 pp.

Smith, J. G., R. B. McCall, and P. K. Chan (1977) Formation of polychlorinated aromatic compounds during aqueous chlorination. *Environ. Pollut.* **14:** 289–296.

Soto, C., J. A. Hellebust, and T. C. Hutchinson (1975) Effect of naphthalene and aqueous crude oil extracts on the green flagellate *Chlamydomonas angulosa* II. Photosynthesis and the uptake and release of naphthalene. *Can. J. Bot.* **53:** 118–126.

Speers, G. C., and E. V. Whitehead (1969) Crude petroleum. pp. 638–675 In: G. Eglinton and M. R. J. Murphy (eds.), *Organic Geochemistry: Methods and Results.* Berlin: Springer-Verlag.

Spencer, W. F., and M. M. Cliath (1972) Volatility of DDT and related compounds. *J. Agr. Food Chem.* **20:** 645–649.

Statham, C. N., C. R. Elcombe, S. P. Szyjka, and J. J. Lech (1978) Effect of polycyclic aromatic hydrocarbons on hepatic microsomal enzymes and disposition of methylnaphthalene in rainbow trout *in vivo. Xenobiotica* **8:** 65–71.

Statham, C. N., M. J. Melancon, Jr., and J. J. Lech (1976) Bioconcentration of xenobiotics in trout bile: a proposed monitoring aid for some waterborne chemicals. *Science* **193:** 680–681.

Stegeman, J. J. (1974) Hydrocarbons in shellfish chronically exposed to low levels of fuel oil. pp. 329–348 In: F. J. Vernberg and W. B. Vernberg (eds.) *Pollution and Physiology of Marine Organisms.* New York: Academic Press.

Stegeman, J. J. (1977) Fate and effects of oil in marine animals. *Oceanus* **20:** 59–66.

Stegeman, J. J., and D. J. Sabo (1976) Aspects of the effects of petroleum hydrocarbons on intermediary metabolism and xenobiotic metabolism in marine fish. pp. 423–436 In: *Sources, Effects and Sinks of Hydrocarbons in the Aquatic Environment.* Washington, D. C.: American Institute of Biological Sciences.

Stegeman, J. J., and J. M. Teal (1973) Accumulation, release and retention of petroleum hydrocarbons by the oyster *Crassostrea virginica. Mar. Biol.* **22:** 37–44.

Stich, H. F., and A. B. Acton (1976) The possible use of fish tumors in monitoring for carcinogens in the marine environment. pp. 44–54 In: C. J. Dawe, D. G. Scarpelli, and S. R. Wellings (eds.) *Tumors in Aquatic Animals. Prog. Exper. Tumor Res.* Vol. 20.

Straughan, D., and D. M. Lawrence (1975) Investigation of ovicell hyperplasia in bryozoans chronically exposed to natural oil seepage. *Water, Air, and Soil Pollut.* **5:** 39–45.

Suess, M. J. (1972a) Laboratory experimentation with 3,4-benzpyrene in aqueous systems and environmental consequences. *Zblt. Bakt. Hyg.* **155:** 541–546.

Suess, M. J. (1972b) Aqueous solutions of 3,4-benzpyrene. *Water Res.* **6:** 981–985.

Suess, M. J. (1976) The environmental load and cycle of polycyclic aromatic hydrocarbons. *Sci. Total Environ.* **6:** 239–250.

Sullivan, J. B. (1974) Marine pollution by carcinogenic hydrocarbons. pp. 261–263 In: *Marine Pollution Monitoring* (*Petroleum*) Washington, D.C.: Natl. Bur. Stand., Special Publ. 409.

Swaisland, A. J., P. L. Glover, and P. Sims (1973) Epoxides in polycyclic aromatic hydrocarbon metabolism and carcinogenesis. *Biochem. Pharmacol.* **22:** 1547–1556.

Swallow, W. H. (1976) Survey of polycyclic aromatic hydrocarbons in selected foods and food additives in New Zealand. *N.Z. J. Sci.* **19:** 407–412.

Thomas, J. F., M. Mukai, and B. D. Tebbens (1968) Fate of airborne benzo[*a*]pyrene. *Environ. Sci. Technol.* **2:** 33–39.

Thomas, R. E., and S. D. Rice (1978) The effect of exposure temperatures on oxygen consumption and opercular breathing rates of pink salmon fry exposed to toluene, naphthalene, and water-soluble fractions of Cook Inlet crude oil and No. 2 fuel oil. In: F. J. Vernberg, W. B. Vernberg, and A. Calabrese (eds.) *Marine Pollution: Functional Processes.* New York: Academic Press.

Thompson, S., and G. Eglinton (1978) Composition and sources of pollutant hydrocarbons in the Severn estuary. *Mar. Pollut. Bull.* **9:** 133–136.

Thomson, R. H. (1965) Quinones: nature, distribution and biosynthesis. pp. 309–332 In: T. W. Goodwin (ed.), *Chemistry and Biochemistry of Plant Pigments.* London, Academic Press.

Thomson, R. H. (1971) *Naturally Occurring Quinones*, 2nd ed. London: Butterworths Scientific Publ. 302 pp.

United States Department of the Interior (1968) The cost of clean water—summary report. *Federal Water Pollution Control Administration* **1:** 1–39.

Vandermeulen, J. H. (ed.) (1978) Symposium on recovery potential of oiled marine northern environments. *J. Fish. Res. Bd. Canada* **35:** 510–795.

Vandermeulen, J. H., and T. P. Ahern (1976) Effect of petroleum hydrocarbons on algal physiology: review and progress report. pp. 107–125 In: A. P. M. Lockwood (ed.) *Effects of Pollutants on Aquatic Organisms.* Cambridge England: Cambridge University Press.

Vandermeulen, J. H., P. D. Keizer, and W. R. Penrose (1977) Persistence of non-alkane components of Bunker C in beach sediments of Chedabucto Bay, and lack of their metabolism by molluscs. pp. 469–474. *Proc.* 1977 *Oil Spill Conference* (*Prevention*, *Behaviour*, *Control*, *Cleanup*). Washington, D.C.: American Petroleum Institute.

Van Overbeek, J., and R. Blondeau (1954) Mode of action of phytotoxic oils. *Weeds* **3:** 55–65.

Varanasi, U., and D. C. Malins (1977) Metabolism of petroleum hydrocarbons: accumulation and biotransformation in marine organisms. pp. 175–270 In: D. C. Malins (ed.) *Effects of Petroleum on Arctic and Subarctic Marine Environments and Organisms. Vol. II. Biological Effects*. New York: Academic Press.

Wakeham, S. G. (1977) Synchronous fluorescence spectroscopy and its application to indigenous and petroleum-derived hydrocarbons in lacustrine sediments. *Environ. Sci. Technol.* **11:** 272–276.

Walker, J. D., J. J. Calomiris, T. L. Herbert, and R. R. Colwell (1976a) Petroleum hydrocarbons: degradation and growth potential for Atlantic Ocean sediment bacteria. *Mar. Biol.* **34:** 1–9.

Walker, J. D., P. A. Seesman, T. L. Herbert, and R. R. Colwell (1976b) Petroleum hydrocarbons: degradation and growth potential of deep-sea sediment bacteria. *Environ. Pollut.* **10:** 89–99.

Wallcave, L., H. Garcia, R. Feldman, W. Lijinsky, and P. Shubik (1971) Skin tumorigenesis in mice by petroleum asphalts and coal-tar pitches of known polynuclear aromatic hydrocarbon content. *Toxicol. Appl. Pharmacol.* **18:** 41–52.

Wallen, I. E., W. C. Greer and R. Lasater (1957) Toxicity to *Gambusia affinis* of certain pure chemicals in turbid waters. *Sew. Indust. Wastes* **29:** 695–711.

Wauchope, R. D., and F. W. Getzen (1972) Temperature dependence of solubilities in water and heats of fusion of solid aromatic hydrocarbons. *J. Chem. Eng. Data* **17:** 38–41.

Weil-Malherbe, H. (1946) The solubilization of polycyclic aromatic hydrocarbons by purines. *Biochem. J.* **40:** 351–363.

Whittle, K. J., J. Murray, P. R. Mackie, R. Hardy, and J. Farmer (1977) Fate of hydrocarbons in fish. pp. 139–142 In: A. D. McIntyre and K. J. Whittle (eds.) *Petroleum Hydrocarbons in the Marine Environment*. Charlottenlund Slot, Denmark: Cons. Intern. Explor. Mer. Vol. 171.

Wilson, R. D., P. H. Monaghan, A. Osanik, L. C. Price, and M. A. Rogers (1974) Natural marine oil seepage. *Science* **184:** 857–865.

Winters, K., J. C. Batterton, and C. van Baalen (1977) Phenalen-1-one: occurrence in a fuel oil and toxicity to microalgae. *Environ. Sci. Technol.* **11:** 270–272.

Winters, K., R. O'Donnell, J. C. Batterton, and C. van Baalen (1976) Water-soluble components of four fuel oils: chemical characterization and effects on growth of microalgae. *Mar. Biol.* **36:** 269–276.

Wiser, W. H. (1973) Some chemical aspects of coal liquification. Paper presented at *Short Course on Coal Characteristics and Coal Conversion Processes* (cited in: J. T. Ensminger, 1977. *Coal: Origin, Classification, and Physical and Chemical Properties*. 104 pp. In: H. M. Braunstein, E. D. Copenhaver, and H. A. Pfuderer (eds.), *Environmental Health and Control Aspects of Coal Conversion: an Information Overview*. Oak Ridge, Tennessee: Oak Ridge National Laboratory ORNL/EIS-94).

Wofford, H. W., and J. M. Neff (1978) Structure–activity relations of organic pollutants: comparative toxicity of fluorene, dibenzofuran, dibenzothiopene, and carbazole to estuarine crustaceans and fish (unpublished manuscript).

Wolfe, D. A. (ed.) (1977) *Fate and Effects of Petroleum Hydrocarbons in Marine Organisms and Ecosystems*. New York: Pergamon Press. 478 pp.

Wong, C. S., W. J. Cretney, P. Christensen, and R. W. Macdonald (1976) Hydrocarbon levels in the marine environment of the southern Beaufort Sea.

*Beaufort Sea Tech. Rept.* 38. Unpublished manuscript. Environment Canada, Victoria, B.C., 113 pp. (cited in: Clark, R. C., Jr., and W. D. MacLeod, Jr., Inputs, transport mechanisms, and observed concentrations of petroleum in the marine environment. pp. 91–223 In: D. C. Malins (ed.) *Effect of Petroleum on Arctic and Subarctic Marine Environments and Organisms. Vol.* 1. *Nature and Fate of Petroleum.* New York: Academic Press.

Woo, C. S., A. P. Dsilva, V. A. Fassel, and G. J. Oestreich (1978) Polynuclear aromatic hydrocarbons in coal-identification by their X-ray excited optical luminescence. *Environ. Sci. Technol.* **12**(2): 173–174.

Woodward, P. W., G. P. Sturm, Jr., J. W. Vogh, S. A. Holmes, and J. E. Dooley (1976) Compositional analyses of synthoil from West Virginia coal. BERC/RI-76/2, January, Bartlesville Energy Research Center, Bartlesville, Oklahoma.

World Health Organisation (1970) European Standards for Drinking Water, 2nd ed., revised. Geneva, Switzerland: WHO. 58 pp.

Wynder, E. L. (1976) Nutrition and cancer. *Fed. Proc.* **35:** 1309–1315.

Yang, S. K., D. W. McCourt, J. C. Leutz, and H. V. Gelboin (1977) Benzo[*a*]pyrene diol epoxides: mechanisms of enzymatic formation and optically active intermediates. *Science* **196:** 1199–1201.

Yang, S. K., P. P. Roller, and H. V. Gelboin (1978) Benzo[*a*]pyrene metabolism: mechanism in the formation of epoxides, phenols, dihydrodiols, and the 7,8-diol-9,10-epoxides. pp. 285–302 In: P. W. Jones and R. I. Freudenthal (eds.), *Carcinogenesis—A Comprehensive Survey. Vol.* 3. *Polynuclear Aromatic Hydrocarbons: Second International Symposium on Analysis, Chemistry, and Biology.* New York: Raven Press.

Yarbrough, J. D., and J. E. Chambers (1977) Crude oil effects on microsomal mixed-function oxidase system components in the striped mullet (*Mugil cephalus*). *Life Sci.* **21:** 1095–1100.

Yevich, P. P., and C. A. Barszcz (1977) Neoplasia in soft-shell clams (*Mya arenaria*) collected from oil-impacted sites. pp. 409–426 In: H. F. Kraybill, C. J. Dawe, C. J. Harshbarger, and R. G. Tardiff (eds.) *Aquatic Pollutants and Biologic Effects with Emphasis on Neoplasia.* New York: Ann. N.Y. Acad. Sci. Vol. 298.

Young, G. P. (1977) Effects of naphthalene and phenanthrene on the grass shrimp *Palaemonetes pugio* (Holthius). Master's thesis. The Graduate College, Texas, A. & M. University, College Station, Texas. 67 pp.

Youngblood, W. W., and M. Blumer (1975) Polycyclic aromatic hydrocarbons in the environment: Homologous series in soils and recent marine sediments. *Geochim Cosmochim. Acta* **39:** 1303–1314.

Zafiriou, O. C. (1977) Marine organic photochemistry previewed. *Mar. Chem.* **5:** 497–522.

Zechmeister, L., and B. K. Koe (1952) The isolation of carcinogenic and other polycyclic aromatic hydrocarbons from barnacles. *Arch. Biochem. Biophys.* **35:** 1–11.

Zitko, V. (1975) Aromatic hydrocarbons in aquatic fauna. *Bull. Environ. Contam. Toxicol.* **14:** 621–631.

ZoBell, C. E. (1971) Sources and biodegradation of carcinogenic hydrocarbons. pp. 441–451. *Proc. Joint Conf. Prevention and Control of Oil Spills.* Washington, D.C.: American Petroleum Institute.

Zsolnay, A. (1977) Inventory of nonvolatile fatty acids and hydrocarbons in the oceans. *Mar. Chem.* **5:** 465–475.

# *Glossary of Abbreviations*

| | |
|---|---|
| AHH | aryl hydrocarbon hydroxylase = mixed-function oxygenase. The enzyme system responsible for oxygenating a polycyclic aromatic hydrocarbon to an epoxide. |
| API | the American Petroleum Institute, Washington, D.C. |
| AR-$C_n$ | an aromatic hydrocarbon with *n* number of alkyl carbons (see Fig. 4). |
| B*a*A | benzo[*a*]anthracene (see Fig. 1E). |
| B*a*P | benzo[*a*]pyrene (see Fig. 1G). |
| B*e*P | benzo[*e*]pyrene (see Fig. 1H). |
| COED | Char-Oil-Energy Development pyrolysis process for coal liquefication developed by the FMC Corporation for the Office of Coal Research, US Department of the Interior. The process converts coal into synthetic crude oil, gas, and char. |
| cpm | radioactive counts per minute as measured by a liquid scintillation counter or other radiation counting device. |
| dpm | radioactive disintegrations per minute. Counts per minute corrected for counting efficiency of the measuring instrument. |
| EPA | US Environmental Protection Agency. |
| gcal/$cm^2$/m | gram-calories per square centimetre per minute. A measure of the intensity of light radiation. |
| GC–MS | gas chromatographic technique in which eluting peaks from the gas chromatogram are injected into a mass spectrometer for identification. |
| GSH | reduced glutathione, an essential biochemical cofactor in several biological oxidation–reduction reactions. |
| IR | infrared spectrometry. |

| | |
|---|---|
| MFO | mixed-function oxygenase. The enzyme system responsible for oxygenating a polycyclic aromatic hydrocarbon or other suitable substrate to an epoxide or phenol. |
| $mW/cm^2$ | milliwatts per square centimeter. A measure of the intensity of incident light radiation. |
| NAD, NADP | nicotinamide adenine dinucleotide and nicotinamide adenine dinucleotide phosphate, oxidized form. The reduced forms are indicated as NADH or NADPH, respectively. Coenzymes involved in many enzymatic oxidation–reduction reactions in living cells. |
| NAS | the US National Academy of Sciences, Washington, D.C. |
| 3-OH-B*a*P | 3-hydroxybenzo[*a*]pyrene. One of the major initial metabolites of benzo[*a*]pyrene produced by the cytochrome P-450–mixed-function oxygenase system. |
| PAH | polycyclic aromatic hydrocarbon(s). Sometimes also called polynuclear aromatic hydrocarbons. Carbon–hydrogen compounds consisting of two or more fused aromatic (benzene) rings. |
| PCB | polychlorinated biphenyls. Industrial chlorinated hydrocarbon formulations sold under a variety of trade names (see review of Peakall, 1975). |
| ppb | parts per billion, micrograms per kilogram, or micrograms per liter. |
| ppm | parts per million, milligrams per kilogram, or milligrams per liter. |
| ‰ salinity | parts per thousand total dissolved solutes. A measure of the concentration of seawater. |
| $t_{1/2}$ | the half-time for a reaction or a rate process. The amount of time required for a rate process to go 50% to completion. |
| TLC | thin-layer chromatography. |
| UV | ultraviolet spectrophotometry. |
| WSF | the water-soluble fraction. The aqueous solution produced by gently mixing water and a poorly soluble substance such as petroleum for a given period of time. |

# *Index*